Management and Industrial Engineering

Series editor

J. Paulo Davim, Aveiro, Portugal

More information about this series at http://www.springer.com/series/11690

Guillermo Cortés-Robles
Jorge Luis García-Alcaraz
Giner Alor-Hernández
Editors

Managing Innovation in Highly Restrictive Environments

Lessons from Latin America and Emerging Markets

Springer

Editors
Guillermo Cortés-Robles
Tecnologico Nacional de Mexico / Instituto Tecnologico de Orizaba
Orizaba, Veracruz
Mexico

Giner Alor-Hernández
Tecnologico Nacional de Mexico / Instituto Tecnologico de Orizaba
Orizaba, Veracruz
Mexico

Jorge Luis García-Alcaraz
Autonomous University of Ciudad Juárez
Ciudad Juárez, Chihuahua
Mexico

ISSN 2365-0532 ISSN 2365-0540 (electronic)
Management and Industrial Engineering
ISBN 978-3-030-06709-0 ISBN 978-3-319-93716-8 (eBook)
https://doi.org/10.1007/978-3-319-93716-8

Printed on acid-free paper

This Springer imprint is published by the registered company Springer Nature Switzerland AG
The registered company address is: Gewerbestrasse 11, 6330 Cham, Switzerland

To Andrea and Gael

Preface

In the past decade, the economies of Latin America and the Caribbean and other emerging economies are trying to mobilize their internal resources to impel prosperity and welfare. A current initiative in this direction is to promote the development of innovation. The transformation of the intellectual capital and available resources into new products, services, processes, or just added value to the society is one of the paths to enhance the prosperity of a country. However, this intention is immersed in a complex context. In the emergent countries, companies achieve performance under challenging conditions, for instance, the technologies are not always accessible, the workforce does not have the required competences or demand constant training not easily available; the lack of techniques to facilitate the assimilation of new technologies and to learn from others; A limited perception about the technical side of innovation, which produces the impression that this process depends on a random creative effort, and no less important, a low assimilation degree of tools for planning the evolution of a product or an innovation system. Despite this complicated circumstances, the internal market and capacities continue to grow. This book explores how the innovation process is carried out under such conditions.

Chapter "Against the Odds. Innovation in Latin American SMEs" offers a broad perspective on how the innovation process is a crucial strategy in Latin America and the Caribbean (LAC). The chapter underlines that most of the innovation theories come from developed countries that have essential differences with the nations in LAC or emerging economies. One of the most important is the enterprise size. In LAC economies, the small- and medium-sized enterprises represent the 99% of the industrial activity. Consequently, the innovation process evolves under conditions not necessarily considered in these theories. Hence, the effort and results in LAC probably ask for an adequate measuring structure. Many pieces of evidence support this argument. In the first place, the Global Competitiveness Report 2016–2017 ranks in the 33rd position the first Latin American country and is Chile who takes this position. Next, country in this rank is Panama in the 42nd position, then Mexico in the 51st position. However, the Happy Planet Index counts a different story. This index evaluates the sustainable well-being of a country. In this ranking, six of the first ten places belong to LAC countries and the first ten places correspond to emerging economies. Probably, it is necessary to transform the concept of innovation and assure the welfare of the country, without losing the sight of happiness. Finally, the first chapter explains that the LAC countries are the laboratory where probably will emerge the next big revolution.

Chapter "Use of Lean-Sigma as a Problem-Solving Method in a Restrictive Environment" describes a common challenging situation in LAC enterprises: the need to solve problems in well-established production systems. In Mexico, the industrial activity related to the assemblage of products is growing, despite the complex international conditions. In this context, a Company needs to coordinate the solving effort along the supply chain. Hence, it is necessary to collaborate with suppliers, to analyze the production system, and to assure that the logistics and distribution will not suffer from any potential solution. Chapter "Use of Lean-Sigma as a Problem-Solving Method in a Restrictive Environment" describes a new solving approach where the rapid improvement of the Lean Manufacturing and the in-depth statistical analysis of Six Sigma get harmonized. The method called "Lean-Sigma" aims to solve problems in the short term. Thus, if a long-term problem arrives, each method takes their original role. Even if this combination is in their initial development stages, results unveil the potential for uncovering a more efficient approach for problem-solving.

Chapter "Creation of Technology-Based Companies: Challenges to Innovate in the Manufacturing Sector of Medical Devices, the Case of Baja California, México" is closely related to the materialization of the triple helix concept. The actors that participate in this complicated relationship are the industry, the universities and research centers, and the government initiatives to foster innovation. Such dynamic conditions could be observed in LAC countries. For instance, the fashion industry in Colombia, the oil technology in Brazil, the Aerospatiale sector in Mexico and Brazil just to mention some examples. Under such conditions, it is a not only a risk activity, but also a complex process to compel the creation of a start-up company. Chapter "Creation of Technology-Based Companies: Challenges to Innovate in the Manufacturing Sector of Medical Devices, the Case of Baja California, México" proposes a framework to deal with this problem. The most relevant contribution is to describe the conditions that facilitate the technology transfer process, which is probably the more significant obstacle to overcome in the commercialization of new developments. The chapter describes some generic guidelines to identify potential allies and some strategies to access financial sources. As a general remark, it is important to notice that LAC entrepreneurs face another problem: the lack of methodologies to design new services or to manage the future transformation of a function.

Chapter "Applying Action Research to Academic Patent-Licensing Path" describes one of the most complex and extenuating situations that a public research center, public laboratory, or university face: the lack of an effective organism or structure to transfer the research results with practical and commercial purposes. Researchers in LAC produce knowledge that frequently debouches into new products or services. Nevertheless, the transfer to this development to the right actor is not an easy task. The problem starts right after there is a patentable solution. LAC countries have not the culture or an efficient organization capable of protecting the intellectual property (IP) at the right rate. Then, the search for the right partner is time demanding and an ineffective process. Besides, the researcher is not aware of all the IP process, a condition that could lead to a failure when asking for a patent. Besides, in some countries in LAC, advanced developments as the software are not patentable, which makes difficult the protection of a potential product. Chapter "Applying Action Research to Academic Patent-Licensing Path" offers some guidelines that certainly will be useful in a different domain.

Chapter "Mass Customization Process in Companies from the Housing Sector in Brazil" deals with a very restrictive market requirement: the need for customized products at the lower possible price. Customization is a definite tendency in the market. It is interesting to notice that a personalized product is not more a product but a service. Thus, the incorporation of customized product in the market is one of the most useful strategies to compete in an environment that changes all the time. Also, it is relevant to note that the customization frequently leads to contradictory requirements that ask for new solving methods. Chapter "Mass Customization Process in Companies from the Housing Sector in Brazil" expose an argument to comprehend better the methodological differences between the recommended customization process and the actual strategy from housing companies from developing economy countries. The return of experiences, in this case, could be useful in a completely different domain. Hence, this is maybe one of the more valuable contributions of this chapter: to propose new questions about how to reutilize the knowledge acquired while customizing a product and transferred to another field.

Chapter "A Talent Management Model for Innovation and Competitiveness in Complex Domains: A Study Case in the Latin American Energy Sector" faces one of the most challenging problems in emerging economies and LAC countries: the need for qualified and highly trained personnel. Specialized human resources are indispensable to compete in the market, to anticipate changes, and the only part of any enterprise with the capacity to create or transform a system. This chapter proposes to integrate new technologies into the Talent Management process and develop a new training process through a Computer-Based Training model. The training model provides online and personalized instruction based on three perspectives: knowledge, skills, and attitudes. The particularity of this system derives from the system capacity to recognize the emotions of the user via an intelligent system, which adapts the training program according to the learning style of the user and their emotions. Emotions are recognized using the facial expressions exposed during a training situation. With this information, the system uses an animated agent to show emotions and empathy with the user. The purpose is to facilitate and reduce stress in the user and thus to improve the learning process. This chapter describes a model that has a considerable potential to mobilize knowledge among the LAC countries and emerging economies.

Chapter "The Role of ICT in Educational Innovation" shares a similar objective with Chapter "A Talent Management Model for Innovation and Competitiveness in Complex Domains: A Study Case in the Latin American Energy Sector": to modify the teaching practices and offer new resources to the learning process through the incorporation of the Information and Communication Technologies (ICT). The Oslo Manual recognizes that the educational system determines the training standards in the workforce and the domestic consumer market. Thereby, an efficient strategy to change the workforce and the perspective of internal costumers is through the university system. However, the benefits of using ICT in the teaching process need a previous validation. The authors propose to use a structural equation model to satisfy this requirement. The case study collects information related to the use of ICT in the learning process. The tool is a survey of 43 items, and the population was a sample of 469 students (at the bachelor degree). The analysis shows enough statistical evidence to state that the use of ICT has a direct and positive impact on the student's performance.

Chapter "A Series of Recommendations for Industrial Design Conceptualizing Based on Emotional Design" proposes the use of Emotional Design as the primary strategy to understand and transform the customer's emotions into design attributes. Research in this field highlights that emotions play a crucial role in the design process. There are even some statements that define design as the art of transforming emotions. Nevertheless, also it seems natural to assimilate this design perspective; the reality follows the opposite direction, at least in LAC where just a few universities have adopted this approach. The chapter suggests some fundamental recommendations to incorporate emotions in the initial design stage: conceptual design, which is one of the most elusive stages. This intention opens a discussion space to explore the Industrial Design Conceptualizing based on Emotional Design. This approach was developed through the analysis of several design emotion-based methodologies, conceptualization theories, and of course, the emotional design. Several cases are the basis for validating the set of recommendation to apply the emotional design. Eight cases, developed by a group of eight industrial design students, implemented the recommendations in academic projects. All resulting products were evaluated to determine the impact of the emotional design in a product. Five out of eight projects show evidence of the correct interpretation of the recommendations. Hence, the set of emotional design guidelines are a useful tool to transform some emotions into product parameters.

Chapter "Indexing and Mapping Examples of Heuristics Compiled from TRIZ": explores the application of the Theory of Inventive Problem Solving (TRIZ) is another approach for problem-solving that attracts the attention of the industry and the academy. One of the reasons is the technological foundation of TRIZ, which offers several resources to mobilize technical knowledge across domains. Despite the broad application of TRIZ in different industrial sectors, in LAC and other emerging economies, the use of TRIZ is in the initial stage of assimilation. TRIZ has several tools, including Inventive Principles, Standard Solutions, and Separation Principles that are abstract recommendations or strategies to solve an inventive problem. In general, these solving strategies are called Inventive

Heuristics, and there are a vast number of them (469 inventive heuristics). Thus, the adoption of TRIZ for a new user demands too much time and effort, and the necessary experience to identify the right application context. Besides, the inventive heuristic overlaps their application with a disagreeable consequence: doubt and confusion in the users. The chapter deals with these issues through the use of an index, and some examples that act as pointers and that makes easier to select one solving strategy to a particular problem. This compilation, organized as a catalog with the goal to facilitate the use of such collection of heuristics.

Chapter "The Use of Affective Computing in the Conceptual Design Stage of New Products" sees innovation as a continual problem-solving process that has several limitations. Some significant drawbacks are, for instance that in the initial stages, the team or an individual need to select what problem should be faced first. Then, through a creative effort, the group or the individual collects some potential solving concepts. In this step of the process, the need for some mechanism to filter the ideas result evident. Consequently, the innovation process asks for efficient approaches to assist decision-making in the evaluation and selection of different potential solutions. Some techniques fulfill this requirement as the Analytic Hierarchic Process or the Technique for Order of Preference by Similarity to Ideal Solution (TOPSIS). Nevertheless, these techniques reveal their limitation when it is necessary to take into account a crucial success factor for the innovation process: emotions. The emotional response to unique product attributes is, in fact, a determinant element to succeed in a market. Thus, it is indispensable to take into account this dimension of the innovation process. This chapter describes the use of the affective computing paradigm to help a team or an individual in the different stages where it is necessary to evaluate technical parameters and emotions.

Chapter "Taking Advantage of the Innovation in Die Service Design" identifies a market need and proposes the design of new services. The metal-mechanic industry is an essential provider or a link in the supply chain. Thus, the performance of this industrial sector has a significant impact on other industrial activities; this is the case of the design of dies. A die is a tool used to cut, punch, or give a particular shape to different material, including metal parts. A dies covers the specific requirements of a process; then it must be designed. In Mexico, national enterprises satisfy only 5% of the total internal market. The rest of this need belongs to external companies and is a market of 2.6b/year. This dependency affects the national industry negatively from different perspectives: the productivity and competitiveness loss dynamism, the innovation process reacts slowly, and maybe a critical factor: the evolution of the skills to satisfy this national need do not evolve. Consequently, the objective of this chapter is demonstrated that the service is feasible. Results show that the market will accept the new service with excellent profitability. Finally, the service design consists of the integration of several technologies. The service was implemented by an SME in Mexico that aims to collaborate with the national welfare.

Chapter "TRIZ Evolution Trends as an Approach for Predicting the Future Development of the Technological Systems in the Food Industry" focuses on a fundamental necessity of the innovation process: to anticipate the future technologic developments and also, the next evolution of the product functionalities, and the food industry is not the exception. The chapter explains that in the food industry there are two innovation drivers: the technological change (technology push) and the market orientation (market pull). Each driver has specific requirements, and thus, it is necessary to coordinate both directions of development under a more comprehensive perspective: to anticipate the future developments of technological food systems. The chapter proposes to use an evolutionary view based on the Theory of Inventive Problem Solving (TRIZ). TRIZ encloses in its toolbox, a set of Evolution Trends (TETs) that has the goal to explain the transition that a system will endure during its development. Thus, to improve the innovation process in the food industry, both drivers are combined with the TRIZ Evolution Trends with the goal to conceive some guidelines of the most favorable transformation to improve or develop technological food systems. A case study depicts the implementation of the guidelines.

Chapter "A Service Design Process Based on the Business Model CANVAS and the C-K Theory" focuses on the service design process. It is important to underline that service design in Latin America is an emergent research topic. Paradoxically, the service domain accounts for more than 60% of the employment in several Latin America economies. The authors propose a combination of the business model CANVAS and the C-K theory. The integration of both approaches produces a design framework, where it is possible to connect the business model with a graphical tool to model processes. Despite the usefulness of the C-K theory, its application in the academic and industrial domain in Latin America is not frequent. On the other hand, the business model CANVAS has a broad use as a tool for business modeling. Several institutions in Mexico recommend the use of this tool to describe business ideas or to model a business. The service design framework in Chapter "A Service Design Process Based on the Business Model CANVAS and the C-K Theory" preserves the advantages of the business model CANVAS and overcome some of their limits by introducing a new dimension: the knowledge space of the C-K theory and the interaction that takes place between concepts and knowledge. As a result, the design process becomes more dynamic and coherent.

Chapter "Integration of Design Thinking and TRIZ Theory to Assist a User in the Formulation of an Innovation Project" explains that the comprehension of the market share and the efficient transformation of some need or customers requirements is a successful strategy for new product development. The Design Thinking approach can produce valuable information about the market need or the customer's requirements. However, it does not have a specific tool to solve the intrinsic problems that the design process invariably unveils. The goal of this chapter is to overcome this limitation by adding an approach capable of producing inventive solutions: the theory of inventive problem solving or the TRIZ theory. In this chapter, the TRIZ theory also obtains some benefits. TRIZ does not have a tool to identify the user's requirements or the market tendencies. Hence, when dealing with an inventive problem that has as objective to determine the market requirements, TRIZ needs the collaboration with other tools. Consequently, the problem-solving process depends on the experience of the solver. The authors present a methodology where the TRIZ theory provides some methods and concepts to model and solve inventive problems, while Design Thinking uses the knowledge extracted from the user to define the design goal.

Probably, the LAC countries and other emerging economies will produce a new perception about innovation, a more open perspective where the technology, economy, and well-being will interact with equilibrium. Next years will reveal if the path must be reformulated or the effort should intensify. Finally, restrictive environments or conditions are not an obstacle to the innovation process, but a force that impels it.

Orizaba, Mexico — Guillermo Cortés-Robles
Ciudad Juárez, Mexico — Jorge Luis García-Alcaraz
Orizaba, Mexico — Giner Alor-Hernández

Acknowledgements

This book crystallizes the effort of several organizations that support research as the National Council of Science and Technology in Mexico, the Tecnologico Nacional de Mexico, and PRODEP. We want to express our gratitude to other organizations that share the same purpose but are not mentioned here. We also appreciate and recognize the trust, effort, and collaboration that all authors endorse us. Finally, we are thankful and appreciate the Springer Publishing experts, especially Petra Jantzen and Mayra Castro for their invaluable help, patience, and advice to materialize this book.

Contents

Against the Odds. Innovation in Latin American SMEs

Jorge Rodriguez-Martinez

Abstract The popular expression "Necessity is the mother of invention," means that it may trigger new ways of doing things, in a more economical, faster or easier way. Most innovation theories in the academic and business literature come from advanced countries and examples usually derived from multinational companies. There is little evidence from emerging economies, such as Latin America. This chapter will focus on how innovation may arise even in highly restrictive environments, especially in small and medium-sized enterprises. For the OECD Oslo Manual 2005, there is the innovation of products, processes, business practices, workplace organization, and social improvements. The data of some international and regional indexes were consulted at the country, city, and personal level. They evaluate competitiveness, creativity, human development, and even happiness. Emerging economies that want to get out of the middle-income trap need to have an innovation-driven economy, despite the current constraints. The situation is of lights and shadows; nevertheless, a brighter future is ahead. Latin America is indeed innovating in its own way, sometimes not captured with the traditional indicators.

Keywords Innovation · Latin America · OECD oslo manual · Constraints

1 Introduction

Creativity is the generation of new ideas, which could be how to solve or look at needs in different ways, such as complete a task with less effort or time, or create a user-friendly product. However, for turning creativity into innovation, an economic decision is necessary; the product launch into the market. For the Cox Review Report (2005), innovation is the successful commercial exploitation of a new product or service. Most countries are increasingly paying more attention to it as an

J. Rodriguez-Martinez (✉)
Universidad Autónoma Metropolitana – Azcapotzalco, CYAD, Procesos. Azcapotzalco, C.P. 02200 México City, Mexico
e-mail: rmj@correo.azc.uam.mx; jorgerodriguez_8@hotmail.co.uk

G. Cortes-Robles et al. (eds.), *Managing Innovation in Highly Restrictive Environments*, Management and Industrial Engineering,
https://doi.org/10.1007/978-3-319-93716-8_1

important source of competitiveness and differentiation in the market place. This section will follow the structure of the OECD Oslo Manual 2005 that defines different types of innovation: product, process, marketing, and organizational.

1.1 How Innovation Appears in the Marketplace. OECD Oslo Manual 2005

The first Oslo Manual is from 1992; its focus was on the different types of improvements at the company level. The second version appeared in 1997. The last version, the current one, appeared in 2005; defines innovation in the product, process, marketing, and organizational level. The OECD strategy offers a wider scope to consider: "It goes far beyond the confines of research labs to users, suppliers and consumers everywhere—in government, business and nonprofit organizations, across borders, across sectors, and across institutions," (OECD 2017).

A superior product or service. "A good or service that is new or significantly improved. The product presents improvements in technical specifications, components, materials, software in the product, user-friendliness or other functional characteristics," (OECD 2005: 11). The innovation process may be incremental as it happened with electronic devices to save data, starting with floppy disks ≫ diskettes 5 ½ ≫ diskettes 3 ½ ≫ CDs ≫ DVD ≫ USB. There is also the radical or disruptive innovation that breaks paradigms, as the radical change of technology in the use of conventional to digital cameras. About 2000, there were more digital cameras sold than conventional ones; however, it was a short-lived victory, as in just over a decade, Smartphones equipped with cameras overtook digital cameras (petapixel.com 2017). The technology used on televisions changed, as well, it went from analog black and white to color TVs; later, they changed to digital TVs ≫ to interactive HDTV.

An upgraded production process. "A new or significantly improved production or delivery method. The benefit may bring time reduction and quality improvement by applying new techniques, equipment or software" (OECD 2005: 12). An example is how the production process evolved in the fast food restaurants. McDonald's® changed the way people in a hurry could have a simple meal, by applying the same concepts of a production line to the restaurant kitchen. Each location is a production cell, with multitasking employees, repeating the same activities and offering a simple menu. Domino's Pizza® reduced costs by simplifying its processes, and the time to make and deliver a pizza in thirty minutes or less, otherwise, it is free.

Original ways to market a product. "A new marketing method involving significant changes in product design or packaging, product placement, product promotion or pricing," (OECD 2005: 12). For example, the on-demand economy based on online platforms that offer goods or services; and according to the Harvard

Business Review (2016), half of the customers are the millennials (age 18–34). New business models respond to the needs of people who demand transportation on a short-notice (Uber® or Cabbify®) or want to buy books and many other goods delivered to their house (Amazon®). Another example of radical changes in the marketplace is the sharing economy, defined as "an economic model in which individuals are able to borrow or rent underused assets owned by someone else… This is the peer-to-peer (P2P) rental market," (Investopedia 2017), like houses or apartments for short-term rent (Airbnb®).

New ways to organize work and achieve the group's objectives. "A new organizational method in business practices, workplace organization or external," (OECD 2005: 12). In Japan, after World War II, the country was devastated and had no resources. The Toyota® Company developed, out of necessity the just in time, a flexible manufacturing system, adapted to a variety of models in small numbers, by reducing all sorts of waste. A most recent example is the Canvas business model®; a planning tool used by entrepreneurs and managers alike that presents an overall view of the nine main elements of a company by highlighting the value propositions and the core value delivered to the consumer (Strategyzer 2017).

New ways to solve challenging social issues. This is the process to answer social needs by the interaction of citizens or organizations to improve the quality of life. It is not part of the original definitions of the OECD, however, is a recurrent topic of interest to policymakers and citizens alike. "Social innovation is the implementation of new ideas (products, services or models) that satisfy needs, create new relationships and offer best results; of such a way that they generate an answer to social demands that affect the interaction process and aim to raise the level of human well-being," (Foro Consultivo 2017).

The chapter has six parts. Section 2 offers a brief overview of the Latin American history and its main socioeconomic characteristics. Section 3 presents Latin American countries in different world rankings; these charts are at the country, city, and personal level. To enrich the comparisons two groups of countries are included: the so-called BRICS—Brazil, Russia, India, China, and South Africa; and the MIST group: Mexico, Indonesia, South Korea, and Turkey. Section 4 has examples of the constraints that limit innovation in the region in economic sectors such as agriculture, manufacturing, social, health sector, and other fields. Section five describes Latin American clusters and how small and medium-sized companies innovate. Finally, the conclusions section discusses some of the challenges the region faces to get out of the middle-income trap to move forward from and efficiency to an innovation-driven economy.

2 Overview of Latin America History and Socioeconomic Characteristics

Latin America is the area of the American continent stretching north to south, which includes most of those countries conquered and colonized by Spain or Portugal. It is a region vast in natural resources: mining, agriculture, oil, precious metals, and wood or cattle breeding. It has a rich and diverse history, populated by important civilizations such as the Incas, Mayan or Aztecs, advanced in astronomy, architecture, mathematics, medicine, agricultural techniques, sculpture, and pottery.

Spain and Portugal's kingdoms retained their colonies for nearly three hundred years; any external trade, mostly minerals, and precious jewels needed approval by the Crown, limiting creative activity. Latin America had universities long before the United States: The University of Santo Domingo (1538), The University of San Marcos in Lima (1551) and The University of Mexico (1553), while Harvard University only opened until 1650. The type of education delivered in the Latin American universities was mainly religious-based and focused on Philosophy, Law, Theology, and Engineering, which limited the creation of new knowledge (Maloney et al. 2002). The actual situation is not that different, a review of the 17 Nobel Prizes awarded to Latin Americans reveals the knowledge areas where the region tends to excel: in Literature, six prizes obtained; Peace, six Prizes; in Physiology or Medicine, four Prizes; and in Chemistry only one Prize (Nobel Prize Org 2017).

The entrance to the Spanish colonies of non-Catholics, foreign companies, businesspersons, intellectuals, or their ideas were tightly controlled, limiting the contact with new European knowledge, therefore, constraining the development of innovation. Maloney et al. (2002: 120) describe that the resource-based growth in Latin America had its origins in Europe: "Furthermore, Spain in the eighteenth century was a resource-rich nation that used its fantastic returns from silver and gold mines in the New World to purchase all that was needed. Thus, developing a rentier mentality rather than one of a nation of hands-on tinkerers such as appeared in Great Britain or the United States…" In other words, Spain preferred to buy goods, weapons, or ships from England, France, or Germany, therefore, financing their industrial revolutions and not developing its own.

Historically, Latin America has not functioned as a region with shared interests. Jonas and McCaughan (1994: 37–38) describe this situation: "The prevailing tendency led Latin American nations not to develop complementary economies, but to separate and isolate themselves, to turn their back on one another while looking toward Europe and, to a lesser degree, the United States." The Latin region during the late nineteenth and early part of the twentieth centuries, underwent a mostly convulse period: wars, foreign invasions, civil wars, struggles between political parties, weak institutions, lack of respect for the law, Coups *d'état*, privileges for a selected few, and instability—not the best environment to innovate.

The initial efforts to industrialize the larger countries of the region depended, almost entirely, on foreign immigrants or international companies investing and exploiting the internal market because there was little local expertise. At some point during the last eight or nine decades, the larger economies (Argentina, Brazil, and Mexico) were in a better economic situation than China, Korea or even Taiwan. They were about to take off and achieve a higher level of development. However, this has been a history of disappointment and missed opportunities. Some characters or small groups of power have put their ambitions above the common good (Reid 2010; Maloney et al. 2002).

In the Postwar period, from about 1945 to the 1980s, several of the larger economies embarked on import-substitution industrialization (ISI) policy led by the State to manufacture the goods that a growing middle class was demanding (Dussel-Peters 2000). The ISI had the positive effect of creating the infrastructure and industrializing the larger economies of the region by imposing high import tariffs on foreign goods, and barriers to trade and investment. However, this protectionist period came at a high cost because it created State and private-owned monopolies, a captive local market, and a lack of incentives to improve quality or develop new ideas (Maloney et al. 2002). In Latin America, the larger economies tried to replicate the strategy followed by Italy, Germany, Japan, and other countries, emulating the production and design of the same products manufactured in the developed countries, such as household products, sound equipment, coaches, and cars that required middle-technology. However, the technical innovation was missing (Bonsiepe 1985; Fernández and Bonsiepe 2008).

The economic emergency in the region during the 1980s, also described as "the lost decade," and early 1990s, consisted of a series of events like the Mexican (tequila) crisis of 1995, the Brazilian (samba) of 1999, and the Argentinean (tango) of 2001 were a watershed. The ingredients were the rising foreign debt, lack of competitiveness of State-owned companies, high inflation rates, and low economic growth among other factors. The unsustainable situation forced the Latin American governments to liberalize their economies, privatize companies, and abandon the ISI model; although there is criticism that it created an increasing economic, social, and even territorial polarization (Dussel-Peters 2000).

A crisis and the collapse of the domestic market forces companies to become innovative out of necessity and either proactively or reactively look for international markets. For example, in 1994, the North American Free Trade Agreement (NAFTA) went into effect between Canada, Mexico, and the United States. The very same year, Mexico suffered a sharp devaluation of its currency. The results were the emergence of "two Mexicos," one that is modern, productive, and outward-looking that took advantage of these new external markets; coexisting with an old-fashioned, unproductive, and inward-looking sector of the economy (McKinsey 2014).

If the European powers were the dominant foreign figures in Latin America from the sixteenth to the nineteenth centuries, during the twentieth century, the United States played the most influential political, economic, and cultural role. However, starting the early twenty-first century, China has become a major player in the region, mainly as an exporter of all sorts of products, from consumer to sophisticated goods (gadgets, autos, etc.) but also as an importer of large amounts of raw material and commodities. It is an important foreign direct investor (Dussel-Peters 2015).

One characteristic of the twentieth and twenty-first centuries are the Multinational Corporations (MNCs) that invest, develop products, and services and sells them in multiple countries. MNCs have traditionally come from the called "triad" economic centers: United States, Japan, and Europe (Germany, France, Italy, the United Kingdom, and others). However, new MNCs may come, although in fewer numbers, from large emerging countries, this is the case of Brazil, China, India, Korea or Mexico (Bartlett and Goshal 2000).

The "Multilatinas" are those large Latin American companies that expanded their operations in the region, the term first used by the *America Economia* magazine at the turn of the new twenty-first century. Casanova et al. (2009: 9) came up with an updated term "Global Latina" for those companies venturing beyond the Latin American region; its definition: "a privately owned, Latin American-based multinational firm with operation on at least one other foreign continent and which generates a minimum of US$500 million in annual revenues." Some of these Global Latinas are Brazilian companies: Vale® (iron), Petrobras® (oil), Embraer® (mid-range planes), Natura® (cosmetics), and Politec® (IT services); Mexican companies: Bimbo® (bread), Cemex® (cement), and America Movil® (communications). Countries with fewer companies are the Chilean company (Concha y Toro®, wine); and from Guatemala, El Pollo Campero® (chicken).

However, the internationalization of companies is not limited to large ones, since the 1980s, the small and medium-sized enterprises (SMEs) started to have a more active role in foreign markets. In the United States, SMEs contribute with around 30% of total exports; among the OECD country members, is about 25% to 35%. In the Asia Pacific region, it reaches 30%. In the European Union, 18% of SMEs are exporting. In Mexico, SMEs contribute with 9% of Mexican exports (Rodríguez 2013).

2.1 Latin American Economic and Social Indicators According to ECLAC

The Economic Commission for Latin America and the Caribbean (ECLAC 2016, CEPAL in Spanish) publishes that the region is the world's most unequal, because of its bad income distribution in the population; the most affected sectors are women, children, and indigenous people. The Gini coefficient measures inequality, with "0" = perfect equality, where everyone has the same income, and "1" = the perfect inequality, where one person has all the income, and the rest have no income. The CEPAL (2016) Gini data for the region reveals that Colombia has 0.530, Brazil 0.515, Mexico 0.507, Costa Rica 0.506, Bolivia 0.493, Chile 0.470, Argentina 0.455, Peru 0.447, and Uruguay 0.403. Some highlights of the current economic and social indicators are as follows:

- Nearly half of the economically active population in cities (46%) works in the informal sector, with low productivity, and limited or none access to social security protection and retirement programs.
- Regarding education, 93% of students in the region complete the first 6 years of basic education, and 73% do so at the secondary level. Most Latin American governments have laws to make education compulsory during the first 12 years of education, but it will take years to make it a reality.
- Girls have made significant progress in access to formal education, mainly at the secondary and tertiary education, when compared with boys.
- The ECLAC's statistics show that in the age bracket of 25–59 years of age, 40% of the population has 0–5 years of study; 18% have between 6–11 years of study. The numbers keep decreasing, only 9% have 12–15 years of study, and even less, just 6% have 16 years of study, the equivalent of a university degree.
- In 2017, the total population is over 626 million people. The region's life expectancy is 76 years, 79 for women, and 72 for men.
- Women still earn just 87% of men's wages. In the region, women hold less than 25% of positions in the branches of state power.
- There are 114.6 mobile telephone subscriptions for every 100 inhabitants, and there are 50.1 internet users per 100 inhabitants.

3 Latin America + BRICS + MIST in World Rankings

This section presents economic and social data from Latin America's largest economies: Brazil, Mexico, Colombia, Argentina, Chile, and Peru, which taken together as a group represent nearly three-quarters of the region's gross domestic product and more than 70% of the population. In some cases, there is data of smaller countries, such as Costa Rica, an outlier who scores high in different rankings such as educational, sustainability and happiness aspects. Other interesting case is Cuba that in spite of having an almost closed off economy for over five decades ranks high in health, educational and sports achievements, and in developing medicines and health treatments, with very few economic resources.

The Latin America data comes from a series of international indexes that offer world rankings with a different focus: country, city and personal level. Two other groups of countries are included in the comparison. The so-called BRICS: Brazil, Russia, India, China, and South Africa, and the MIST group: Mexico, Indonesia, South Korea, and Turkey. These two acronyms come from the Goldman Sachs consultancy firm (2003). Brazil and Mexico, the region's largest economies, appear, as members of BRICS and MIST groups, but along with Argentina are also parts of the G20, the top 20 largest economies. Nevertheless, new classifications keep emerging; the A. T. Kearney Global Business Policy Council (GBPC 2015) analyzed the seven emerging markets with more potential for growth for the 2015–2020 periods. The 2020-Seven: Chile, China, Malaysia, Mexico, Peru, Philippines, and Poland; but this list does not appear in this study.

The methodology consisted in reviewing databases of international and regional organizations: United Nations (UN), United Nations Educational, Scientific and Cultural Organization (UNESCO), World Trade Organization (WTO), World Intellectual Property Organization (WIPO), and the Organization for Economic Co-operation and Development (OECD). In the Latin American region: ECLAC, CEPAL, and the Network for Science and Technology Indicators (in Spanish RICYT), among others. Academic databases and journals published in the Web of Science, Scopus, Springer publications, Redalyc, Latindex, Dialnet, Scholar Google, Scimago, and others. Some data comes from additional searches in international and regional economic and financial publications, either academic or practitioner-oriented, and visits to the web pages of relevant companies and organizations.

3.1 Rankings for LATAM, BRICS, and MIST, at the Country Level

Total Trade in Relation to Gross Domestic Product (GDP) and Percentage Invested in R and D.

The World Bank (2015) reports that in all the Latin American, BRICS and MIST groups, the external trade activity increased during the 1960–2016 period. The percentages represent the sum of imports and exports of to the gross domestic product (GDP). The most open large countries are Korea and Mexico, both with 78%, Costa Rica 64%, South Africa 60%, Chile 56%, Turkey 47%, and Russia with 46%. The total trade activity of Brazil of 25% is weak for international standards, reflecting a closed and protected market.

The investment in research and development (R and D), as a percentage of the Gross Domestic Product, is one of the most common indicators used to measure the innovation of an economy. In the 2000–2014 period, the overwhelming majority of analyzed countries have increased their invested percentage. The two outstanding of the rest are Korea that almost duplicated the investment from 2.18 to 4.29%, and China that more than doubled the amount from 0.89 to 2.02%. Thus, it does not come as a surprise when WIPO (2016) reports that Korea and China, have two of the top five offices, along with the United States Patent and Trademark Office (USPTO), the European Union and Japan. The world's average of the top 40 countries is 1.70%, and none of the Latin Americans is even close. The leading country is Brazil with only 1.21%, and the rest are a long distance away from the group average.

It is possible to observe in Table 1 that Korea and China are ahead in this ranking, in position number 26 and 28, respectively; this puts them very close to the top 25 developed countries. Of the Latin American countries, Chile is the best

Table 1 The Global Competitiveness Index 2016–2017 for LATAM and some selected countries (WEF 2017)

	Overall index Out of 138 Economies			Basic Requirements –Institutions –Infrastructure –Macroeconomics –Health and education		Efficiency enhancers –Higher educations and training –Infrastructure –Macroeconomics –Health and education		Innovation–sophistication –Business sophistication. –Innovation	
Country	Rank of 138	Score (1–7)		Rank of 138	Score (1–7)	Rank of 138	Score (1–7)	Rank of 138	Score (1–7)
Korea	26	5.03		19	5.71	26	4.88	22	4.81
China	28	4.95		30	5.34	30	4.79	29	4.22
Chile	33	4.64		37	5.08	31	4.77	56	3.73
India	39	4.52		63	4.62	46	4.41	30	4.22
Indonesia	41	4.52		52	4.78	49	4.38	32	4.16
Panama	42	4.51		34	5.15	51	4.36	44	3.93
Russia	43	4.51		59	4.68	38	4.56	66	3.62
South Africa	47	4.47		84	4.37	35	4.62	31	4.18
Mexico	51	4.41		71	4.56	45	4.41	50	3.83
Costa Rica	54	4.40		57	4.70	52	4.34	45	3.93
Turkey	55	4.39		56	4.70	53	4.32	65	3.63
Colombia	61	4.30		85	4.35	48	4.38	63	3.65
Peru	67	4.23		77	4.43	57	4.26	108	3.30
Brazil	81	4.10		103	4.0	61	4.2	72	3.60

Source Jorge Rodríguez-Martínez with information from Global Competitiveness Index 2016–2017(WEF)

ranked, in the 33 position, doing well in the Basic Requirements and Efficiency Enhancers section, but needs to improve in the Innovation section. However, the same applies to all the economies of the region. The challenge for these efficiency-driven countries is to move up the economic ladder with unique products and services.

The Global Innovation Index (GII) started in 2007, uses about 80 key indicators, divided into two parts. Innovation Input: Institutions and its Environment, Human Capital and Research, Infrastructure, Market and Business Sophistication. The second section deals with the Innovation Output: Knowledge and Technology Outputs and Creative Outputs. In this study, the higher ranked country is Korea in 11th place out of 128 countries, scores high in all, but the Institutions and its Environment criteria. Korea already has all the ingredients of a developed country. China comes in 25th place; its highest score is in the Innovation Output section, precisely the weakness of all the Latin American countries. The objective of China to move from manufacturer to innovator is in the making. In 2013 China overtook the US as the more active country in registering and obtaining patents, trademarks, and industrial designs (WIPO 2013). The Latin countries start to appear at the bottom of the upper half, with Chile in 44th place, followed by Mexico in 61st, Uruguay 62nd, and Colombia in 63rd. Also, Brazil 69th and Argentina 81th are in the lower half (Table 2).

The Global Creativity Index (GCI) measures economic growth and sustainable prosperity based on the so-called 3Ts: Talent, Technology, and Tolerance. The hypotheses of Richard Florida, dean of the Martin Prosperity Institute, at the University of Toronto, is that the countries that score higher have more balanced societies, where the wealth is better distributed. Florida states that creativity, as measured by the GCI, influence the economic development, competitiveness, and its prosperity. And besides, it has an impact on productivity, entrepreneurship and human development. From about 30% to 40% of the population of the advanced economies, fall in the category of the creative class, comprised by knowledge workers, who may be professionals in technology, education, health service, business, or science. In Table 3, two South American nations score higher than even Korea or China. Uruguay is in 26th, and Argentina 27th thanks to the very high scores in the tolerance index; however, they score low in the Global Talent and Global Technology Indexes. Some countries have low scores in the first index, linked to educational achievements and lack of good university graduates, related to economic growth. This is the case with Brazil, China, and Mexico, revealing a disadvantage if they aspire to an innovation-led economy (Table 3).

Table 4 lists a classification by the stage of development:

The three classes are given as follows:

- **Factor-driven economies**. It is composed of the poorest countries. Its focus is on agriculture and extraction business as mining. It depends on low-paid unskilled labor.

Table 2 The global innovation index 2016, LATAM, BRICS and MIST counties (WIPO,2016), (GII 2016)

		INNOVATION INPUT Score (1–100) rank out of 128 countries					INNOVATION OUTPUT Score (1–100) rank out of 128 countries	
Rank out of 140	Country/ economy score (0–100)	Institutions and its environment –Political –Regulatory –Business	Human capital/ research –Educational –Tertiary education –Research and development	Infrastructure –ICT –General infrastructure –Ecological sustainability	Market sophistication –Credit –Investment –Trade –Competition –Market scale	Business sophistication –Knowledge workers –Innovation linkages –Knowledge absorption	Knowledge and technology outputs –Knowledge creation –Knowledge impact –Knowledge diffusion	Creative outputs –Intangible assets –Creative goods and services –Online creativity
11	Korea-57.1	31	3	9	14	13	5	21
25	China-50.6	79	29	36	21	7	6	30
42	Turkey-39.0	82	43	62	46	86	45	31
43	Russia-38.5	73	23	60	63	37	40	66
44	Chile-38.4	36	62	38	47	41	59	55
54	South Africa-34.9	46	55	85	17	56	63	77
61	Mexico-34.6	65	53	67	51	77	70	62
62	Uruguay-34.3	47	78	37	104	95	80	56
63	Colombia-34.2	71	80	35	40	70	82	68
66	India-33.6	96	63	87	33	57	43	94
69	Brazil-33.2	78	60	59	57	39	67	90
71	Peru-32.5	66	81	57	35	54	109	73
81	Argentina-30.2	106	47	65	106	69	97	83
82	Indonesia-29.1	122	92	80	62	106	71	85

Table 3 The Global Creativity Index (GCI) 2015/LATAM, BRICS and MIST countries (GCI 2015)

Total/Rank out of 139 countries	Global Technology Index –R and D investment –Patents/per capita –Technology index	Global Talent Index –Creative class –Educational Attainment –Talent Index	Tolerance –Racial and ethnic minorities –Gays and lesbians –Tolerance index	Global creativity index
26- Uruguay	48	45	7	0.688
27-Argentina	48	35	19	0.681
29-Brazil	27	68	15	0.667
31-South Korea	1	50	70	0.660
34-Chile	56	39	31	0.611
36-Costa Rica	47	61	20	0.607
38-Russia	30	62	57	0.579
39-South Africa	30	62	22	0.564
41-Cuba	94	17	37	0.556
62-China	14	87	96	0.462
69-Peru	62	79	60	0.418
71-Colombia	89	75	36	0.410
73-Mexico	54	94	56	0.407
88-Turkey	58	53	123	0.348
99.India	52	92	108	0.292
115.Indonesia	67	108	115	0.202

Table 4 Classification by each stage of development: LATAM, BRICS, and MIST countries (WEF 2017)

Stage 1 Factor-driven (35 economies) <$2,000 US dollars	Transition from Stage 1 > Stage 2 Factor-driven $2,000-$2,999	Stage 2 Efficiency-driven (30 economies) $3,000-$8,999	Transition from Stage 2 to Stage 3 (19 economies) $9,000-$17,000	Stage 3 Innovation-driven (37 economies) >$17,000
Nicaragua	Bolivia	Brazil	Argentina	Canada
India	Honduras	Colombia	Chile	Germany
	Venezuela	Peru	Costa Rica	Japan
	Russia	China	Mexico	Korea
		Indonesia	Uruguay South Africa Turkey	United Kingdom United States

- **Efficiency-driven economies**. The economy is more competitive, the production processes are more efficient, and with good quality. The country is a platform for manufacturing; they may be tier 1 suppliers for Original Equipment Manufacturers (OEM). They are under constant pressure from other countries that may offer similar products at a lower cost.
- **Innovation-driven economies**. The economy moves up to higher value-added activities, which are more knowledge-intensive. Their activities of research + development + innovation links with the market to develop products and services that have value added. Thus, knowledge-based services are technologic and economic factors in the innovation-driven economies.

The Shih Smile Curve in Fig. 1 presents the value distribution along the global chain. The highest value is on the left and right extremes. Their focus is on intangible activities with high value-added and economic benefits, such as R and D, design, logistics, marketing, sale, and post-sale. On the other hand, the efficiency-driven companies are in the center; they concentrate on tangible activities such as manufacturing, where the economic benefit is lower.

One example is the maquiladora industry (in-bond) located mainly on the Mexican side along the US border. They face international challenges from other low-cost countries. The companies that survive go through an industrial upgrading. The first maquila generation consisted of simple manual assembly activities. The second generation had a higher technological level and automation. The third generation includes not only manufacturing but also design, research, and development of products. Finally, the fourth generation coordinates the logistics and manufacturing chain of factories not only in Mexico but also in other countries (Carrillo and Gomis 2003; Carrillo and Lara 2003).

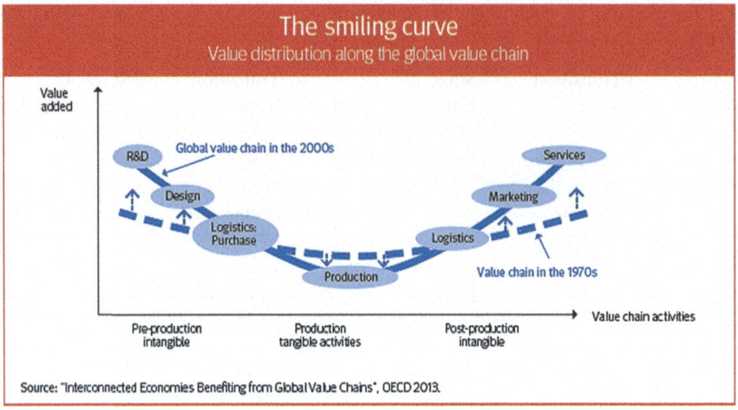

Fig. 1 Stan Shih Smile Curve. The value distribution along the global value chain (OECD 2013)

3.2 Data from Other Indexes, but Information not Displayed in Tables, Key Findings

Global Manufacturing Competitiveness Index (GMCI) (2016) The Deloitte Touche Ltd. publishes the GMCI. Out of a total score of 100, there are four Latin American countries among the top 40. Mexico appears in the 8th place, scores 69.5, in the 2020 forecast it will occupy the 7th place with 75.9 points. Mexico is the largest exporter in the region and the fifteenth globally; it is climbing up the economic ladder exporting technologically complex products. Brazil is ranked in the GMCI 2016 in the 29th place with 46.2 points; the prediction for 2020 is that it will ascend to 23rd place with 52.9 points. Colombia in 2016 is in 36th place with 35.7 points, for 2020 it will move up to 34th place with 40.9 points. Argentina in 2016 occupies the 39th place, with 22.9 points, and for 2020, it will remain in the same 39th place but with 24.6 points.

The Economic Complexity Index, **Harvard-MIT Media Lab** Cimoli et al. (2009) describe the technological specialization index (TSI) that helps to explain why the Asian economies have consistently increased their TSI since the 1980s. It analyzes the manufacturing value added; in this period, the Latin countries only achieve marginal increments. The exception is Mexico, which obtained a better performance than the other Latin American countries, as the bulk of its exports are high and medium complexity manufactures. The Atlas of Economic Complexity Index measures the quantity of knowledge that each country uses. It includes data from 128 countries. In the 2015 edition, the best performers of the analyzed countries in this article are Korea, 4th; Mexico is in 21st; China, 26th; Turkey, 42nd; Russia, 45th; Brazil, 47th; Costa Rica, 51st; Uruguay, 55th; Colombia, 60th; and South Africa is in 64th.

3.3 *Rankings for LATAM + BRICS + MIST: City and Natural Heritage*

UNESCO, World Heritage List

The United Nations Organization for Education, Science, and Culture (UNESCO) has compiled the World Heritage List (UNESCO 2017). The criteria are that the site must represent a masterpiece, exhibit, and be an outstanding example of a type of a building, or a natural setting of exceptional beauty or importance. UNESCO has to date 1,073 locations; the Latin America region contributes with 139 locations, which represent 13% of the total. The top countries are Italy, 51; China, 48; Spain, 44; France, 41; Germany, 41; India, 35; and Mexico with 34 sites. In Latin America, Mexico is the country with more world heritage positions, followed by Brazil 21, Cuba, 9; Colombia, 8; Chile, 6; and Costa Rica with 4. Some countries in the region (Brazil, Colombia, Costa Rica, Ecuador, Mexico, and Peru) are megadiverse for the richness of the animal, plant variety, and cultural diversity. The UNESCO Universal Declaration of 2001 proclaims the respect for biological diversity and cultural diversity as sources of innovation, creativity, and exchange.

UNESCO Creative Cities Network (UCCN) 2017

The UCCN network (2004) "aims to strengthen cooperation with and among cities that have recognized creativity as a strategic factor of sustainable development as regards economic, social, cultural and environmental aspects," (Unesco.org 2017). There are 116 city members, from 54 countries (see Table 5). It covers seven

Table 5 Creative cities network—seven creative fields (UNESCO 2017)

Creative field	Quantity	City	Country	Member Since:
Crafts and folk art	4	Nassau San Cristóbal Duran Jacmet	Bahamas México Ecuador Haiti	2014 2015 2015 2014
Design	3	Buenos Aires Curitiba Puebla	Argentina Brazil Mexico	2005 2014 2015
Film	1	Santos	Brazil	2015
Gastronomy	4	Belem Ensenada Popayan Florianopolis	Brazil Mexico Colombia Brazil	2015 2015 2014 2014
Literature	1	Montevideo	Uruguay	2015
Music	4	Bogotá Medellín Kingston Salvador	Colombia Colombia Jamaica Brazil	2012 2015 2015 2015
Media Arts	0	–	–	–

creative fields: crafts and folk art, design, film, gastronomy, literature, music, and the media arts. There are currently 17 Latin American cities in the UCCN. Brazil is the most active country with five cities, followed by Mexico and Colombia, both with three; Argentina, Bahamas, Ecuador, Haiti, Uruguay, and Jamaica all have one metropolis.

The list of the different creative fields is composed of Crafts and folk art: four; Gastronomy: four; Music: four; Design: three; Film: one; Literature: one; and none for Media Arts (UNESCO 2017). Nevertheless, one natural candidate is Guadalajara in Mexico, a digital city. The media city has its headquarters at *Mexico Innovación y Diseño* (MIND 2017). However, as this chapter is about to go to press, UNESCO announced on October 31, 2017, that 64 new cities join the UCCN NETWORK, and eleven are from Latin America. Gastronomy, four; music, two; crafts, two; design, two; and media, one city: Guadalajara. This will make a total of 28 Latin American cities on the list (UNESCO 2017).

UNESCO Intangible Cultural Heritage of Humanity

The term "cultural heritage" goes beyond monuments and collection of objects. "It also includes traditions or living expressions inherited from our ancestors and passed on to our descendants, such as oral traditions, performing arts, social practices, rituals, festive events, and practices concerning nature and the universe or the knowledge and skills to produce traditional crafts," (UNESCO 2017). The intangible cultural heritage has certain characteristics: traditional and living, inclusive, representative and community-based. This program started in 2008, to date there are 429 intangible heritage initiatives. The larger countries have the most registers: Mexico, and Colombia, both with nine; Brazil, eight; the Dominican Republic and Ecuador with three; Argentina, Guatemala, Cuba, and Chile they all have two; while Costa Rica only has one. The Latin American countries are protecting their heritage, such as traditional music and dance: Tango (Argentina), Capoeira (Brazil), Mariachi (Mexico), Marimba (Ecuador), Rumba (Cuba) or the Merengue dance (Dominican Republic). The intangible heritage is an important source of innovation.

3.4 Rankings for LATAM, BRICs, and MIST at the Person Level

Human Development Index (HDI) (2014), **United Nations** (see Table 6) The HDI is part of the United Nations Development Program (UNDP 2017); formed in 1990; it includes data of 188 countries. "It was created to emphasize that people and their capabilities should be the ultimate criteria for assessing the development of a country, not economic growth alone." The human development index consists of life expectancy at birth, expected number of the schooling years, and Gross National Income (GNI) per capita in US dollars. In the rank of "very high human development," there are three countries in the Table: Korea in 18th,

Table 6 Human development index 2014, LATAM, BRICS and MIST countries (UN 2014)

HDI Rank	Country	Human development index	Life expectancy at birth	Expected years of schooling (years)	Mean years of schooling	GNI Gross national income per capita	GNI per capita rank minus HDI rank 2015	2014
Very high human development								
18	Korea	0.901	82.1	16.6	12.9	34,541	12	18
38	Chile	0.847	82.0	16.3	9.9	21,665	16	38
45	Argentina	0.827	76.5	17.3	9.9	20,945	12	45
High human development								
54	Uruguay	0.795	77.4	15.5	8.6	19,148	8	54
66	Costa Rica	0.776	79.6	14.2	8.7	14,006	14	66
68	Cuba	0.775	79.6	13.9	11.8	7,455	48	69
71	Turkey	0.767	75.5	14.6	7.9	18,705	-7	72
77	Mexico	0.762	77.0	13.3	8.6	16,383	-9	77
79	Brazil	0.754	74.7	15.2	7.8	14,145	-1	79
87	Peru	0.740	74.8	13.4	9.0	11,295	6	89
90	China	0.738	76.0	13.5	7.6	13,345	-7	91
95	Colombia	0.727	74.2	13.6	7.6	12,762	-10	95
Medium human development								
113	Indonesia	0.689	69.1	12.9	7.9	10.053	-8	113

followed by Chile, 38 and Argentina, 45. In the following section "High human development," the best ranked is Uruguay, 54; Costa Rica, 66; Cuba, 68; Turkey, 71; Mexico, 77; Brazil, 79; and Peru, 87. In the rank of "medium human development," there is only one country, at a great distance from the leaders, Indonesia, 113.

Happy Planet Index (HPI) 2016, New Economics Foundation The New Economics Foundation has the slogan: "economics as if people and the planet mattered." The HPI evaluates 140 countries. The data comes from assessing how efficient are countries at using environmental resources, and if their citizens may lead long and happy lives. The four elements are well-being, life expectancy, inequality of outcomes, and ecological footprint. Most of the advanced and wealthy countries tend not to score well in the HPI. Instead, the leaders are countries from Latin America and the Asia Pacific, as they have a more balanced life and the ecological footprint that leaves behind is smaller. Latin American countries, tend not perform very well in the competitiveness, innovation, and manufacturing indexes; appear in mid-table, or in the bottom half. They start to improve in the creative cities, world heritage, and intangible heritage. However, in this ranking, they are the absolute leader. The first overall three out of 140 countries are from Latin America: Costa Rica, first; Mexico, second; and Colombia, third; followed at a distance by Uruguay in 14th place. Indonesia comes in 16th position. The next four countries are also from the Latin region: Argentina, 19; Peru, 21; Brazil, 23; and Chile in the 25th place. It is ironic that in this data the roles of China and Korea change, previously in all other Tables they had the top positions and here are the laggards; China occupies the 72nd place and Korea the 80th position.

4 Constraints that Limit Innovation in Latin America

A highly restrictive environment limits innovation in Latin American. It is a region formed by emerging economies, with complex political, economic, social, and technological issues. This situation limits the capacity of governments, companies, and individuals to improve products or services at the local scene. However, due to neoliberal policies, most Latin countries have their domestic markets open to foreign competitors and are part of Free Trade Agreements. This section will attempt to highlight some of the main problems and constraints.

The Corruption. Transparency International (2017), a Non-Governmental Organization, present in over 100 countries, defines: "Corruption is the abuse of entrusted power for private gain," (Transparency.org. 2017). Their research reveals there is a strong connection between corruption and inequality. The Latin America region scored very low in the 2016 Corruption Perceptions Index, as it only obtained 44 points out of a maximum score of a 100. It calls the attention that the larger economies are the ones that scored the lowest in a ranking of 176 countries: Uruguay (21), Chile (24), Costa Rica (41), Cuba (60), Brazil (79), Colombia (90),

Argentina (95), Peru (101), México (123), and Venezuela (166). The challenge for the governments is not only to implement transparency but also to attack the impunity of those who break the law. A current problem with corrupt companies is that they prefer to take shortcuts instead of investing and innovating for the long term.

The Poverty. The ECLAC (2016) reported that 12% of the Latin American population, or about 70 million, live in extreme poverty. In rural areas, the situation is worse as it affects one-third of the people. High levels of inequality may affect growth negatively by causing a lack of investment in human capital among low-income families.

The Most Relevant Worries for Latin American Citizens. One study revealed the perceptions of Latin American citizens (Latinbarometro 2016), their biggest concern, for 37%, were economic (general economy and employment). For 25%, relates to violence (drug trafficking and urban gangs). Other problems are corruption with 7%, politics 5%, education 4%, and health-related issues with 4%.

The Informality. The problem is so common that about 55% of workers in the region are in the informal sector. In Mexico, the second largest Latin economy, the situation is very similar. The National Institute of Statistic (INEGI 2017) published that about 29 million people, more than half of the labor force, are in the informal sector. Nevertheless, despite its low level of productivity, they contribute with 26% of the country's GDP. A drawback is that employees in the informal sector are vulnerable to any work-injury or illness and do not have social security.

The Preoccupant Low levels of Education. The United Nations (2011) reported that in the age bracket 15–29, there were 156 million young people, of which only 9% completed university education. Most countries in Latin America have passed laws that the expected years of schooling for their young people are twelve years; however, the reality is different. The country with the highest mean years of schooling is Cuba with 11.8, followed by Argentina and Chile both with 9.9; Peru, 9; Costa Rica, 8.7; and Mexico and Uruguay both with 8.6 years.

The weak rule of law and the lack of compliance. The International Property Rights Index (IPRI 2017) measures the protection of property rights in 127 countries, either physical or intellectual (patents, trademarks, and industrial designs); it also evaluates the legal and political environment. In the 2017 edition, Latin America as a region increased its IPRI score from 4.747 to 5.234, on a scale of 10. The rankings come in five quintiles. In the first quintile, there are the top 17 countries, but none of the countries in this study appears there. The second quintile has 22 countries, of the Latin: Chile (28) and Uruguay (36), and from Asia and Africa: South Africa (26), and Korea (34). In the third quintile, there are 25 countries: Costa Rica (45), Brazil (58), Colombia (62), China (52), and India (54). In the fourth quintile, there are 29 countries: Peru (65), Mexico (67), Indonesia (68), Turkey (78), and Argentina (97).

The Protectionism. There is a growing level of tariff protection in areas of export interest to the region, Europe, and United States, preventing exporters from taking full advantage of tariff-free access. There are nontariff barriers as well that may go from sanitation criteria to standardization requirements.

The Demographics and Shifts in the Demand in International Markets. There are two trends, on one extreme there is a declining fertility rate, and in the other, the growth of the aging population. There is a shift in the international demand that brings an end of the commodities boom during the first decade of the twenty-first century (McKinsey 2017).

5 Latin American Clusters and Types of Innovation

This section presents two perspectives of innovation, from an overall view of clusters, and to the company level. A cluster is a place where companies, universities, research centers, public institutions, and finance centers are in proximity to create new products or processes. Some of the Latin American most representative Clusters appear in Table 7 (Giulani, Petrobelli, and Rabellotti 2005). Traditional manufacturing centers, such as textiles, shoes, furniture, tiles or jewelry. Natural resource-based locations, such as wine, sugar, salmon, and milk-dairy; other products are mangoes, grapes, melons, and apples. Complex products may come

Table 7 Overall view of some Latin America's Clusters: traditional, manufacturing and others (Giulani, Petrobelli, and Rabellotti 2005), PROMEXICO Clusters (2017), Bortogaray (2000)

Traditional manufacturing clusters	Natural resource-based clusters, wine, sugar, copper, salmon, fruit, etc.	Complex products automobile assembly, aircraft, consumer electronics	Specialized suppliers software
Textiles	**Wine**	**Aircraft**	**Software videogames**
Medellín (Col) Itaji, Santa Catarina (Br)	Colchagua (Ch)	Sao Paolo (Br)	Joinville (Br)
Itaji, Santa Catarina (Br)	Serra Gaucha RGS (Br)	Querétaro (Mx)	México City (MX)
Central American Countries (CR, Gt, Hon)	Valle Guadalupe, BC (Mx)	Chihuahua (Mx) Baja California (Mx)	Guadalajara (Mx) Monterrey (Mx)
Shoes	**Sugar**	**Automotive**	Buenos Aires (Ar)
Sinos valley (Br)	Valle del Cauca (Co)	Nova Serrana (Br)	**Creative industries**
Campina Grande (Br)	**Marble**	Caixa do Soul (Br)	México City (Mx)
León (Mx)	Espíritu Santo (Br)	Ciudad Juárez (Mx)	Guadalajara (Mx)
Guadalajara (Mx)	**Salmon**	Ramos Arizpe (Mx)	Buenos Aires (Ar)

(continued)

Table 7 (continued)

Traditional manufacturing clusters	Natural resource-based clusters, wine, sugar, copper, salmon, fruit, etc.	Complex products automobile assembly, aircraft, consumer electronics	Specialized suppliers software
Furniture	Region Austral (Ch)	Guanajuato (Mx)	**Medical devices**
Serra Gaucha (Br)	**Milk-dairy**	Aguascalientes (Mx)	Baja California (Mx)
Uba, Minas Gerais (Br)	Boaco-Chontales (Nic)	Puebla (Mx)	Chihuahua (Mx)
Espíritu Santo (Br)	**Mangoes/Grapes**	**Pharmaceutical**	**Electronics**
Sao Bento do Soul (Br)	Petrolina-Juazeiro (Br)	México City (Mx)	San Jose (Costa Rica)
Segusino-Chipilo (Mx)	Quinta Región (Ch)	Mar del Plata (Ar)	Guadalajara (Mx)
Tiles	**Melons/Grapes**	**Household appliances**	**Biotechnology**
Santa Catarina (Br)	Rio Grande Norte (Br)	Nuevo León (Mx)	La Habana (Cu)
Jewelry	**Apples**	**Processed food**	Montevideo (Uy)
Guadalajara (Mx)	Santa Catarina (Br)	México City, Puebla, Oaxaca-Veracruz (Mx)	**Film making**
Taxco (Mx)	Chihuaha (Mx)	**Metal-mechanical**	México City (MX)
		Monterrey (Mx)	Buenos Aires (Ar)

from industrial areas, such as auto-parts and automobile assembly, aircraft, pharmaceutical, or household products. Specialized suppliers compose the last one; some examples are software, video games, medical devices, electronics or biotechnology.

Latin America, with its growing middle classes, is an important market for multinational companies that establish manufacturing facilities, but lately, its growing well-educated workforce and relatively low costs have made it attractive for research and development activities (Boutellier, Gassmann, and von Zedtwitz 2000). The policymakers are paying attention to clusters, for example, the Mexican Government through the National Statistics Institute and the National Institute of the Entrepreneur (INADEM 2017) created a cluster mapping of the country to attract foreign direct investment (FDI).

At the company level, there are different types of innovation in Latin America. The criteria to choose the cases was to show that in Latin America, in spite of hurdles, restrictions, and lack of human and material resources, there are success stories, most not well known. They are the fruit of talented individuals or companies who persevered against the odds. The traditional academic literature tends to have an American or European focus, thus, neglecting the rich, different and diverse experiences that come from emerging economies, such as Latin America. The examples follow the OECD classifications from the 2005 Oslo Manual. That is a new product, process, marketing method, and organizational methods. In addition, due to its increasing importance, the topic of social improvements is part of the analysis. The objective is to offer a better perspective of how improvements take place in the region.

5.1 New or Significantly Improved Product (Good or Services)

- Tridilosa (MX), Heberto Castillo created a tridimensional construction system that uses a combination of iron and concrete. The innovation consisted in the lightness of the structure, savings of about 66% in the material. These types of structures appear in roofs that cover large areas and even bridges. He obtained a patent for this invention in the seventies (Conacyt 2017).
- Algramo® (Ch) vending machine that sells items such as beans, lentils, rice or sugar in bulk at lower prices. The product targets consumers that belong to the bottom of the pyramid who buy small quantities of food but end up paying up to 40% more. Local stores offer consumers the option of the Algramo vending machines, profits are split with store owners (Allchile 2017).
- The contraceptive pill (Mx), Luis E. Miramontes coinvented the first birth control pill in 1956. The pill made possible the sexual revolution of the 1960s allowing couples to plan how many children they wanted (WEForum 2016).
- Color TV (MX), Guillermo Gonzalez-Camarena filed in 1942 for a patent in Mexico and in the United States; although color TVs became popular until the 1960s (WEForum 2016).
- The ballpoint pen (Ar), the Hungarian Lazlo Biro, designed the pen that used fast drying oil-based ink and a tungsten ball that allow the ink to flow into the paper. He fled to Argentina just before the World War II, where he obtained a patent and during several years he was the leading producer of pens (WEForum 2016).
- The automatic tortilla machine (Mx), the tortillas were handmade until the first machines appeared in the 1940s; they utilize rollers to shape the tortilla and wires to separate them. The tortilla is a basic ingredient of Mexican food, either flat, folded or roll up as in tacos (Prodiamex 2017).

5.2 New Process

• The Colombian Association of Flower Exporters (Asocolflores®) represent Colombian-owned farms that grow flowers. It is the country's second-leading agriculture export. The logistics include cutting flowers, dethroning, refrigerating, packaging, and sending them by plane to 89 international destinations (PMA 2015).
• (Mx) Juan Celada is the inventor of the direct reduction system for the iron, or sponge iron, patented by Mexican corporation Hylsa, Monterrey, in 1957. The company used imported scrap iron from the USA; however, in the 1950s, due to the Korean War, this was no longer possible. Therefore, Hylsa had to change its technology process, the resulting Hyl-Process, improved quality and reduced costs. It is Mexico's number one tech contribution to the world steel industry (Siyanbola, et al. 2012).
• (Mx) The International Agricultural Research Center (CIMMYT 2017) it grew out of a pilot program sponsored during the 1940s and 1950s period, with the objective of raising farm productivity in Mexico. It is an important research institution, which contributes to the Green Revolution and to an increase in the production of maize and wheat (Innovalatino 2011).
• (Br) The agricultural research institute, EMBRAPA®, played an important role in the green revolution and it contributed to an increase in productivity. Brazil is the largest exporter of coffee, sugar, orange juice, and tobacco. And it is the second biggest producer of soya (Innovalatino 2011).

5.3 New Marketing Method

• (Br) Natura Cosméticos® it is "a pioneering manufacturer of eco-friendly personal cosmetics, fragrance and personal hygiene products." The ingredients for the sustainable products of the company come from local communities. It has a direct sales network with over one million sales representatives or consultants (Natura 2017).
• (Col) Colombia's Federation of Coffee Producers, instead of competing in international markets on its own, they teamed together and agreed on having one premium brand, Juan Valdez®, for its coffee products. The brand represents 500,000 coffee producers' families; they have successfully applied marketing, and diversified into retailing with its international chain of coffee shops (Juan Valdez 2017).
• (Mx) the retail unit Elektra®, and Banco Azteca® (store, and bank), target customers from the bottom of the pyramid (BOP), most of them work in the informal economy and are not eligible to commercial credit. They may buy household products in small installments. Elektra offers a foreign remittance service with

money transfers from the USA, paid in their facilities, which are open seven days a week. They have stores in Mexico, Central, and South America (Elektra 2017).
• (Mx) OXXO® Convenience stores, they are small and open 24/7. The customer may buy dairy products, snacks, drinks and pay for city services, telephone or credit cards. In 2016, Oxxo had 15,525 stores in Mexico and Latin America, making it the leading retailer in the region by number of units (Femsa 2017).

5.3.1 New Organizational Methods in Business Practice

• (Br) Doctor Solution®, offers the formal service of plumbers, electricians and construction workers, with quality, warranty, and even an invoice. These types of employees tend to work in the informal sector and are usually unreliable. Brazil is the fourth largest market in the world in number of franchises and it is now exporting business models that respond to the specific needs of Latin American countries (Doctor Solution 2017).
• (Ar) Guerra Creativa®, provides design services by crowdsourcing. If a client wants a new Web page, they will host a design contest for a few days, better design, quick responses and reduced costs. A participative digital community offers a variety of options and the client selects the one that better fits his purpose (Guerra Creativa 2017).
• (Ar) Mercado Libre®, it was the Latin answer to eBay®. It is the largest and most visited e-commerce Web site in the region, serving 15 countries and over 30 million customers that buy all sorts of goods, sell and pay using this platform (Mercado Libre 2017).
• (Mx) Insitum®, is a leading innovation-consulting firm in Latin America that help companies to solve complex problems. It uses user-centered methodologies such as Design Thinking, User Experience Design (UX) and lean innovation. It specializes in strategic design, service design, organizational design, and product design (Insitum 2017; Fastcompany 2017).

5.4 New Workplace Organization or External Organization

- (Ar) Los Grobos®, it is one of the largest grain producers and agricultural services providers in the world, even though it does not own any land, or tractors or harvesters. It is a global company, which provides food and energy production, risk management, human and social capital (Los Grobo 2017).

- (Co) Colombian Parquesoft® is a Technology Incubator in Medellin's Innovation District. A cluster that works as an ecosystem; it has more than 400 companies of digital art, science, and ITC. There are 12 technology centers, in Cali, Medellin, and other cities, providing ITC service to 42 countries (Parquesoft 2017).
- (Ar) Centro Metropolitano de Diseño® (The Metropolitan Design Center), is a public institution part of the General Direction of Creative Industries in Buenos Aires. "This is the place where design and innovation come together; it promotes meetings of design professionals, companies, institutions and a system to generate value". Its vision is to become the main public promoter of the social, economic, and cultural importance of design (Buenos Aires gob 2017).

5.5 New Social Innovation—Social Benefits

- (Br) The Curitiba Bus Rapid Transit (BRT) System. This is a low-cost mass transit system that is an attractive alternative to the most expensive subway option. It has a dedicated lane in the avenues, use bus stations for boarding, and a smart card for payments. There are now more than 50 BRT systems in North and South America, Europe, and Asia (Hensher and Golob 2008).
- (Br) The Center for Digital Inclusion® (CDI). Its key concept is helping low-income communities to help themselves. "CDI community centers are: self-managed, self-sustainable, and implement the CDI pedagogy." There are 842 schools in Brazil and 15 countries and about 1,640,000 certified people (Innovalatino 2011).
- (Ch) Un Techo para mi País® (A roof for my country), founded in Chile by a group of young people that build houses together with the families living in slums. There are about 104 million people living in these conditions in Latin America. TECHO has expanded to 15 countries in the region. The program has involved 723,178 volunteers that work with the family to build in a couple of days a simple, but decent house, by promoting social development, fostering social awareness and local action (Techo 2017).
- (Co) The urban transformation of Medellin. In 1992, it was one of the most dangerous cities in the world. Today, this city of 2 1/2 million people is a laboratory of progressive architectural and urban interventions, for social integration. The city has undergone a spatial, social, economic and cultural transformation. In 2012, it was recognized as one of the world's most innovative cities (WEF 2016).
- (Ch) Recycla Chile®, it is the first and only electronic recycling company in the country. It disarms appliances by extracting and separating the raw materials. The employees are people with criminal records. In 2008, it obtained the Energy Globe Award (Recycla 2017; Innovalatino 2011).

6 Conclusions and Final Remarks

This essay attempts to present an overall view of Latin America's innovation, a region of promises, with a young population that aspires to obtain a better quality of life. The task is overwhelming but fascinating, as information is scattered. Innovation may arise even in highly restrictive environments, in a region plagued by economic, social, and technological problems. This is in spite of high informality, bureaucracy, corruption, low-skilled work, weak productivity, and out-of-date technologies that characterize large parts of the regional economy. In this group of countries, small and medium-sized enterprises (SMEs) represent nearly 99% of the total number of companies, the majority fighting for their own survival, therefore, improvement activities are not their chief concern. However, there is a small group of SMEs, albeit growing that are the unsung heroes, nimble, creative, smart, and close to the market. Their target may be the bottom of the pyramid or the growing middle local classes. They may also target niches in international markets with demanding consumers seeking products of high-quality and design, attractive price, reliable service, and delivery.

There is a need for regional public policies that may create a shift from natural resources to a knowledge-society. Nevertheless, this will not happen overnight, as it is necessary to improve human capital through investments in quality education, R and D, ICT, innovation systems, better infrastructure and enforcing the rule of law and protection of Intellectual Property (de Ferranti et al. 2002). The aim of most Latin American countries is to move out from the so-called middle-income trap, based on efficiency to an innovation-driven economy. And to move away from commodity-type markets that are price-sensitive to market-niches that demand design, novelty, technology, service and a better experience. Nevertheless, the race is on, as several other countries, from Asia or East Europe, with a comparable level of development, have similar aspirations.

Of the regional studies of innovation, the best known, more complete and probably the best, is the European Innovation Scoreboard. In Latin America, the OECD Oslo Manual, already described in a previous section, was the reference to come up with and adapted version that better reflects the reality of the region. The Network for Science and Technology Indicators (RICYT in Spanish) published the Bogotá Manual in 2001, as a guide to the national innovation surveys in the region.

Some key recommendations from Casanova et al. (2016: 82) are: "… A national vision and a drive for partnerships, innovation related to natural resources, human capital as a catalyst of innovation, supporting SMEs through cluster policies, innovation information ecosystems beyond traditional measures, and support for innovation leading to social inclusion and sustainability […] Latin America is innovating, albeit in its unique way. It is redefining innovation, giving new paradigms and facets to innovation and, thus, charting its own growth story, tailor-made to tackle its weaknesses and build on its core strengths. Thus, the economies are becoming better equipped to seize their opportunities, even with the backdrop of the scarcity of resources and low R&D expenditure."

Cassoni and Ramada-Sarasola (2012) argued that standard innovation output indicators are unable to account for new procedures and the great variety of improvements and novelties in the Latin American market, as there are imperfections and high degrees of uncertainty. Lourdes Casanova (2011) has a similar point of view: "We [tend to] measure innovation in very traditional ways. We measure regarding investments in R&D, patents registered, the number of Ph.D. holders, and in the quantity and quality of published work." The InnovaLatino report demonstrates that the use of those indicators is not adequate to create an accurate picture of the actual potential of Latin America. And in spite of the fact of occupying low positions, there is a high level of improvements that do not appear in the rankings. "Capturing innovation [in Latin America] does not come from traditional sources," (Innovation Excellence 2011).

There is a growing use of Information Communication Technologies in Latin America, among the younger generations. There is a link between higher education, socioeconomic level, age, and the more intense use of social networks (Latin Barometer 2016). Facebook is the most popular option with 50%, followed by WhatsApp with 49%, and Youtube with 28%. These are potential tools not only for irrelevant social issues but also for work and innovation-related activities; according to ECLAC, the Latin region is already the top user of social networks.

The job situation, especially among Latin American young people is precarious, as result of several factors, such as the continuous downsizing of companies that try to reduce their fixed costs by outsourcing activities. However, there are opportunities as well with the surge of ICT technologies, social networks, Smartphones, and the restless attitude of millennials with new ideas, such as the movement called "the makers" that use 3D printing in small-scale laboratories.

The combinations of these factors contribute to the creation of a start-up culture. Johnson, Christensen, and Kagermann (2008) suggest by starting to detect an unfulfilled market opportunity to satisfy a customer who needs to have a job done, which is to solve a problem that needs a solution, through a better customer value proposition. One example is the Startup Mexico initiative opened a few years ago to coach entrepreneurs through the first period of incubation of their business. Campus Party attracts university students for a weekend to present new ideas, take workshops, attend conferences, and create networks, etc. Another example is Innovation Match MX that aims to attract and "be an innovative meeting point where companies, researchers, and students will be able to share knowledge, experiences, and explore opportunities of technological-based products or services. The forum aims to reunite Mexican talent; regardless of in which country they live, with Mexican companies and institutions to arrange technological projects that boost the innovation in the country (Innovationmatch 2017).

In a knowledge-society, advanced learning plays a central role. The OECD (2017) points out that "the potential for higher education remains unrealized in Latin America." Although enrolment of university students has improved, graduation rates are still low; these institutions face quality and adequacy problems. On average, only 14% of the population aged 15–64 completes the tertiary education (2017: 152). There is an insufficient number of Latin American students interested in science, only 6%; on the other extreme, the most popular programs are social sciences, law, and business with an average of 39%. In the field of engineering, manufacturing, and construction, the Latin America data of 17% is similar to the OECD countries with 15%. Mexico with 115,000 graduates per year is the country in the region with the largest number of alumnae in engineering and technology. The same OECD in his book *Skills for Innovation and Research* (2011) defined two sets of skills: hard skills are necessary for science, mathematics, ability to learn a new language. On the other hand, soft skills facilitate multicultural collaboration between people and companies from different countries, ability to solve problems, to communicate, and to lead, and complete a project.

International students and migrant networks are reshaping the world; they have increased mobility and intense flow of information thanks to ICT, which helps to disseminate new concepts. Many of the emerging world's brightest minds attend Western universities. As opportunities arise in their native countries, students go back home with the knowledge acquired and the contacts made. "Diasporas create connections that help people with good ideas to collaborate with each other…" (Economist 2011: 13 and 72). There is no lack of talent in Latin America. The Small Business Administration (SBA) publishes that Hispanic-Americans are the fastest growing population group in the USA, with nearly 55 million people; they have opened 3.3 million businesses. Foreign-born American starts businesses at a ratio of three to one when compared with non-foreign born. The Hispanic entrepreneurs have acquired a new name "Latinovators" (Small Business Administration 2016).

The World Economic Forum 2017 on Latin America revealed that there is an acute skills shortage. Companies in the region complained that about half (50%) of formal firms could not find the candidates fulfilling the skills they need; especially in the most advanced sectors of the economy, such as automotive or aerospace. (WEForum 2017). The ironic situation is that there are groups of young people who are neither studying nor working; thus, the region is wasting its more precious resource.

In a speech delivered by the former president of Ecuador, Alfredo Palacio (AFESE.com, May 22, 2006), he presents the economic dependency of a country when its future is subject to the exploitation of its natural resources and has no other option if only sells crude oil at market prices: "…This is how the Ecuador's Budget is composed, it comes fundamentally from oil, with no value added, there are no neurons, no work, the oil that we obtain from over there; we send it abroad regardless of what they pay, we send it, because we live out of that …"

Another example is the speech by the former Brazilian Ministry of Industry, Commerce, and Trade, Mr. Fernando Pimentel (August 30, 2011), who illustrates the gravity of not doing anything to change the actual situation of a Latin America, selling raw materials and importing finished goods: "If the government let the markets act alone, South America will become the farm and the mine of the world, while Asia, the (manufacturing) plant of the world (Canacero.org.mx 2011)."

This work attempts to present the myriad examples and initiatives of how innovation is happening in Latin America, it is ubiquitous, adapting to changing circumstances; however, the actual official and private indicators are not considering. New products, services, and processes are emerging in the region, despite the lack, or insufficiency, of financial, human, or technological resources. The essential elements are already there. As paradigms are changing, new opportunities arise with the 3D printing technology, or with the on-demand economy and sharing economy. A different type of consumption model that is inclusive and with a better distribution of wealth is urgently necessary. The stakeholders are in place: private firms, universities, government research institutions, ministries, and the citizens; however, it is essential to work together, cocreate and add not rest.

However, new ideas may flourish in a culture that is tolerant of failure, promote entrepreneurship at an early age, and disseminate the achievements, but also the lessons learned of those that initiate new things and fail. Innovation also requires resources, therefore, is advisable to create a support networking with venture capital funding, angel investors and risk capital. Niccolo Machiavelli wrote in *The Prince*, "It ought to be remembered that there is nothing more difficult to take in hand, more perilous to conduct, or more uncertain in its success, than to take the lead in the introduction of a new order of things."

The Latin American countries, with their rich cultural history, have an international opportunity in the Creative Economy. As is the case of design, as mentioned earlier, the first schools in Latin America opened during the 1960s during the import-substitution period. It is a discipline that applies to different sectors, such as transportation, education, manufacturing, health and domestic appliances. More than half a century later, the design has started to mature, although still should achieve recognition among company owners who may use it as a strategic tool; on the other side, consumers deserve well-designed products regardless of the price they pay.

There are hundreds of schools and thousands of graduates working in the region, the USA, Europe or Asia. Latin American designers have received many international design prizes and awards. The store of the Museum of Modern Art (MoMA) in New York City celebrated a Destination series (2005–2012 period), to discover design concepts from around the world. It showed the work of Argentinean, Brazilian, and Mexican designers. In Mexico, during the 1970–1976 presidential periods, there was a national design prize geared to the international markets. A new design prize started in 2016. In Brazil, they have a design award since 1988, and several other countries followed the example. Three Latin America's cities are in the UNESCO Design category: Buenos Aires, Curitiba, and Puebla. As this chapter is about to go to press, the UNESCO announced in 2017 the designation of two new design cities: Brasilia and Mexico City. Also, Mexico City was designated World Design Capital© for 2018.

There is a touristic, social, and economic opportunity in the creation of collective marks to identify the world heritage sites, the immaterial heritage, or the title of creative cities that some countries have obtained from UNESCO. A pending subject is the creation of a culture of registration of intellectual property of patents, utility models, trademarks, and industrial designs. Katelhöhn and Ogliastri (2013) concluded that innovation in Latin America is geographically more concentrated than economic activity; the evidence is that even though Latin America contributed in 2012 with 8.7% of the world's Gross Domestic Product, it only generated 0.19% of the patent production registered in the USPTO.

It is a daunting job, but if Latin America wants to control its destiny, it must develop ideas and turn them into new business models, and not wait to receive it from elsewhere. Innovation is probably the best option for the region to improve the standards of living and level of education, reduce the income disparity, and compete in the world markets with environmentally friendly products. The World Economic Forum 2017 declared the three biggest challenges for the region: corruption, education and skills development, and reducing inequality. The Latin American dream envisions products and services that proudly will bear not just the "Made in Latin America" mark but rather the "Innovated" or "Created in Latin America" brand.

Acknowledgements The work and technical support in the making of the Tables of Ivonne Arellano, research assistant, is appreciated. Thanks to the editors and referees for their comments and suggestions to improve this article, nevertheless, the author is the only responsible for what is written.

References

Bartlett, C. A., & Ghoshal, S. (2000). Going Global: Lessons from late movers. *Harvard Business Review, 78*(2), 132–142.

Bonsiepe, G. (1985). *El diseño de la periferia.* Gustavo Gili, México: Debates y experiencias.

Boutellier, R., Gassman, O., & von Zedwitz, M. (2000). *Managing Global Innovation.* Berlin: Springer.

Bortogaray, I. (2000). Innovation clusters in Latin America. 4th International Conference on Technology Policy and Innovation, Curitiva, Brazil. 28–31 August 2000.

Carrillo, J., & Gomis, R. (2003). Los retos de las maquiladoras ante la pérdida de competitividad. *Comercio Exterior, 53*(4), 318–327.

Carrillo, J., & Lara, A. (2003). Maquiladoras de cuarta generación y coordinación centralizada. *Revista Ciencias Administrativas, Fortaleza, 9*(2), 161–171.

Casanova, L., Golstein, A., Almeida, A., Fraser, M., Molina, R., Hoeber, H., & Arruda, C. (2009) Multilatinas to Global Latinas: The new Latin American Multinationals. Banco Interamericano de Desarrollo (BID) Washington.

Casanova, L., Dayton-Johnson, J., Olaya-Fonstad, N., & Jain, S. (2016) Innovation in emerging markets: The case of Latin America. In: J. Haar et al. (Ed) *Innovation in emerging markets, Palgrave Houndmills*, Basingstoke:Macmillan.

Cassoni, A., & Ramada-Sarasola, M. (2012). The return to Innovation in Latin America: Inexistent or Mismeasured? *Latin American Business Review, 13*(2), 141–169.

Cimoli, M., Ferraz, J. C., & Primi, A. (2009). Science. *Technology and Innovation Policies in Global Open Economies: Reflections from Latin America and the Caribbean, Globalization, Competitiveness & Governability Journal, 3*(1), 1–29.

CEPAL. (2016). *Panorama Social de América Latina.* Naciones Unidas: Santiago.

Conacyt. (2014). *Programa Especial de Ciencia, Tecnología e Innovación,* México, Consejo Nacional de Ciencia y Tecnología.

Deloitte Touche Tohmatsu Ltd. (2016). *Global Manufacturing Competitiveness Index*

Dussel-Peters, E. (2000). *Polarizing Mexico.* The impact of liberalization strategy: Boulder-Colorado, Lynne Rienner Publishers.

Dussel-Peters, E. (Ed.). (2015). *Economía, Comercio e Inversión 2015*. México, Red ALC-China: América Latina y el Caribe y China.
Economist. (2011). The magic of Diasporas. How migrant business networks are reshaping the world. *The Economist, 13*, 72–74.
Fernández, S., & Bonsiepe, G. (2008). *Historia del diseño en América Latina y el Caribe*. Sao Paulo, Blucher: Industrialización y comunicación visual para la autonomía.
de Ferranti, D., Perry, G. E., Lederman, D., & Maloney, W. E. (2002). *From natural resources to the Knowledge economy*. Washington: World Bank.
Foro Consultivo Científico y Tecnológico. (2017). *Ecosistema de Innovación Social en México,* México, Foro Consultivo Científico y Tecnológico.
Giuliani, E., Pietrobelli, C., & Rabellotti, R. (2005). Upgrading in global value chains: Lessons from Latin American Clusters. *World Development, 33*(4), 549–573.
Hensher, D. E., Golob, T. F. (2008). *Bus Rapid Transit Systems. A Comparative Assessment, Transportation*, Vol. 35, July 4, 2008 (pp. 501–518).
Jonas, S., & McCaughan, E. (1994). *Reconstructing a social justice agenda*. Boulder-Colorado: Westview Press.
Johnson M.W., Christensen, C. M., Kagermann (2008) Reinventing your business model. *Harvard Business Review, 86*, 51–59.
Katelhöhn, N., & Ogliastri, E. (2013). Introduction. *Innovation in Latin America, Academia Revista Latinoamericana de Administración, 26*(1), 12–32.
Martin Prosperity Institute. (2011). *Creativity and Prosperity: The Global Creativity Index.*
Marini, R. M. (1994). The path to Latin American integration. In S. Jonas & E. McCaughan (Eds.), *Reconstructing a Social Justice Agenda* (pp. 37–50). Boulder Co: Westview.
Martinez, A., de Souza, I., & Liu, F. (2003). Multinationals vs. Multilatinas: Latin America's great race. *Strategy + Business,* Fall, Issue 32.
Maloney, W. F., Manzano, O., & Warner, A. (2002) Missed Opportunities and Resource-Based Growth in Latin America, *Economía, 3*(1), 111–167.
McKinsey Global Institute. (2014). *A Tale of Two Mexicos: Growth and Prosperity in a Two-Speed Economy*: McKinsey Global Institute.
Martinez, A., de Souza, I., & Liu, F. (2003): Multinationals vs. Multilatinas: Latin America's great race, *Strategy and Business*, Fall, Issue 32. https://www.strategy-business.com/article/03307?gko=85c85
OECD. (2012). Better Skills, Better Jobs, Better Lives. A Strategic Approach to Skills Policies, OECD Paris.
OECD. (2005). Oslo Manual. Guidelines for Collecting and Interpreting Innovation Data, OECD Paris.
OECD. (2016). Regulatory Policy in Latin America and the Caribbean, OECD Paris.
OECD. (2017). OECD Skills Strategy. Diagnóstico de la OCDE sobre la Estrategia de Competencias, Destrezas y Habilidades de México, OECD Paris.
OECD. (2017). Latin American Economic Outlook. Youth, Skills and Entrepreneurship, OECD Paris.
Ricyt/OEA/Cyted. (2001). Manual de Bogotá. Normalización de indicadores de innovación tecnológica en América Latina y el Caribe. Bogotá, Red Iberoamericana de Indicadores de Ciencia y Tecnología (RICYT).
Reid, M. (2010). So near and yet so far. A special report on Latin America. *The Economist, 11*, 1–18.
Rodríguez-Martínez J. (2013). Estrategias de internacionalización de las PyMEs basadas en la información e innovación. El panorama internacional y el caso de México, UAM-A, México.
Rodríguez-Martínez, J. (2009) Las actividades de innovación según el Manual de Oslo (2005) de la OCDE: La innovación de productos. Anuario Administración y Tecnología para el Diseño 2009, UAM-A. México (pp. 145–166).
Siyanbola, W., Egbetokun, A., Abiola-Adebowale, B., & Olamade, O. (2012). *Innovation systems and capabilities in developing regions. Concepts issues and cases.* Farnham, UK: Gower Publishing LTD.

United Nations/United Nations Development Programme/UNESCO. (2013).*Creative Economy Report. 2013 Special Edition, Widening local development pathways,* Paris, UNESCO.
World Economic Forum. (2017). In K. Schwab, (Ed.), *The Global Competitiveness Report* 2016–2017, World Economic Forum, Geneva.

Online References and Publications

AllChile. (2017). http://www.allchile.net/viewtopic.php?t=13094. Accessed August 2017.
At Kearney Consultancy Firm. Retrieved August, 2017, from https://www.atkearney.com.
AFESE Revista, 2006. https://afese.com/img/revistas/revista44/discursopalacio.pdf
Canacero.org. (2011). http://www.canacero.org.mx/Es/assets/19-estudio-ilafa-germano-mendes.pdf
Centro Metropolitano de Diseño, Buenos Aires. http://www.buenosaires.gob.ar/cmd/institucional/que-es.Accessed August 2017.
CIMMYT, Mexico, International Agricultural Research Center. http://www.cimmyt.org/organization/. Accessed August 2017.
Conacyt Prensa. (2017). Retrieved August, 2017 from http://www.conacytprensa.mx.
Cox Review Report. (2005). Retrieved August, 2017, from http://webarchive.nationalarchives.gov.uk/20120704143146/, http://www.hm-treasury.gov.uk/d/Cox_review-foreword-definition-terms-exec-summary.pdf.
Doctor Solucion franchises. Retrieved August, 2017, from https://doctorsolucion.com.mx/.
Economic Commission for Latin America and the Caribbean (ECLAC), CEPAL. Retrieved August, 2017, from https://www.cepal.org/en/articles/2016-latin-america-worlds-most-unequal-region-heres-how-fix-it.
Elektra Stores. Retrieved August, 2017, from https://www.elektra.com.mx/.
Embrapa, Brazil, International Agricultural Research Center. Retrieved August, 2017, from https://www.embrapa.br/en/international.
Fast Company. Retrieved August, 2017, from www.fastcompany.com.
Foro Consultivo México. Retrieved August, 2017, from http://www.foroconsultivo.org.mx/. http://www.foroconsultivo.org.mx/FCCyT/boletines-de-prensa/presentan-casos-de-%C3%A9xito-de-innovaci%C3%B3n-social-en-m%C3%A9xico.
Global Manufacturing Competitiveness Index (GMCI), 2016. Retrieved August, 2017, from https://www2.deloitte.com/content/dam/Deloitte/global/Documents/Manufacturing/gx-global-mfg-competitiveness-index-2016.pdf.
Goldman Sachs Consultancy Firm. Global Economic Paper No. 99, Path to 2050. Retrieved August, 2017, from http://www.goldmansachs.com/our-thinking/archive/BRICs-and-Beyond.html.
Guerra Creativa, Design Studio. Retrieved August, 2017, from https://www.guerra-creativa.com/es/.
Harvard Business Review, 2016. The on-demand economy is growing. Retrieved August, 2017, from https://hbr.org/2016/04/.
Human Development Index (HDI) 2014, United Nations Development Programme. http://hdr.undp.org/sites/default/files/2016_human_development_report.pdf.Accessed August 2017.
Instituto Nacional del Emprendedor (INADEM). (2017). Cluster map of Mexico. Retrieved August, 2017, from http://www.icluster.inadem.gob.mx/index.php?idioma=ing.
Innovalatino. (2011). Fundación Telefónica, España. Retrieved August, 2017, from http://www.fundaciontelefonica.com.mx/publicaciones-listado/pagina-item-publicaciones/itempubli/85/.
Innovation Match. Retrieved August, 2017, from https://innovationmatchmx.com/innovative-mexicans/.
Insitum, Consulting Firm. Retrieved August, 2017, from http://www.insitum.com.
International Property Rights Index, (IPRI 2017). Retrieved August, 2017, from https://www.internationalpropertyrightsindex.org/countries.

Investopedia. Retrieved August, 2017, from http://www.investopedia.com.
Juan Valdez, Coffee Stores. Retrieved August, 2017, from https://www.juanvaldezcafestore.com/es/.
Latinbarometro Corporation, Main worries for Latin America, 2016 study. Retrieved August, 2017, from http://www.latinobarometro.org/lat.jsp.
Los Grobo. Retrieved August, 2017, from http://www.losgrobo.com/en/a-global-company/argentina/products-and-services.html.
Mercado Libre, e-commerce. Retrieved August, 2017, from https://www.mercadolibre.com/.
Nobel Prize Organization. Retrieved August, 2017, from https://www.nobelprize.org/nobel_prizes/lists/countries.html.
Natura Cosmetics. Retrieved August, 2017, from http://www.natura.com.
OECD, Innovation Strategy. Retrieved August, 2017, from https://www.oecd.org/site/innovationstrategy/defininginnovation.htm.
FEMSA, OXXO Convenience stores. Retrieved August, 2017, from http://www.annualreport.femsa.com/femsa-comercio.html.
Parquesoft Colombia. Retrieved August, 2017, from http://www.parquesoft.com/nosotros.
Petapixel. Retrieved August, 2017, from https://petapixel.com/2015/04/09/this-is-what-the-history-of-camera-sales-looks-like-with-smartphones-included/.
PMA, Colombia. Retrieved August, 2017, from https://www.pma.com/content/articles/2015/04/colombian-floral-industry-overview.
Prodiamex, Automatic Tortilla Machine. Retrieved August, 2017, from http://prodiamex.com/productos/historiadelatortilla/historiadelatortillaUK.htm.
Strategyzer. Retrieved August, 2017, from https://strategyzer.com/.
Techo para mi País (A roof for my country). Retrieved August, 2017, from http://www.techo.org/.
Transparency International. (2017). https://www.transparency.org.
UNESCO. Accessed August 2017. Intangible Cultural Heritage and World Heritage Convention, http://whc.unesco.org/en/list/ Creative Cities Network, https://en.unesco.org/creative-cities/
United Nations. (2011). Retrieved August, 2017, from http://www.un.org/esa/socdev/documents/youth/fact-sheets/youth-regional-eclac.pdf.
World Economic Forum. (2016). Retrieved August, 2017, from www.weforum.org/agenda/2016.
World Intellectual Property Organization. WIPO 2014 Indicators. Retrieved August, 2017, from http://www.wipo.int/edocs/pubdocs/en/wipo_pub_941_2014.pdf.

Use of Lean-Sigma as a Problem-Solving Method in a Restrictive Environment

Noé Alba-Baena, Francisco Estrada-Orantes and Christian Valenzuela-Reyes

Abstract This chapter presents the use of Lean-Sigma methodology and tools in a Latin American restrictive environment where time, economic, and technological limitations are common. These conditions limits the use of solutions such as reengineering and major technological investments. The mentioned restrictive characteristics are the perfect scenario for the use of fast and efficient solutions for customer-focused challenges. Lean-Sigma is a fast responder methodology focused on problem-solving using the effective approach of Lean Manufacturing combined with the statistical analysis of Six-Sigma. The use of Lean-Sigma methodology has been previously reported, and its value proved as an efficient solving methodology in highly restrictive environments. However, this chapter describes the inclusion of Design for Six Sigma (DFSS). After the initial considerations, an example related to a production problem in an exterior-home products, is described from a manufacturing facility at Juarez, Mexico. The project is focused on a brass component for outdoor lighting products. The customer complaints shows that this component comes from the supplier with differences in color and tones, through the combination of Lean-Sigma and DFSS a solution is given. The solution includes the design of a qualitative measuring device, the modification of a cleaning process, and the standardization of the chemical coloring process (Patina). Once validated, the designed device was also used for determining the acceptance levels at the incoming inspection station and was incorporated into the quality control station. Finally, the results and knowledge were successfully transferred to the supplier. The results shows that the combination of Lean-Sigma with the DFSS methodologies and their tools can satisfactorily be used in a restrictive environment.

N. Alba-Baena (✉) · F. Estrada-Orantes · C. Valenzuela-Reyes
Department of Industrial and Manufacturing Engineering, Autonomous University of Ciudad Juarez, Av. Plutarco Elías Calles #1210, Juárez, Chihuahua 32310, México
e-mail: nalba@uacj.mx

F. Estrada-Orantes
e-mail: frestrad@uacj.mx

C. Valenzuela-Reyes
e-mail: christian.valenzuela.r87@gmail.com

G. Cortes-Robles et al. (eds.), *Managing Innovation in Highly Restrictive Environments*, Management and Industrial Engineering,
https://doi.org/10.1007/978-3-319-93716-8_2

1 Introduction

The factories established in Mexico to assemble final products and supplied them to the USA market (since the 1960s) are characterized by their diversity of origin: from the USA, Japan, and European ownerships among others. These facilities are dedicated to producing electronic equipment, clothing, furniture appliances, auto-parts, and more products. During the last six decades, this manufacturing business in Mexico coined and developed the term "maquiladora" production system (derives from the Spanish word maquilar, "to process"). A maquiladora is an industrial plant that assembles imported components into products for export and owned by foreign or domestic entities (Wilson 2010). By taking advantage of the well-qualified and cheap labor most of these business uses as a main competitive strategy, generating large profits, measuring the efficiency of the management mainly by cost reduction. The competitiveness and advantages of Mexico as a destiny for establishing manufacturing business reached its maximum in 2001 when 3700 factories employed more than 1,300,000 employees. According to Contreras et al. (2006), after this peak, the pressure for finding more efficient solutions and stronger strategies increases because the global competition is growing mainly from China and Central American countries. These conditions created, in the first place, the need for using new strategies, including the use and the experimentation of different methodologies. And later, to use combinations of tools and best practices adapted to the reality of the new scenario. Managers in this environment now have to compete not only in regarding the classical production triad: time, quality, and costs measurements, but also to compete for giving better service and to reacting faster than their competitors to the needs of the customers.

In this environment, the managerial efficiency is highlighted and measured by the capacity to respond to the challenges that the system is offering. Making the "right" decisions is the main responsibility and for this, the manager has to quickly acquire and compile a set of tools and techniques that have helped him in the past or by taking the risk to experience a new set of tools. Some tools are simple and straightforward, others are complex and sophisticated, but the selection is based on the well-proven ones as first options. For instance, we can decide to use a simple control chart or a value stream map (VSM) to decide at the time in a critical meeting. Preparation and practice are the key factors for gaining experience in this decision-making process, especially for a restrictive environment like the one occurring in Latin America (LA). It is necessary to gain such abilities and respond to the daily challenges by improving the use of advanced tools, increase the experience, and (with the time) become the experts in obtaining the maximum benefit with the most useful and fast solution. If we consider that in the LA environment "one solution is valid only for one set of conditions and, every day a manager faces dynamic conditions where it is not possible to use plain recipes" and "must have the most of a set of tools available and have the habit of including the more efficient ones". Then, it is possible to be more efficient managers by using more data, more statistical analysis, and less of speculation in our budgets, projects, and programs.

Historically, Henry Ford was who promoted the use of mass production, through the production in assembly lines to drive forward the competitiveness using as a basis the mass production. However, his contributions go further, was Ford the promoter of the principle of "keep everything moving" or "keep the flow of production" principles (Hopp and Spearman 2011). Everything has to keep moving at a steady pace and continue without disruptions: raw materials, work in process, and finished products must keep the same pace. For achieving such goals during the expansion and bonanza of the twentieth century, large corporations (mostly United States) focus their efforts on reaching technological innovation, high productivity rates and the costs reduction, this until the large economic crisis in the 1970s when it was necessary to find alternatives for achieving the same economic goals and their stabilization. An alternative was, to change the production processes and to use strategies for adapting the existing technologies. These conditions asked the large corporations and strategists to look into the continuous improvement paths. Schönberger (Schonberger 2008) points out that, since then, the improvement programs are not on the daily agenda but, when there are problems, and there is a risk in the financial health, a company looks for the initiatives of continuous improvement as a potential solution to get out of the risks.

Currently, continuous improvement strategies can be identified in two ways of thinking and structuring solutions: the path focused on quality and the path of deliveries on time. The quality path proposes to eliminate the root cause through a statistical analysis that measures, analyzes, and monitors the root cause of a problem, and by solving the root cause, generate a chain reaction for eliminating the problem (Deming 1986). This path has its best representative in the Six-Sigma methodology (Pande et al. 2000), which include a variety and extent set of tools and supporting manuals (Rath 2000). Six Sigma is based on a logical sequence known as DMAIC: Define, Measure, Analyze, Implementation, and Control.

In the other hand, the path for delivery on time is better represented by the methodology of Lean Manufacturing. Initially designed by Taiishi Onho in 1978 (Onho 1988) is based on the Toyota production system (TPS) but took its actual structure in the 1990s after Womack et al. (1990) publications. And it is based on the principles of "just do it" and "keep it simple" which makes the Lean concept a reactive methodology and an effective solution in critical conditions. Lean manufacturing is based on five principles: (1) Specify the added value of a product, (2) Identify the critical path, (3) Keep the flow, (4) implement a "pull" system, and (5) seek for perfection (Womack et al. 1990; Womack and Jones 2010). Based on these principles, the Lean Manufacturing uses tools to identify and eliminate waste, keep the production flow, reducing work times, and reduce costs to the extent that it is possible.

Recently, there have been proposals for integrating these strategic paths. George (2002) (George and George 2003; George et al. 2004) first proposed a robust model known as Lean-Six Sigma approach. This path uses the methodological DMAIC structure of Six Sigma with an approach to Lean when is necessary to propose some solutions. To combine them, there are a broad and robust set of tools that are available in each phase of the methodology. Lean-Six Sigma emphasizes on the completion of each phase before starting the next one. Also, it uses a validation review to ensure that all components of Lean and Six Sigma have been implemented. It is orientated to medium and long-term projects, focusing mainly on the improvement processes rather than solving short-term problems.

A different approach has been recently reported Alba-Baena and Estrada Orantes (Alba-Baena et al. 2016; Estrada-Orantes and Alba-Baena 2014), where the methodology of Lean-Sigma plays a crucial role. Lean-Sigma is based on the strategy of "do it at the speed of Lean with the depth of Sigma," includes and uses the same methodologies, but it is focused on solving situations in the short term. Lean-Sigma as used by these authors is based on the rapid identification and fast solution to a problem that is affecting the process to keep the flow of the process. Worth is to mention that in general, these solutions may not achieve the statistical level of Six Sigma (or 3.4 ppm) but to bring back the process to a stable state and to give a quick respond to the changes in the process, or to achieve the main goal as projected.

The Latin American restrictive environment has been characterized by restrictions such as time, economic, and technological limitations. These conditions limits the use of solutions such as reengineering and other major technological investments. These restrictive characteristics sets the perfect scenario for the use of fast and efficient solutions to customer-focused challenges. Managers in LA have proved that in a restrictive environment it is possible to successfully implement state of the art methodologies for problem-solving and continuous improvements. Examples in continuous improvement implementations have been published by several authors such as from Coy et al. (2016) and Camacho et al. (2016), also in the product design (López et al. 2016) and Romo et al. (Romo et al. 2016) and optimization methodologies (Pérez et al. 2016). Implementations in this restrictive environment have been reported also for solving problems using Lean-Sigma as a fast response and effective methodology, applications in the automotive industry (de la Cruz Rodríguez et al. 2016), goods assembly (Garcia et al. 2016), and air conditioning filters assembly (Sifuentes et al. 2016) have successfully proven its efficiency.

Also Lean-Sigma methodology in the automotive industry has been previously reported by the authors as an effective problem-solving methodology in highly restrictive environments. Estrada-Orantes and Alba-Baena (2014), reported that the Lean-Sigma could be used to propose effective solutions, but also towards the improvement of the solving process itself. The Lean-Sigma approach, as used by the authors in their research as a problem-solving oriented methodology, more than an improvement project-oriented methodology. The efforts, using this approach, are focused on eliminating a waste or an obstacle to the continuous manufacturing flow, rather than concentrating on achieving annualized savings. Then is possible to extend the previous definition. Then, the objective of the Lean-Sigma methodology is to solve a problem in the shortest time possible, based on five rapid-improvement steps:

1. Identify and measure the problem. What and how big is the problem?
2. Root Cause Analysis. What is the root cause of the problem?
3. Develop Solution Alternatives. Identify the alternative that best solves the problem.

4. Verify the Solution. Make sure that the problem is eliminated by the proposed solution.
5. Control Plan. Make a quick and effective plan so that the problematic condition does not come back.

As an Example, in Estrada-Orantes and Alba-Baena (2014), the authors' report describes a situation where a condition identified as "flash" is occurring in plastic components coming from an injection molding process. Data from a specific machine reports that approximately 99% of its production is exhibiting the "flash." After applying Lean-Sigma, the results demonstrate that the process moved from a not stable and non-predictable performance with 99.62% defective to a state of statistical control, which is stable, and predictable, with an overall performance of 218 ppm, achieving such state in seven days. The same authors reported the use of Lean-Sigma during the period of ramp-up for an ink cartridge production process (Alba-Baena et al. 2016). In this case, two families of products were produced on separate production lines. The high-runner uses an automated process and is forecasted to reduce its production volumes gradually −12.52%/yr. The other family is assembled using a manual process and has a promising sales forecast of +12.53%/yr. The project integrated both family products into the existing automated process making any necessary adjustments, while keeping the productivity and quality levels the same. A total of 24 adjustment activities were used for the assembly, and 18 activities were added for the ink filling processes. The implementation functionality was measured combining the average productivity rate (103,490/day) which results slightly higher than the initial and, the quality levels that were measured below 6600 ppm for both Families, which is an improvement from the initial values of 7,957 ppm and 37,305 ppm for Families A and B respectively. Also, the results show that the changeover times reduced from 57 min/setup to 30 min/setup. Based on the described results, the authors claim that the Lean-Sigma methodology and its tools helped to achieve the quality and productivity goals in a short period with a strong statistical foundation.

Moreover, under restrictive conditions, the use of resources and time are critical and limited. Therefore, the initial stages of the solution process gain even more importance. The initial diagnosis will help to clarify the alarm values and conditions (customer complaints, requirement changes, or goals), followed by the alarm evaluation. A comparison between the alarm values and the product specifications will help in defining the actions to take. Also, a comparison of the data from the process at the time of the occurrence and from the actual process will reduce the possibility of overworking a solution. Such evaluation also helps in determining the degree of importance and steps to follow towards the solution. It will also give feedback to the customer or determine the feasibility of a proposal. In the initial analysis of a problem, it is necessary to determine the response variable that will be used for measuring the progress and target values for a giving problem. The input variables and data collection also have to be characterized for preparing the set of statistical procedures to use. Statistical tools such as control charts and descriptive statistics will help to understand the process behavior and historical tendencies.

Data analysis includes the measuring and evaluation of the input variables and its effect on the response variable. Also, it is necessary to measure the effect of the relationship between them. These data have to be accompanied by a set of tools. Therefore, the manager will have the ability to modify the response variable and the validation process.

The initial diagnosis steps can be summarized as follows:

1. Clarify and understand the alarm.
2. Measure and compare the alarm data to the specification and process data.
3. Identify the response variable(s) to measure (productivity, quality, costs, etc.).
4. Identify the data type(s) and measure them.
5. Plot the collected data, while comparing to the specifications of the variable.
6. Analyze and understand the measured variations and tendencies.
7. Measure and track the effect of the changes in the response variables.
8. Identify the relationship between the response variable(s) and the input variables and those from the process.
9. Find the set of tools for modifying the process and the response variables' levels.
10. Determine the target values and validation methods.

As mentioned before and summarizing, Lean-Sigma methodology embraces the rapid-improvement approach of Lean Manufacturing and the deep statistical analysis of Six Sigma. Additionally, Lean-Sigma may incorporate any needed tool from other methodologies, such as Design for Six Sigma (DFSS) and still maintain the fast solving pace, and the high statistical accuracy. This chapter presents a case study where the Lean-Sigma methodology is used to achieve a fast solution to a cosmetic quality problem occurring in the manufacturing process of exterior-home products located in Juarez, Mexico. The solution also includes the design of a qualitative measuring device, which was used for determining the acceptance parameter levels at the incoming inspection station, and later as a quality control device.

2 Case Study

2.1 Context of the Case

The case describes a productivity problem in the final assembly process of outdoor light fixtures at the Juarez (Mexico) facility. The company, for strategical reasons, decided to have the final assembly process in the Juarez facility, while the fabrication process of the components was assigned to several Asian facilities and different vendors overseas. The final assembly process started to experience color variations in the incoming components, and some customers filled several quality complaints.

One year after the startup, the customer service area recorded several customers' complains, mainly about variation in the color tone and several quality issues such as stains, discolorations, broken, and malfunctioning products. Most of the complaints involved a part known as the chassis. This component is fabricated using a polymer, copper or brass material. 90% of the production uses the brass chassis. Brass components are colored by using a chemical process, called Patina. Coloring using the Patina process is susceptible to the chemical composition of the reactants to their concentrations, exposure time, and the neutralization process. The variability of the Patina process and the mentioned factors added to the customer complaints makes the brass products the target of this project.

The inspection of an incoming container shows that approximately 50% of the received chassis lots are acceptable according to the quality standards used for this purpose. Also, this data shows that the three main reasons for rejecting the chassis components are pieces with different tones, or discolored or stained. Figure 1b shows an example of a nominal colored chassis. Figure 1a shows the representative samples of a clear piece and Fig. 1c a dark piece.

2.2 *Initial Statement*

This case presents a common scenario faced by producers when combining different vendors located around the world. At arrival, the components are inspected for quality purposes using an AQL process at the incoming inspection area. For some of these components, the patinated chassis fixture is inspected for dimensions and visually for color. The levels of scrap observed on this component are around 30 to 50% of the received fixture-lots. Figure 2 shows details of the common defects found in the initial inspection, which includes cosmetic defects such as

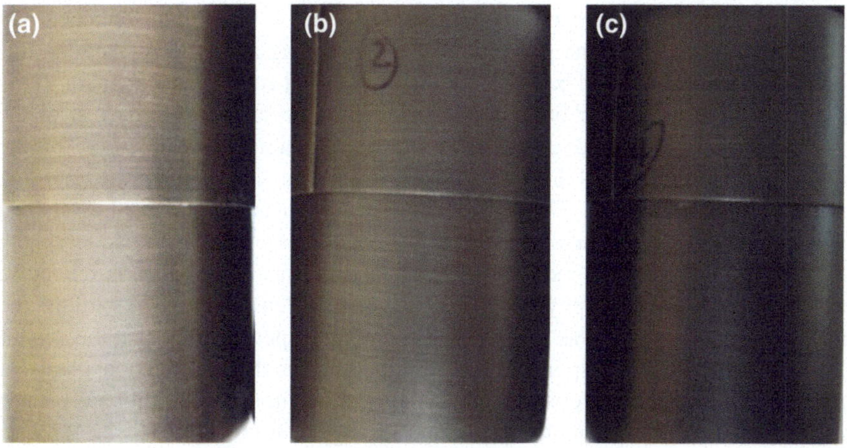

Fig. 1 Brass chassis tones **a** light, **b** nominal, and **c** dark pieces

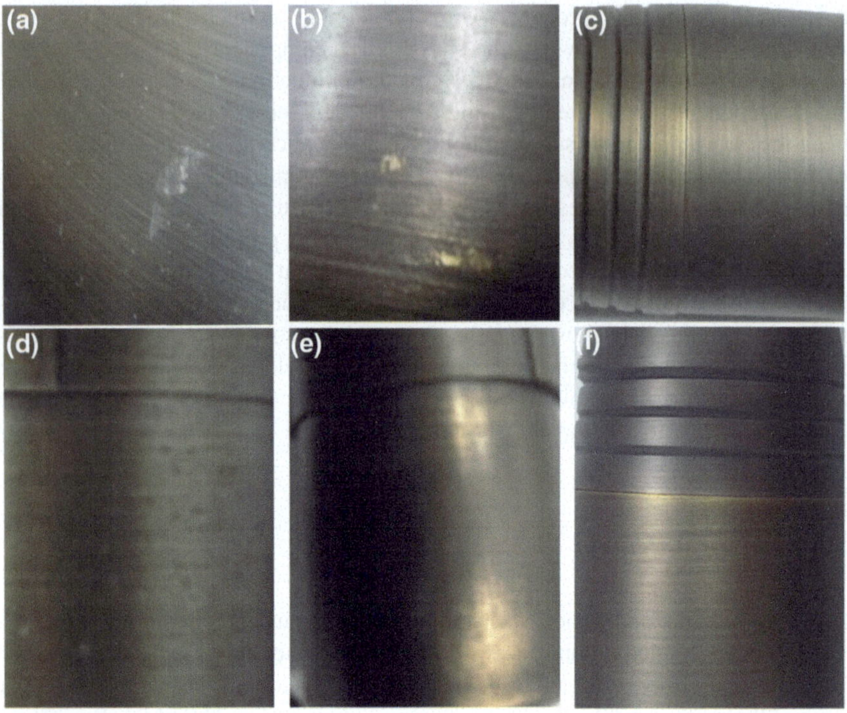

Fig. 2 Rejected samples at the incoming inspection station: **a**, **b** stained sample; **c**, **d** discolored areas and **e**, **f** different colorations

staining spots (Figs. 2a, b), non-patinated areas in pieces (discolored in Figs. 2c, d), and components with different coloring tones (Figs. 2e, f).

Figure 3a shows the results of a selected container at the incoming inspection area. The acceptable components represent around 47% of the container, the components showing patina process defects are ranging around 38% (see the different tones and discolored bars at Fig. 3a), and stained components are about 12% and incomplete fixtures or other quality related issues account for the rest of the components. On the other hand, data from the quality control station of the assembly process recorded during a full year shows that approximately 20% of the product has some defects. 77% of the defective products are classified as stained (see Fig. 3b), 10% of them are incomplete products, 7% are the functionally defective ones (electrical, mechanical, etc.), and the rest is scrap parts related to non-repairable products (physical damages, broken lenses, etc.). Regarding the defective components, it was found at the incoming inspection area, that the vendor (located overseas) has responded with a temporary containment-action plan that includes the increment of lot sizes and the number of shipments. Additionally, he is absorbing the inspection, replacement, and management costs.

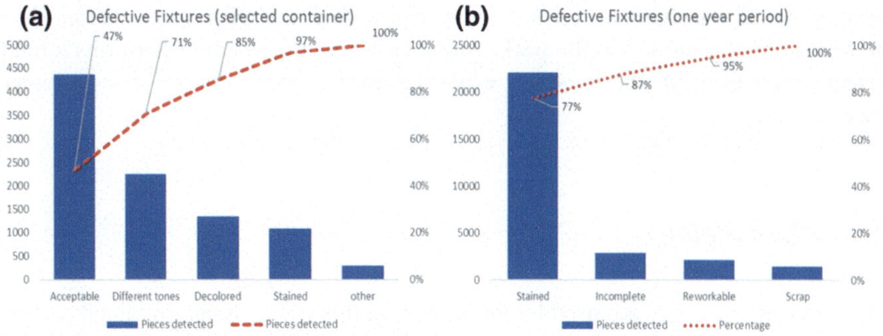

Fig. 3 Defective-fixtures Pareto charts from **a** A specific container at the incoming inspection station, and **b** the collected by the inspection station at the assembly line

2.3 Problem Description

Figure 4 shows a Suppliers, Inputs, Process, Outputs, Customers or SIPOC diagram representing the assembly process at the Juarez facility. For confidentiality, the Suppliers and customers' names from the original diagram were changed for generic ones. However, the playing roles are identified. In the "Inputs" column, the diagram shows only the six components needed for the assembly of the described product. The "customers" are classified by size (retail, wholesale, and special orders) and type (internal or external).

For simplification, in this document, the assembly process is represented with five general stages (see Fig. 4): chassis assembly, wiring, optical assembly, inspection, and packaging. In the chassis assembly step, the chassis is prepared and inspected the mechanical functionality of the components, then, are assembled. The assembled components are then placed in a holding fixture that is transported to the wiring station, where the wiring and soldering are performed. The lens, O-rings and light bulb (or LEDs, depending on the model) are assembled to the chassis, then the product is tested for electrical safety. A final visual inspection and functionality test

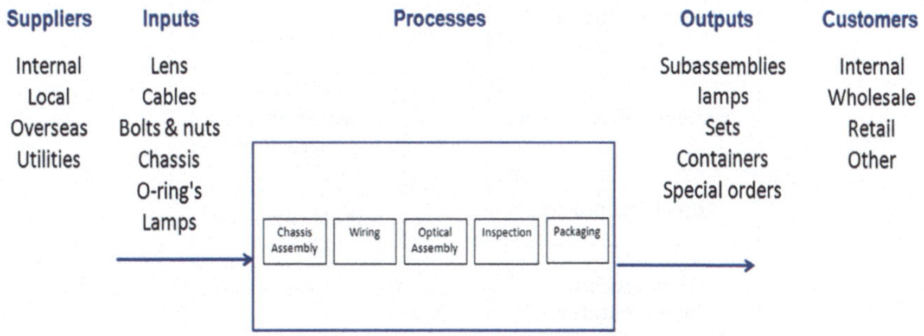

Fig. 4 Graphical representation of the SIPOC and the outdoor lighting assembly process steps

(electrical and mechanical testing) are performed at the inspection station. The product is finally moved to the packaging area where the operator performs a final visual inspection of the product while he packs according to the customer specifications.

2.4 Methodology

The Lean-Sigma approach used by the authors in this research, as explained earlier, is based on five rapid-improvement steps (depicted in Fig. 5): identify and measure the problem, perform the root cause analysis, develop a solution, verify the solution, and make a control plan. Also in the solution stage, DFSS steps are shown as used for this case. For the design of the solution, first, the key considerations for having the design specifications are based in the voice of the costumer (VoC) that in this case combines the final customer requirements and the quality control user's needs.

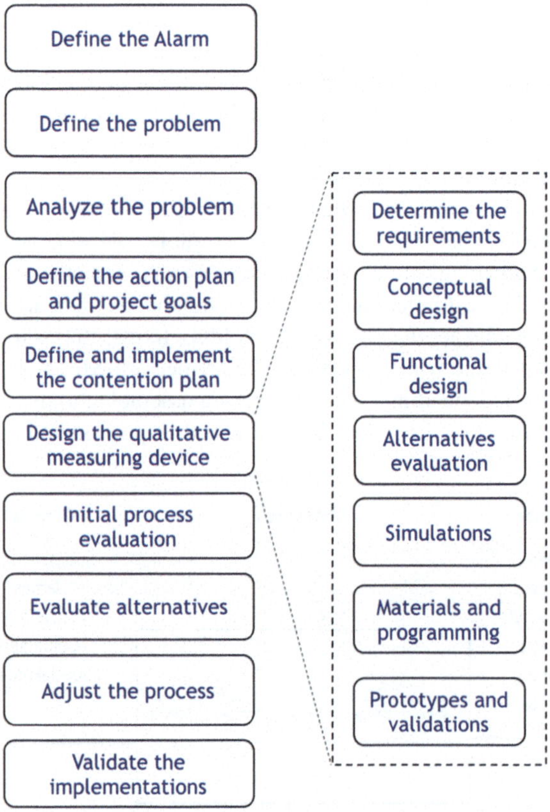

Fig. 5 The methodology used for solving the problem presented in this case study

The specifications for designing the color testing device are based in the Quality Function Deployment (QFD) technique. The QFD is used for defining the conceptual and functional designs and for refining and defining the product specifications. In this case, the QFD data was used for listing the specifications for the color testing device. A series of simulations and some partial prototypes were developed with the purpose to validate the final design in the next step of the Lean-sigma sequence.

2.5 Results

The methodology presented in Fig. 5 is useful to dealing with the problem requirements and finding a solution as shown in the next points which describes the most relevant results.

2.5.1 Step One. Identify and Measure the Problem

The step one is a careful inspection of the incoming chassis components, but it is conducted very quickly. The collected data are useful to identify the problem as a color variation of the brass chassis component reaching the customer site. Using initial measuring methodology, it is determined that 38% of the parts have color variation issues.

2.5.2 Step 2. Root Cause Analysis

Step number two is conducted using a brainstorm session. After considering a large list of ideas, the causes were reduced to two, namely:

1. Color variation at the customer site caused by the inability to effectively separate the parts in ranges of color intensity. This cause was directly related to the current visual measuring system used.
2. Color variation caused by the patina process at the supplier facility.

2.5.3 Step 3. Develop a solution

The first solution was directed towards the measuring system inside the manufacturing facility. A proposed solution was to design a measuring device with the ability to quantify any color deviation from the customer specified color. Six DFSS tools are used to design the device, as depicted in Fig. 5, and is described as follow:

Step 3.1 Voice of the Customer (VoC)

The customer is using the outdoor-light product in batches of six and more parts. The customer is expecting to have a standardized color pattern for all of the parts in a batch, and stay like that for the warranted life of the product. Additionally, if it is needed to replace a component at a later date, the customer expects to have a chassis with the same color as the installed ones.

On the other hand, the user of the measuring system identifies the following characteristics as requirements for the device: The device must show the data to the user (inspection); it has the capability to connect and transfer data to a computer and data center; it has to control the tested component and the surrounding illumination; it must be heavy duty and reliable for long therm use; it must have quick response and reading signals; it must use a quick testing procedure; and it must be easy to operate.

Step 3.2 Conceptual and functional design

The purpose of these two steps is to have the design specifications that will be used as guidelines for having the product requirements, in this case, the testing device characteristics. The goal of the conceptual design is to determine the features and characteristics as the customer envisions them. First, make a list and rank the characteristics according to the customer's appreciation and his determined importance, and give a weight for each of the listed characteristics. Then, a conceptual drawing and integration are presented to the customers for review. After the conceptual design step is concluded, the functional design requires the team to define the engineering parameters and to define the tolerances of each of the measured parameter. The main parameter is that the device has to measure the color of the piece. To accomplish this purpose, three components of the color have to be sensed or measured: hue, saturation, and brightness. Such data has to be used for defining the color identification (based on a predefined code). Data are also useful for identify and discriminate the acceptable, light, from the dark pieces. Also, in the functional design step, it is necessary to define the working conditions for the device to perform with robustness. Here, and for an accurate measurement of the color of the piece, the signal reaching the measuring device has to become as clean as possible. Then, other sources of color must be reduced or eliminated. The device also must have robust components for the working load at the factory and flexible to be integrated into the production process in their current state or with minor modifications. Last, in this step it is important to analyze the relationship among the different functions for establishing restrictions and functional requirements.

Finally, a QFD diagram (quality function deployment) is used to merge and use the gathered information, and to rank the main product characteristics as given by the customers. Also, the information considers to enlist and rank the engineering functions of the color measurement device. The ranges, units, and values also have been determined, then, the requirements and relationships are used for defining the

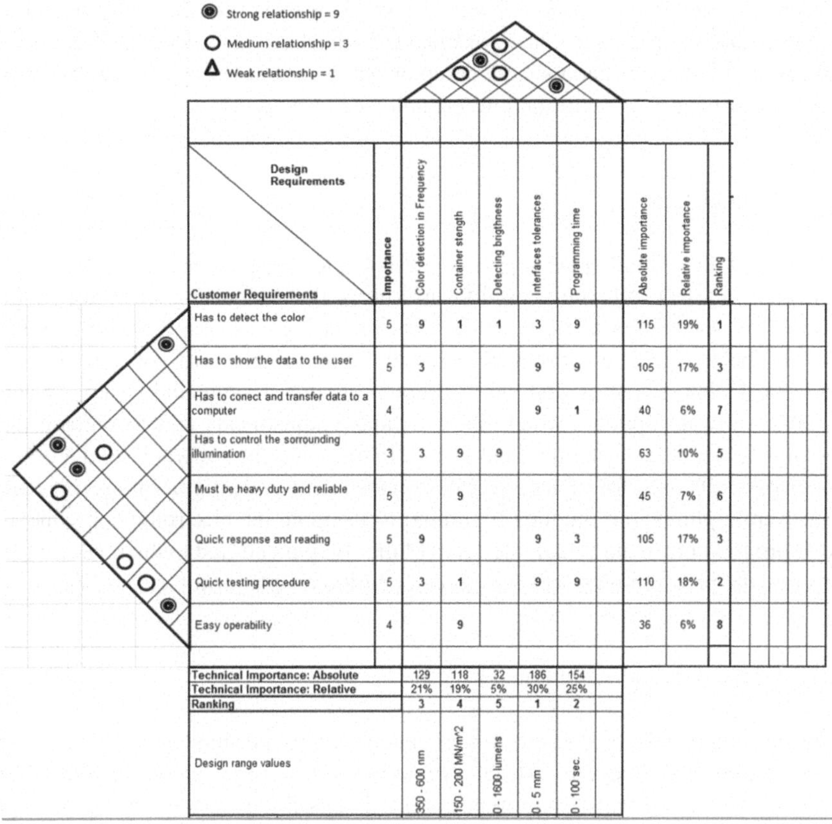

Fig. 6 QFD or the design of the color measuring equipment to determine chassis coloration

product design specifications or engineering requirements. The resultant values and considerations are seen in the bottom of Fig. 6 where is possible to see the product design specifications and requirements for designing the Color Measuring Equipment (CME). Along with this exercise, the team benchmarks the market for the available devices (not shown in Fig. 6) for making a feasibility study and determine the more efficient components as used.

Step 3.3 Evaluation of Alternatives

The goal of this evaluation is to determine the more suitable components and materials for the defined specifications. For this task, there are several decision tools, such as the CES EduPack® that helps in choosing the materials (see Ashby 2008), also for the selection the mechanical and physical characteristics, CO2 profiles, and the costs among others parameters. Also for this step, other weighing tools are possible to use the data from the buying system at the company or to use the advice from the suppliers and consultants for making the selection. From such evaluation, a list of materials and components is generated including the programming software and information technologies to be used.

Step 3.4 Simulations

The goal of this simulation step is to design and integrate the components of the device. A CAD software is used for making scaled to real size drawings of the components and pieces as designed and shown in Fig. 7b. Some features of the software helped in the simulated assembly and functionality of the device. Mechanical characteristics and other properties can be simulated using several available software options, which allow to make adjustments and to prepare the drawings for manufacturing the final pieces. In this case, the software SolidWorks® was successfully used for such tasks. For the electronics and programming, there are also several options, but the idea is similar, to integrate the electronic components and to program the features and interfaces of the components and to the final user. In this case, the programming was based on an Arduino® platform (see Fig. 7a).

Step 3.5 Materials and programming

Once the components, pieces, electronics, and programs are simulated, the next step is to acquire and integrate partial prototypes for a physical evaluation. The advantage of having partial prototypes is that is possible to make corrections and adjustments for a specific function or program without compromise the rest of the functions of the final prototype. After testing the different functions and prototypes, it is possible to move to the next step.

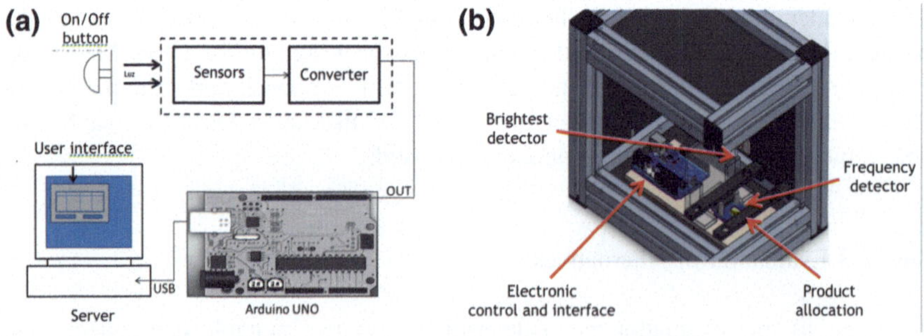

Fig. 7 **a** Scheme of the device's system and **b** Prototype for detecting the color components of the receiving product at the incoming inspection station

Step 3.6 a Prototype description

The final prototype integrates all the functions, and the challenge in this step, is to make them work together. Figure 7 shows the integration of the color measuring device; Fig. 7a shows the basic schema of the integration. The functionality of the prototype is described as follow: the prototype is based on a series of sensors that measures frequency for the hue component of the color readings. A lux-meter is used for measuring the brightest and saturation color components and, finally, a micro-controller board (Arduino UNO) is used to calculate and record the data. Later, with the recorded, the data is displaying using a portable touch screen where the user can interact with the system using the same screen. In the prototype, the sensors are located in an enclosed chamber that insulates the system from external light (see Fig. 7b). The prototype includes a fixture for holding and moves the component; the device is also connected to an external monitor (not included in the Figure) as a visual interface to the user.

The operation and use of the device shown in Fig. 7 is described as follows: in the dark chamber, after placing a chassis for evaluation, the white light is received by the sensors. Then, the detected readings and data variations of the RGB-W colors (Red, Green, Blue, and White) are converted to colorimetric values: hue (in degrees), saturation (in percentage), and brightness (in percentage). The color is identified and the data is stored, and used for statistics then displayed to the user. Then, the station opens the chamber for exchanging the chassis for the next inspection. The next step in the solution process is to consider the vendors and customers for determining the characteristics of the chassis color in the final product. The required color is agreed with the customers and vendors to be defined as antique brass color that is determined by a selection of masterpieces that serves as reference for a qualitative classification. Thus, along with the nominal masterpieces, the team agreed to classify the "clear" pieces to be lighter than the nominal antique brass and the "dark" pieces to be darker than the nominal ones.

Step 3.6 b Prototype validation

Gage Repeatability & Reproducibility (Gage R&R) studies were used during the calibration and validation process. Three factors were identified as the main source of variation: Geometry of the piece to measure, the use of a trigger button attached to the prototype, and a combination of coding pins for transmitting the collected data to the Arduino. For the calibration, the team used 100 selected masterpieces matching the antique brass color as defined and described previously. The measuring process required at least six recordings for each piece and position. After several interactions the changes in the fixtures and adjustments in the sensors' position, the variation was measured in less than 10% for each of the color variables (8.67% as seen in the example for the hue in Fig. 8), which was considered as acceptable for these prototypes. After the calibration process, the team approved the

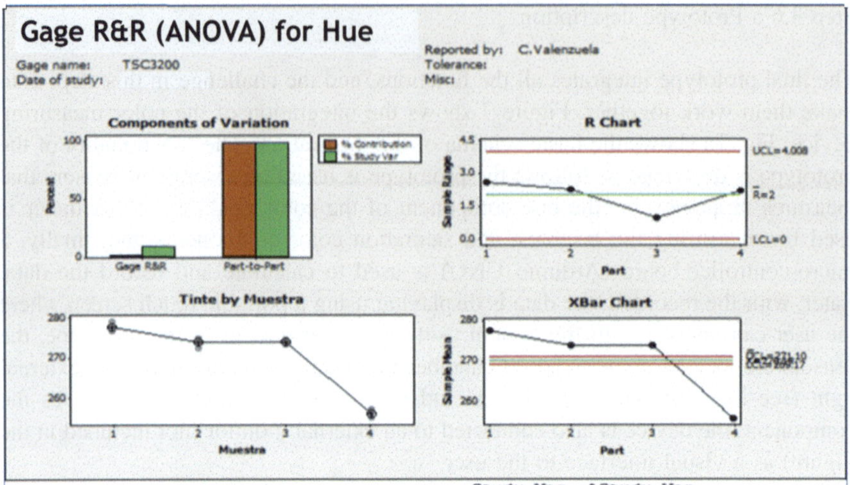

Source	StdDev (SD)	Study Var (6 * SD)	%Study Var (%SV)
Total Gage R&R	0.84114	5.0468	8.67
Repeatability	0.84114	5.0468	8.67
Part-To-Part	9.67078	58.0247	99.62
Total Variation	9.70729	58.2437	100.00

Number of Distinct Categories = 16

Fig. 8 Gage R&R study for the Hue during the calibration process

prototype for its integration and replication for the final inspection stations as they were adapted to the inspection operation in the production process.

2.6 *Step 4. Verify Solution*

After the calibration of the prototypes and by measuring the nominal masterpieces, a nominal value of 272.5 Hz was presented as the target for the Hue component, with 8.35% and 32% for the saturation and brightness, respectively. Also, the data from the accepted pieces (labeled as nominal) or acceptable were used for establishing the range and limit values for determining saturation, brightness, and hue and analyze the capability of the process. For achieving this purpose, the collected data in Fig. 9 shows that it is possible to use the given groups and limits for differentiating the colored pieces and their groups. Then, the next step is to measure the possibility of having a piece in an in-between zone.

The central group has a mean of 275 Hz for the Hue values, which is close to the targeted (272.5 Hz) but is also close to the darker ones which mean is far as ~7 Hz. But seems to be different enough from the light-colored pieces which mean shows a difference of ~15 Hz. Figures 9b and c shows the values and basic statistics for the saturation (Fig. 9b) and brightness (see Fig. 9c). The normal pieces show saturation values with a mean of ~8%, with a variation close to 1% value

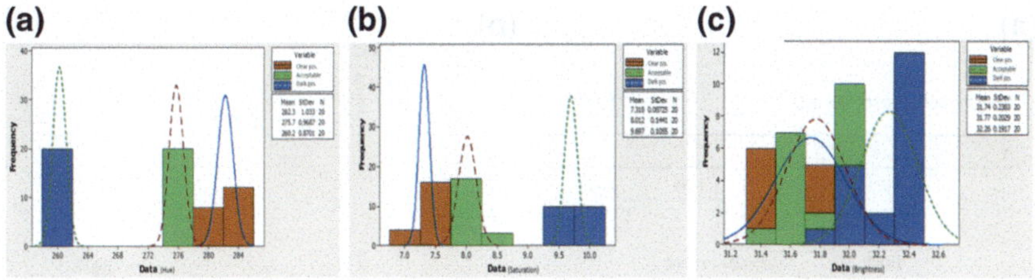

Fig. 9 Masterpieces data used for determining the coloring components and ranges of variation: **a** Saturation, **b** Brightness and **c** Hue data

difference between group means. Here, the dark pieces group seems to be well defined by the other two groups while the light and normal groups may have an overlapping at the extreme points. The nominal (or normal) percentage values for the brightness shown to be around 31.7% (see Fig. 9c). Here, the data is overlapping, the clear and acceptable groups' data shown to be part of the same population, overlapping and interacting with the dark pieces data. For determining the degree of overlapping and variation of the color components in each of the groups the team used an ANOVA. Using the shown data and understanding the behavior of the color components variation, the team decided to establish the nominal limits and present the initial standardized values. For the hue values, the nominal frequency was determined in 272.5 Hz with a tolerance of ±10 Hz. The saturation limits were delimitated between 6.97 and 9.71%, and the brightness between 31.74 and 32.26%. Then, the limits and range values for the color components were added to the programming of the controller program of the CME for labeling and identifying the color of the chassis in testing. The screen image shown in Fig. 10a exemplifies the user interface screen and the readings where the values for each component of the color are displayed to the user. The same Fig. 10a shows the suggested identification for one testing piece, where is displayed the choosing from one of the three color groups. Also, the screen shows the options for saving the data to the database or for deleting the displayed results.

After the calibration and establishing the limits, the team exercises ANOVA in the three coloring components' data (Hue, Saturation, and Brightness). This data is used for calculating the confidence level in differentiating the nominal groups and pieces. Figures 10b and d shows the boxplots accompanying the ANOVA results: Fig. 10b shows the boxplots for the Hue data, where are compared the light and dark masterpieces data. From the ANOVA results is possible to establish a statistical difference to the nominal color data with a confidence level of 98.95%. Comparable results are shown for the saturation factor, as seen in Fig. 10c. Here it is possible to differentiate the saturation values from the masterpiece groups and could be used for identifying the group where a chassis belongs. For this data the confidence level in detecting statistical differences among the masterpiece groups is calculated in 96.72%. However, in the case of the data plotted from the brightness groups (Fig. 10d), data shows similar distribution values for the clear and nominal

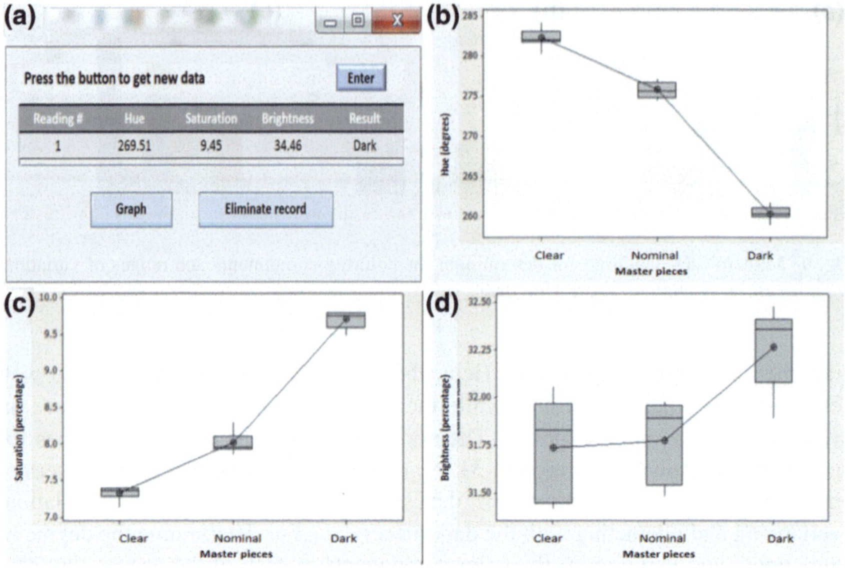

Fig. 10 **a** screen that serves as the interface with the user and boxplots after the variance testing for the selected masterpieces **b** Hue, **c** Saturation and **d** Brightness

groups and values separated from the dark pieces' box. The confidence level for using the brightness as a color identifier factor has been calculated in 56.05%, which is expected and acceptable due to the polishing process is the same for all the pieces. Nevertheless, the team has determined that in the decision process and for classifying a piece to be nominal or not, the brightness is less critical and will be considered as a reference value. Then the brightness is a parameter to consider for identifying a difference in the color and quality of an incoming chassis.

Figure 11 shows the initial capability analysis (as seen in Fig. 11a), at the incoming inspection station for each of the measured variables. Hue data shows that the received lots have an average of 50% of conformant materials, 47.5% are dark colored materials, and light pieces represents 7.5% of the samples. In consequence,

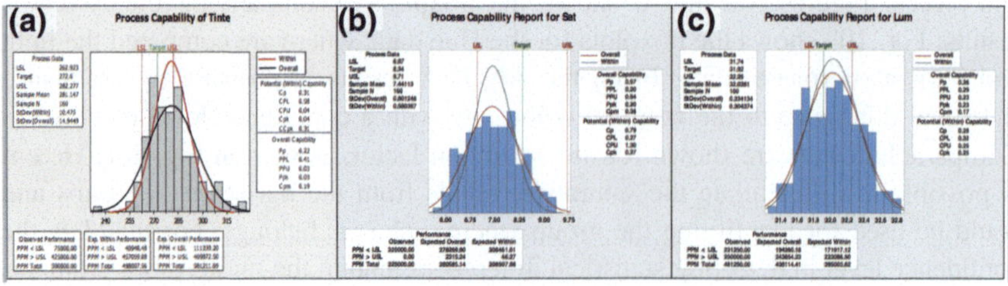

Fig. 11 Initial process capability analysis (Cpk) using the calibrated equipment for data of each of the color components **a** Hue, **b** saturation and **c** brightness from a selected batch

the process capability (Cpk) calculated for this study is in 0.04, and the out-layer proportions are expected to be 11.3% below the lower limit. The dark pieces are expected to be 46.9% and the pieces in the nominal range in of 58.2% of the received materials. For the saturation data shown in Fig. 11b, the results gives a calculated Cpk of 0.27 with 28% of the sample materials out of the nominal values mainly at the lower side expecting for the population a total of 32.5%. However, from the study, there is expected to have a 27.8% of the received chassis to have lower saturation percentages and 0.2% pieces with over saturation. In the same way, the brightness Cpk is calculated in 0.25, and the nominal brightness accounts for 56.2%. Data show to be centered with a slight tendency (24.4%) to be brighter than the nominal (see Fig. 11c). but the calculated percentages out limiting data is expected to be 23.1% and 25% at the opaque and over bright sides respectively.

2.7 *Step 5. Control Plan*

Changes in the process have been proposed for including the findings of this research. The received chassis lots are now inspected and separated into three groups: Light, Nominal and Dark colored. After, for the non-comforming components a reworking process starts with a sandblasting step. This is followed by a second Patina chemical processing which is described as follow: The prepared mixture (described above) is heated to 95 °C, the cleaned pieces are then hanged for 1 h, while the mixture is stirred every 5 min. After, a visual inspection and a wax polishing finish the reworking step is ended. Without considering data of the reworked pieces, the efficiency in the use of the color testing device is monitored and a learning curve is plotted. Once the users of the CME in the production area reaches an 85% efficiency in their learning curve in the use of the testing device, the team collects data samples of the receiving lots. Then, for consider that the Lean-Sigma process is completed, the team uses this data for comparing to the initial receiving data (at the time of starting this project). The comparison of the conditions of the chassis batches and their color distribution, the behavior of the mean, and the variance values for the hue are shown in Fig. 12. For the comparison to the actual process, Fig. 12 shows the two-sample test results (Fig. 12a) and a representative boxplot graphs with the established USL and LSL limits marked for reference (Fig. 12b). The test (p-value == 0.00) shows that there is a significant difference between the initial mean and the actual (final) data. The boxplot shows that the initial data has a median and mean closer to the USL, while the same reference points for the after-filtering data are close each other with the tendency to the LSL.

With the limits and controlled process, the use of I-MR chart is proposed for monitoring the process. The I-MR chart in Fig. 13 plots the individual color observations and moving ranges for the initial and after the implementation of the solution. As seen in the Figure, the initial data shows a clear dispersion of the chassis' color data, observing that most of them are out of the specified limits.

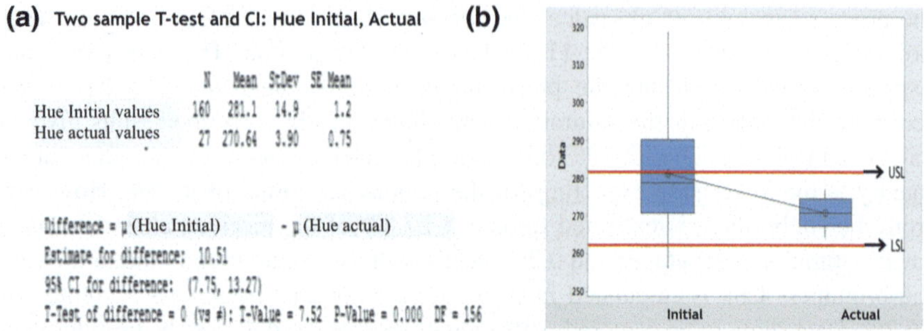

Fig. 12 **a** Two-sample T test results for the initial recorded Hue values vs. data after using the color testing device and **b** the corresponding boxplot comparison

However, the data after implementation of the solution shows a controlled color data means and moving ranges as seen at the right part of the chart in Fig. 13. The control acquired by using the color testing device is observed in the boxplot comparison in Fig. 14a. The quartiles showed to be more equilibrated as compared to the initial data plot. In the same Figure, the comparison shows that the median for the hue values is closer to the targeted value in the after implementation data. Also, for determine and compare the capability of the actual process data to the initial capability, Fig. 14b shows the capability analysis and comparisons. From this Figure (14b) it can be noticed that the Cpk moved from a 0.036 to 0.65, this value is low as compared to the conventionally acceptable Cpk (1.33). However, the use of the color testing device helped in moving the outcome of the nominal colored pieces from 50 to 97.24% and provides the initial value for the next steps in the continuous improvement process.

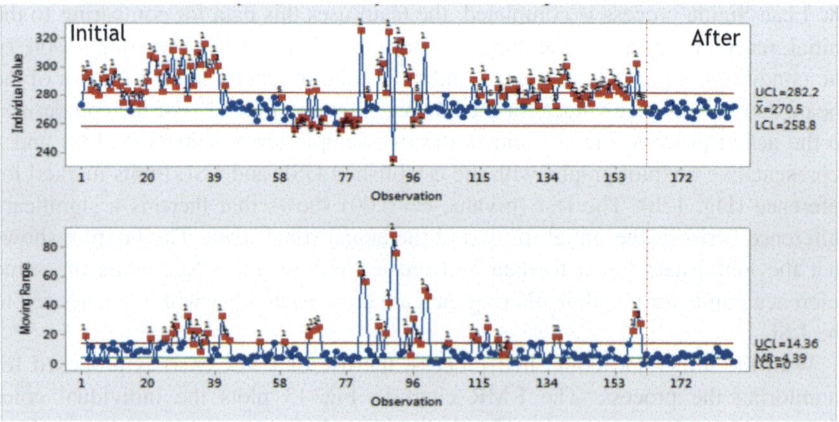

Fig. 13 I-MR chart is showing for comparison the initial and data after implementation of the inspection for the Hue color component

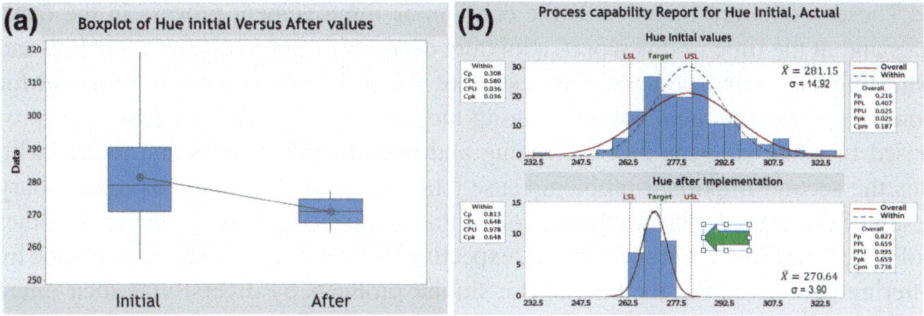

Fig. 14 **a** Boxplot and **b** process capability graphs comparing the initial and after implementation data

Finally, the next step in this project is to analyze and adjust and optimize the Patina formula and its process to achieve a reduced variation and more controlled color values (target) as established by the design specifications. Followed by the analysis of the cleaning process after the Patina processing, for reducing the stains and cosmetic marks in the final product.

3 Conclusions

The case presents the solution procedure in a final assembly facility located in Juarez, Mexico, with the described characteristics of a LA restrictive conditions. These solution efforts are focused in the component "chassis" which is the main component of an outdoor lighting fixture. The problem was detected from a customer's complaint and is described as a difference in color between the received products and the later replacing fixtures. As a containment plan, the components received from an overseas supplier are subject to a 100% incoming inspection. For finding a solution, the team decided to combine Lean-Sigma methodology and DFSS tools for designing and implementing the use of a color testing device in the incoming and final inspection stations. The inspection apparatus was designed and validated by the team using the qualitative inspection data for determining a standardized target and limit values for the color components. As result, the qualitative color-tone inspection, based on a visual separation of the chassis is substituted by a quantitative inspection process using the designed measuring device. The measuring is based in the three color components (hue, saturation, and brightness) and are used for determining the color values. The obtained readings are later compare to the ranges calculated and determined for accepting the incoming chassis. After implementation, the process capability moves the outcome from a 50% probability of sending to the customer mismatching colored products to a 97.2% probability of sending chassis and products colored in the acceptable color range.

These results are included in the continuous improvement process in the facilities and at the time of this report, a mixture design was proposed for correcting and adjusting the patina process. The results of the mixture design will be useful for controlling the chemical coloring during the reworking procedure. Also, it is proposed to make changes in the cleaning and neutralizing process and reduce with this the second important problem cause (stained chassis). Finally, as a secondary result of this project, the administration of this company ventures in the use of other coloring tones' for the same product, expecting to increase the business outcome by offering options during the bidding for similar products by diversifying their offers to the market.

4 Final Remarks

Managers in a restrictive environment have the challenge to solve problems at hi-speed but with limited resources. For such conditions, experts have developed tools and methods which combine different elements of efficient methodologies. Such is the case of Lean-Sigma approach that has been proven to be one an efficient way to solve problems under these conditions. The case study shows that it is possible to use Lean-Sigma in combination with other methodologies such as DFSS, for problem-solving under the mentioned conditions by taking advantage of the fast response of Lean while having the deep statistical analysis of Six-Sigma. In this case, the use of Lean-Sigma with Design for Six Sigma boost the solution process and complement the improvement, and corrective actions as the process move from 50% defective to $\sim$2% at the end of this report.

References

Alba-Baena, N., & Estrada F. J., Sierra Torres O. O. (2016). Using lean-sigma for the integration of two products during a ramp-up event. In: A.-H. Giner, S.-R. Cuauhtémoc, G.-A. Jorge Luis (Eds.) *Handbook of research on managerial strategies for achieving optimal performance in industrial processes* (pp. 405–427). Hershey, PA, USA: IGI Global. https://doi.org/10.4018/978-1-5225-0130-5.ch019.

Ashby, M. (2008). *The CES EduPack database of natural and man-made materials*. Cambridge, UK: Cambridge University and Granta Design.

Camacho, J., Baena, N. A., Hernández, A., Molina, J., Barrera, M. F. R., & de León, D. Z. (2016). Mejora de productividad de una línea de producción a través de la implementación de un programa especializado a la captura del métrico de eficiencia general de los equipos "OEE". CULCyT (55).

Contreras, Ó. F., Carrillo, J., Garcia, H., & Olea, M. J. (2006). Desempeño laboral de las maquiladoras: Una evaluación de la seguridad en el trabajo. *Frontera norte, 18*(35), 55–86.

Coy, M. R., Salazar, J. M., Martínez, L. F. F., & Baena, N. A. (2016) Aplicación de la Metodología Seis Sigma para controlar los parámetros significativos en la producción de un guante de Nylon con recubrimiento de Látex. CULCyT (58).

de la Cruz Rodríguez, M. I., Orantes, F. J. E., Mendoza, M. D., Saldaña, J. F. E., & Rodríguez, R. R. (2016). Metodología para el mejoramiento continuo de procesos de manufactura, basado en lean sigma y aplicada al proceso de elaboración de arneses automotrices. CULCyT (56).

Deming, W. E. (1986) Out of the crisis. Massachusetts Institute of Technology (p. 510). Cambridge, MA: Center for advanced engineering study.

Estrada-Orantes, F. J., & Alba-Baena, N.G. (2014) Creating the Lean-Sigma Synergy. In: J. L. García-Alcaraz., A. A. Maldonado-Macías, & G. Cortes-Robles (Eds.) *Lean manufacturing in the developing world: methodology, case studies and trends from latin America* (pp. 117–134). Cham: Springer International Publishing. https://doi.org/10.1007/978-3-319-04951-9_6.

George, M., Maxey, J., Rowlands, D., & Upton, M. (2004). *The lean six sigma pocket toolbook: A quick reference guide to 70 tools for improving quality and speed.* Mcgraw-hill.

George, M. L. (2002) *Lean Six sigma: combining six sigma quality with lean production speed.* McGraw-Hill Education.

George, M. L., & George, M. (2003). *Lean six sigma for service.* New York, NY: McGraw-Hill.

Gracia, O. C., Orantes, F. J. E., & Pérez, F. H. (2016). Aplicación de la metodología Lean-Sigma en la solución de problemas en procesos de manufactura: Caso de Estudio. CULCyT (57).

Hopp, W. J., & Spearman, M. L. (2011) *Factory physics.* Waveland Press.

López, D. R., Gaytán, M. C. C., Baena, N. A.,& Orantes, F. E. (2016). Diseño de una estación de medición y análisis del proceso de engastado de cables. CULCyT (58).

Ohno, T. (1988). *Toyota production system: beyond large-scale production.* CRC Press.

Pande, P. S., Neuman, R. P., & Cavanagh, R. R. (2000). *The six sigma way.* McGraw-Hill.

Pérez, A. H. G., Orantes, F. J. E., & Pérez F. H. (2016). Diseño de parámetros robusto utilizando el método Taguchi aplicado a un equipo de crimpado de fusibles tipo midget. CULCyT (57).

Rath, S. (2000). *Six Sigma Pocket Guide Rath and Strong management consultants.* MA, USA: Lexinton.

Romo, J. S., Baena, N. A, González, J. R., Pérez, D. M. A., & Orantes, F. E. (2016). Rediseño y caracterización de un equipo de prueba eléctrica en el ensamble del producto Jack Bam. CULCyT (59).

Schonberger, R. J. (2008). *World class manufacturing.* Simon and Schuster.

Sifuentes, A. M., Orantes, F. J. E., Baena, N. A., Urdiales, M.G. V., Mandujano, H. A. T., & Salcido, M. H. T. (2016). Implementación de la metodología lean-sigma para la reducción de tiempos de preparación de maquinaria. CULCyT (57).

Wilson PA (2010) Maquiladoras. Texas State Historical Association. Retrieved October 26, 2017, from http://www.tshaonline.org/handbook/online/articles/dzm02.

Womack, J. P., & Jones, D. T. (2010). *Lean thinking: banish waste and create wealth in your corporation.* Simon and Schuster.

Womack, J. P., Jones, D. T., & Roos, D. (1990). *Machine that changed the world.* Simon and Schuster.

Creation of Technology-Based Companies: Challenges to Innovate in the Manufacturing Sector of Medical Devices, the Case of Baja California, México

Karina C. Arredondo-Soto, Teresa Carrillo-Gutiérrez, Marcela Solís-Quinteros and Luis A. Ávila-López

Abstract This chapter presents the challenges in the start-up of companies in Mexico, considering as the primary requirement that these companies apply innovative technologies in their processes, and these are willing to participate in the Technology Transfer Process (TTP). First, it analyzed the importance of Higher Education Institutions (HEIs) and Research Centers (RCs) as support agencies for the creation of new hi-tech companies. Also, the programs with public funds are listed in response to the need for technical and scientific support in the TTP. Considering this scenario, for the particular case study in Baja California, the dimensions of the participatory Medical Devices Cluster (MDC) and their expectations of growth were identified, followed by the definition of the needs and the challenges for TTP.

1 Introduction

According to Oxford Business Group (2017), Mexico is the leading exporter of medical devices in Latin America and is the leading supplier of these products to the United States. Baja California shows an advanced industrial cluster, organized

K. C. Arredondo-Soto (✉) · T. Carrillo-Gutiérrez · M. Solís-Quinteros · L. A. Ávila-López
Universidad Autónoma de Baja California, Calzada Universidad,
#14418 Parque Industrial Internacional Tijuana, C.P. 22390 Tijuana, BC, Mexico
e-mail: karina.arredondo@uabc.edu.mx

T. Carrillo-Gutiérrez
e-mail: tcarrillo@uabc.edu.mx

M. Solís-Quinteros
e-mail: marcela.solis@uabc.edu.mx

L. A. Ávila-López
e-mail: avila.luis@uabc.edu.mx

G. Cortes-Robles et al. (eds.), *Managing Innovation in Highly Restrictive Environments*, Management and Industrial Engineering,
https://doi.org/10.1007/978-3-319-93716-8_3

by conglomerates where the manufacture of medical products stands out at a national level for its economic importance, accounting for 79 companies by 2015. In this context and due to the characteristics of the sector, it is necessary to create Technology-Based Companies (TBC). The technological transfer is understood from two aspects, one is the transfer between companies, and the second the transfer between the generators of knowledge (universities) toward the companies. The Mexican standard of terminology in the management of technology NMX-GT-001-IMNC-2007 published by the Diario Oficial de la Federacion defines the transfer of technology as: "the process by which the transfer or licensing of rights is negotiated on intellectual capital. Understanding by intellectual capital, the intangible goods product of the human intellect, which constitutes the sum of all knowledge of an organization."

The case study includes technology transfer from (a) business to business, from Canada to Mexico; and (b) from university to company, the Autonomous University of Baja California (AUBC) and the company Avantti Medi Clear (AMC) participated. It is interesting to note that as a result of the documentary research and analysis of the context, it was identified as a crucial need: to have plants with the technological capacity to provide medical device sterilization services. This service was offered only abroad, with implications of time and money for logistics. When considering the technical and economic feasibility of setting up a company in the region, it was estimated that logistics costs derived from the sterilization process could be reduced by up to 40% to producers of medical devices using the services locally.

There is a lack of flexibility for the transfer of technology, summarized in obstacles for a normative and industrial culture. The research centers go at a pace and companies go at another, the government's response is to generate incentive programs for innovation. According to CONACYT (2014), the Impulse to Innovation Program (IIP) financed with public resources proposes three participation modalities:

- INNOVAPYME: exclusively dedicated to projects whose proponent is a company with 100 or fewer employees. It can be individual or linked with HEI or RC.
- INNOVATEC: exclusively dedicated to projects whose proponent is a company with more than 100 employees. It can be individual or linked.
- PROINNOVA: dedicated exclusively to projects presented in link with at least two HEIs or two RCs or one of each.

It can be concluded that in Mexico, efforts have been made to increase the creation of companies based on technology. However, these are considered insufficient from the economic perspective, and it is destined less than 1% of the Gross Domestic Product (GDP) to innovation (Solís 2016). From a strategic standpoint, it is mandatory to define what needs are in the world and what we are going to specialize.

This chapter presents the challenges to innovate in the manufacturing sector of medical devices, which is characterized by the need for creation of TBC. The applied methodology appears in Sect. 2, and it consists of four steps: (1) work starting, (2) preparation for the project, (3) realization of the project, and (4) setting up the study. Section 3 presents the Mexican context related to TBC, the challenges in their start-up, the financial programs to strengthen of the TTP, and the role of the HEI and the RC in this field. Sections 4–6 analyze the case study, the learned lessons, and finally, the conclusions, respectively.

2 Methodology

According to the methodology proposed by Scott and Ramil (2014), the development of a case study is a useful research model to gather information in real-life contexts. Case studies, when carefully designed, can contribute to significant changes in different scenarios. Therefore, this study was performed under this approach, which includes the steps described below.

Step 1: Work starting
First, it is necessary to define the objectives of the study to select the cases and the team working in the research. Among the inclusion criteria are: the selected case study is a new technology-based company with at least one project linked to an HEI and financed with public funds. It is also important that the company agrees to spread its gained experience about the technology transfer process.

Step 2 Preparation for the project
Second, it is required to perform a document review to identify the key issues, establish relationships with strategic contacts, identify consultants in the field to support the research, and the preparation of the project methodology including the scheduling.

Step 3 Realization of the project
Third, put the process into context. It must obtain a systematic overview to determine the scope, focus, and use of the process; work with local contacts or consultants and record information and hold a stakeholder workshop.

Step 4 Setting up the study
Set criteria and determine outcomes that will be used to assess the process and its results.

This methodology is applied in a company with the characteristic of being located on the northern border of Mexico; then, it is necessary to know the particularities of the environment concerning TBC. The next section analyzed the challenges to creating a TBC in the international, national, and regional level. Then, it identified the national programs to support technology transfer and innovation. Finally, it reviewed the importance of HEI and RC in the creation of new technology-based companies.

3 Technology-Based Companies: The Mexican Context

In the last 40 years, the Mexican industrial sector in the Northern Border Region (NBR) has excelled due to its strategic geographical location with the United States, which has a large number of companies located in the country. Another advantage is the abundant and skilled labor, as well as personnel trained at technical and engineering level. Then, the number of companies carrying out high-technology activities has increased, and not only manual and semiautomatic assembly operations. The industry has been specializing, and growth projections are positive. Therefore, the Mexican context for the materialization of projects in which they are involved is identified in detail: private investment, public funds, and universities.

3.1 Challenges to Create Technology-Based Companies at the International, National, and Local Level

Today, in the world population, there is a rising ecological awareness demanding more and better products that are friendly to the environment. Young people are looking for new and varied technology, as well as consumers who want to be better informed. Hence, there are the right context, opportunities, and challenges to create and develop new products to meet this demand. Therefore, it is necessary to strengthen the innovative capacity to enter credible international markets.

The creation of technology-based companies requires a process of technological innovation that can originate new products, processes, and methods of organization and marketing. When substantial changes occur as a result of improvement, these transformations are classified as incremental innovation (small changes or perfections in products and processes) and radical innovation (application of technologies that generate emerging sectors). In fact, these are the upper and lower limits in the classification; however, there are many other classes in the middle (OECD 2005). Whatever the case, the success or failure to implement new technologies depends on variables such as economic, attitudinal, educational, cultural, organizational, competitiveness, infrastructure, and government policies (Leal 2012).

According to Rojas-Pescio and Roa-Petrasic (2016), technology-based companies must have the following characteristics: innovative products and services, based on technologies and scientific knowledge, with intensive R&O (Research & Operations) activities, with high and advanced specialization of their workers. Other characteristics of technology-based companies emphasize the rapid introduction of changes in product and process design, flexibility and specialization of equipment, greater technological dynamism, easy adaptation of production to demand, and finally, an organizational scheme that facilitates the integration of their processes (Elorz 2003).

It should be noted that the creation of a technology-based company is not easy. It is necessary to obtain information about several features, for instance: (1) to know the profile of the entrepreneur and (2) to know with certainty what are their business opportunities in the market. This information facilitates the management and development of the project to identify the type of technology that is most appropriate to develop. (3) The entrepreneur should have the capacity and professional experience to evaluate potential opportunities with common sense, judicious, and above all, an entrepreneurial team spirit.

Challenges at the international level: According to Jiménez (2016), research centers play a valuable role in the generation of innovation in industrialized countries, where the university–company relationship is functional since universities have the technological infrastructure to create and develop new products, favoring new enterprises with a solid structure for impelling innovation. Unfortunately, in Latin America, the situation is different because the link between the universities and the industry is an opportunity area. SMEs do not hire advanced human capital, and their infrastructure is not capable of implementing activities focused on innovation. Table 1 presents the characteristics of the developed countries compared to the countries of Latin America.

Developed countries are distinguished by their innovation systems, which, in coordination with the growth of their companies' internal capacities, have achieved improvement processes that allow them to remain at the forefront of technological and innovation developments. Among the most relevant characteristics of innovation are the labor flexibility, a high industrial specialization, research internationalization, a sophisticated and low-risk venture capital, a technological and innovation orientation, the concentration on the management of industrial production in R&D (Research & Development), etc.

Table 1 Characteristics of innovation between developed and Latin American countries Elorz (2003)

Developed countries	Latin America
Workforce flexibility	There are low rates of participation and training of human resources
High industrial specialization	Innovations are generally at the adaptive and incremental level
University–enterprise relationship is functional since universities have the technological infrastructure to create and develop new products	The HEI is disconnected from the industry and does not have the infrastructure to create activities focused on innovation
High growth of internal capabilities in their companies	Poor growth in internal capabilities
Technological and innovation orientation	Technological development is sometimes individual and isolated limiting its competitiveness
Concentration in the management of industrial production in R&D	They cannot strengthen their R&D systems

In contrast, Latin America and other countries have not yet consolidated their R&D systems, in relation with industrialized countries. Innovations are generally at the adaptive and incremental level. At the entrepreneurial level, technological development is sometimes individual and isolated, limiting its competitiveness. There are few cooperation networks among the same companies, as well as low participation rates and human resource formation (Morales et al. 2012).

Challenges at the national level, Mexico's outlook: A study performed by Pérez Hernández et al. (2017) used the technique of multivariate cluster analysis on the evolution of technological capacity in Mexico. Authors found that there are different levels of scientific capabilities related to the ability of absorption, innovation, and infrastructure, where the gap between the Mexico's states is broadly concentrated mainly in the Mexico City, Nuevo Leon, Queretaro, Jalisco, and Estado de Mexico.

In 2003, CONACYT launched the AVANCE program for the creation and expansion of new companies based on the exploitation of scientific or technological developments. The program has favored the interest of investors in different areas: (a) technological innovation projects with a portfolio of implemented support schemes, (b) national patents, (c) technology transfer offices, (d) business school program, (e) strategic alliances, and (f) innovation networks for competitiveness and seed capital fund. Unfortunately, the resources of the AVANCE program are limited, and the higher number of projects that it supports directs to projects that are in the last stages of its development, displacing the new firms that are usually excluded (Malkin 2012).

The Ministry of Economy has established different programs and instruments of support for the creation of new companies based on technologies that seek their development or consolidation of the market (Malkin 2012). Although these programs are attractive because it is a means for interested companies to access public resources to introduce innovation processes, there are still shortcomings, since resources are limited and lack of coordination among government agencies to create mechanisms to facilitate the support to the companies in the different stages of the innovation process.

Regional challenges in Baja California: The Baja California Research, Innovation and Technological Development System, was created with the purpose of grouping higher education institutions, research centers, and business clusters. These organizations have a common objective: they are capable of strengthening and enhancing the training of high-level human resources level in projects of scientific research, innovation, and technological development. Among the sectors that have been formed are aerospace, information technologies, automotive, electronics, hospital medical, medical products, and logistics. The challenge for these groups is not only to develop the expansion of technological knowledge. They must deal with a lack of sensitivity for identifying the customer's needs, translating the adequate supply into a weakness in achieving innovative products. There are organizational and market-related obstacles that prevent effective coordination between the

Table 2 Areas and specialization niches for the State innovation agenda in Baja California

Aerospace	Advanced manufacturing	Information and communication technologies
Supplier certification	Medical devices	Medical devices
Advanced materials	Automotive	Medical devices, aerospace

different phases of the production process in the value chain (Special Program on Science, Technology and Innovation 2015–2019).

The innovation agenda for Baja California promotes a greater private sector investment in technological development and innovation. The objective is to improve the characterization of strategic infrastructures, as well as the promotion of programs focused on the development and growth of specialized talent.

The innovation agenda aims to create the conditions to create a synergy between sectors and regions as shown in Table 2. Areas of specialization were selected to carry out such strategies as food agroindustry, advanced manufacturing (mainly for the automotive and medical devices industry), aerospace, biotechnology, and alternative energy (Special Program for Science, Technology and Innovation 2015–2019).

3.2 National Programs to Support Technology Transfer and Innovation

The importance of increasing the level of participation of companies in technological innovation projects is continually discussed, but who is responsible for having a culture of innovation and technology transfer? Although the government has an important role in promoting innovation through policies, regulatory exemptions, and special programs, among other initiatives, but are not the only actor (López 2008). Innovation also requires overcoming the inability of many companies to meet technical progress and incorporate new knowledge and technology. The intervention of the state is justified as a regulator of the life of the organizations. This institutional framework assumes that the consequences of market failures on the performance of the economy are greater if the state does not get involved than if it did.

Efforts of the Mexican Government: Mexico through the CONACYT counts until 2017 with the IIP, which focuses on encouraging innovation within the business sector. The general objective is to promote, at the national level, the investment of companies in activities and projects related to research, technological development, and innovation through the granting of complementary financial incentives. In such a way, these supports have the greatest possible impact on the national economic competitiveness. The IIP is relatively new. It emerged in 2009, and for 2017 had supported more than 5,900 projects and 2,900 companies, belonging to 34 different industrial sectors. During this period, it has been possible to support only one-third of the proposals. It is clear that available resources are insufficient to cover the national demand, but the good news is the interest from companies to participate in these types of programs. This is the way to foment and promote the growth and competitiveness of the company (Farias-Zuñiga 2012). Its specific objectives are as follows:

- Encourage the annual growth of investment in the national productive sector in Research, Technological Development and Innovation (RTDI). It is important to note that the program provides complementary economic support, without this means the replacement of the investment that companies make in RTDI activities during the fiscal year.
- Promote the linking of companies in the "education-science-technology-innovation" knowledge chain and its articulation with the productive chain of the strategic sector in question.
- Train and incorporate specialized human resources in RTDI activities in companies.
- Generate new products, processes, and services with high added value, and contribute to the competitiveness of companies.
- Contribute to the generation of intellectual property in the country and to the strategy that ensures its appropriation and protection.
- Expand the coverage base of support to national companies from a decentralized perspective.

This program aims at Mexican companies enrolled in the National Registry of Scientific and Technological Institutions and Companies (RENIECYT), carrying out Research, Technological Development and Innovation (RTDI) activities in the country. An enterprise can participate individually or in connection with Higher Educational Institutions National Public or Private Superiors (HEIs) and National Public Research Centers and Institutes (NPRC&I).

According to Farias-Zuñiga (2012), the PEI is the instrument through which the CONACYT allocates economic resources to encourage companies to invest in innovations that translate into business opportunities. The modalities of the PEI are presented in Table 3.

Table 3 Modalities of the PEI (CONACYT 2014)

Modality	Receiver	Objective
Innovapyme	SME	Projects that generate a high added value and promote the link with other companies
Innovatec	All kind of enterprises	Plans that invest in research and development infrastructure
Proinnova	All kind of enterprises	Proposals in fields that are precursors to knowledge and developed in connection with institutions of higher education or research centers

In Mexico, there is an institution called the National System of Researchers (NSR), which according to the Presidential Agreement published in the Diario Oficial de la Federacion on July 26, 1984, the system aims to recognize the work of people dedicated to the production of scientific and technological knowledge. The NSR has played a very formal role in the constitution and development of a community of qualified, selected, promoted, and rewarded researchers. Without this system, managed centrally by the CONACYT and financed with its budget, the level of excellence of Mexico's research activities and the number and diversity of internationally recognized researchers would not be what they are today.

Statewide efforts in Baja California: At the state level, some studies (Rivera and Valladares 2016; Góngora et al. 2010; Mungaray et al. 2013a, b) suggest that in Mexico, state agencies support innovation by almost 90%, with an incremental budgetary trend. The agencies mainly support training (94.4%), incubation (88.9%), direct technological development (88.9%), through consultancy (83.3%), and productive investment and SME promotion events (83.3% each). Indirectly, the emphasis has also been placed on the development of researchers, who can carry out the transfer of classrooms to companies.

The Network of Technology Transfer Offices in Mexico: The network of Technology Transfer Offices (TTO) in Mexico is an inclusive project since it is made up of public and private educational institutions, companies, and government. This network supports innovation, commercialization, and technology transfer. The TTO network has the purpose of facilitating the interaction between academia and business, and it provides a forum for the exchange of best practices from technology transfer offices. The training of its members supports this effort.

Unfortunately, according to Jalife (2017), after the creation of the TTO infrastructure, CONACYT has stopped the calls that involved them, condemning many of them to inactivity and ostracism. Despite the flaws and dislikes, the plan to give life to the TTO is virtuous and feasible, and we cannot, as a country, give up without persevering or basing the decision on the lack of commitment of a few. The TTO had a very promising background, which was intended to carry out the triple helix and achieve an academic revolution to promote the transfer of technology (Etzkowitz 1993). The process did not lead to an adequate link between the institutions; perhaps, it was a little ambitious to start with so many programs, which, despite belonging to similar sectors, present a wide diversity.

Rivas-Rodríguez and Lino (2014) comment that another obstacle to the operation of TTO is a cultural, related to intellectual property. The research centers go at one pace, and the companies go to another, so the authorities have been forced to generate incentive programs for innovation that require as a requirement that within a project work at least two research institutions and one business. The TTO is not exclusively an institution of CONACYT. However, the presence is essential for the project to survive.

3.3 Importance of HEIs and Research Centers in the Creation of New Technology-Based Companies

The World Bank (2003) has identified four pillars that allow observing the level of development of a knowledge economy. A pillar is educated and skilled labor, which consists of having a well-educated and qualified population, as it is essential for the creation, acquisition, dissemination, and efficient use of knowledge. Another pillar is the system of effective innovation. It is the public and private development of research and development, which results in new products or goods, new processes, and new knowledge.

A third pillar is the appropriate information and communication infrastructure, which consists of the installed capacities that enable the development of innovative, scientific, and technological activities. It is also a pillar where the economic and institutional regime gains knowledge: it refers to the network of institutions rules and procedures that influence the way a country acquires, creates, disseminates, and uses information.

The new theories of growth and trade assume the existence of a solid link between the increase of the knowledge base and the rise of the rate of productivity. Also, the knowledge economy is a pillar for long-term economic growth in Mexico. It is argued that sustained investment in education, innovation, and information technology will lead to an increase in the use and creation of knowledge in economic production, leading to help economic growth in Mexico's 32 states (Sánchez and Ríos 2011).

According to the Global Competitiveness Report (GCR) (2017), economies fall into three categories: (1) factor-based, (2) based on efficiency, and (3) based on innovation. Mexico is in stage two, which refers to an economy based on efficiency, ranking in the 51st position of 139 comparative economies. However, in the 2014–2015 Global Competitiveness Report, Mexico falls to position number 61. Innovation, human capital development, and institutional strengthening continue to play a decisive role in determining which the most competitive economies in the world are.

The objectives, strategies, and lines of action of the Special Program of Science, Technology and Innovation 2014–2018 (PECiTI) are aligned with Goal III: Mexico with Quality Education of the National Development Plan (2013–2018). The Objective 3.5 says: "*To make scientific, technological and innovation development pillars for sustainable economic and social progress.*" (CONACYT 2014; Diputados 2014). This goal is based on the existing empirical evidence that shows that societies that put knowledge at the base of their transformation and development have access to better levels of well-being. To achieve the mentioned objective, five strategies are followed:

- Contribute to national investment in scientific research and technological development to grow annually and reach a level of 1% of GDP.

- Promote the development of local scientific, technological, and innovation vocations and capacities to strengthen sustainable and inclusive regional development.
- Contribute to the strengthening of the country's scientific and technological infrastructure.

From the perspective of the economy of innovation, an evolutionist proposal of technological innovation has been developed, in which different actors that are within the structure of a system recognized as National Innovation System come into play. These are as follows: (1) universities, companies, and productive sectors in general, research centers, financial system, technical HEI, and intermediary organizations to support business activity (Gutiérrez 2004; Dutrénit et al. 2001); (2) service providers, product designers, and consumers (Gutiérrez 2004; Nelson, 1993); and (3) government or the government structure.

According to Gutiérrez (2004), state governments, companies, and the academy turned out to be the fundamental elements, but not the only ones. The production and transfer of scientific and technological knowledge constitute the essential exchange, learning, and interaction between the actors of this system.

The Law on Science and Technology of Mexico (Congreso de la Unión 2015) for the NSSTI (National System of Science, Technology, and Innovation) consolidates the State policy in carrying out scientific, technological, and innovation activities and promotes coordination and cooperation in this area. It is made up of the following actors and elements:

- The State policy in Science, Technology, and Innovation (STI) defined by the General Council of Scientific Research, Technological Development and Innovation.
- The Special Program in Science, Technology, and Innovation, as well as sectorial and regional programs in science, technology, and innovation.
- Guiding principles and legal, administrative, and economic instruments to support scientific research, technological development, and innovation established in the regulations.
- The dependencies and entities of the Federal Public Administration that carry out activities of scientific research, technological development, and innovation or support to them, as well as the institutions of the social, private sectors, and governments of the federative entities, through the procedures coordination, participation, and linkage in accordance with the Law and the applicable regulatory framework.
- The National Network of Groups and Research Centers and the scientific research activities of universities and higher education institutions.

The NSSTI is composed of the instruments of government, public policy and planning, and a set of actors for whom, due to their diversity, the articulation work is complicated: the public sector in their three levels, the academic sector and research, and the set of companies with STI activities. According to Robles-Peiro (2002), it should be noted that innovation in the new economy is based on the link between the different economic and social actors. The generation and application of knowledge occur mainly through the interaction of a three-horn propeller: government, entrepreneurs, and universities (Leydesdorff and Etzkowitz 2001; Robles-Peiro 2002).

Currently, CONACYT (2014) is located as the coordinator and articulating axis of the NSSTI. The system has robust links with HEI and RC. The three pillars of a successful regional innovation system are as follows: (i) enterprises organized in functional clusters; (ii) the academic sector, through timely and relevant educational provision and effective technology transfer mechanisms; and (iii) government through of a proactive policy with targeted and strategic actions (Mungaray et al. 2013a, b).

Higher education, also called tertiary education, contributes to the improvement of the institutional system by training the competent and responsible professionals required for sound management of the economy and the public sector. Its academic and research activities provide support to the national innovation system (The World Bank 2003). According to Mungaray et al. (2013a, b), in the knowledge economy, the first task of regional development policies is to promote processes of learning and interaction among systems, subsystems, organizations, and individuals that allow sustainable economic and social development. In the knowledge economy, the more actors participate in their exploitation, and the more linkage between them is much better (Robles-Peiro 2002).

Universities are the chief centers of research, both basic and applied. For several reasons, it is important to maintain advanced training and research programs at the graduate level. Graduate students are required to work in research and development institutes, both public and private, as well as in high-tech manufacturing companies. Graduate programs are essential to train university teachers and improve the quality of tertiary education, not only current but future generations (The World Bank 2003). According to Mungaray et al. (2013a, b), HEI that carries out research plays an important role as a source of information and fundamental knowledge. More occasionally, they are a source of relevant industrial technology to support a knowledge-based economy.

The participation of university researchers in the national industry contributes to the generation of a stable scientific and technological research capacity in the HEI, as well as to the creation of a research and development ability in the industry itself, which proves to be an indispensable condition for the construction of an integrated system of science, technology, and production (Acuña 1993). Torres (2010) mentions that the HEIs have the imperative commitment to confront their curricula and adjust them regarding competitiveness, relevance, and quality, to make them plausible with the new educational, cultural, economic, and policies that come from the process of economic globalization.

The links between universities and the demands and needs of a concrete city, region, or territory have historically not been evident. In society, there is a growing request for a closer and more direct link between higher education institutions and other actors in economic, social, cultural, and technological life. Also in developing countries, the university is increasingly viewed as an agent for the development of potential, resources, and values for society itself, even in a complex reality with an uncertain future such as that of emerging economies (Petrillo et al. 2013). Today's universities must assume a leading responsibility in shaping the new reality, which is now more directed toward collective well-being, and social development, promoting its link with society and the transfer of technology (Petrillo et al. 2013). However, the problem is that there cannot be an effective cooperation if there is no real link between the school and the company, which may apply to real-life processes (Martinez et al. 2010).

Moreno Zagal and Maggi Yáñez (2011), mention that linking is a fundamental area in HEI, which provides them with the necessary tools to meet the requirements of all social actors through relevant professional training, following these needs. Also, it establishes the necessary bridges of communication and exchange between sectors, which facilitates the students of the higher level their insertion in the social and productive sectors; so they can contribute to their dynamism and strengthening in an active and committed way. Gutiérrez (2004) also mentions that new conceptions regarding linkage have been presented, which in turn have given rise to different policies and programs of action:

1. The linkage as a possibility to support the industrial development of Mexico under an economic model of technological independence.
2. As a possibility to attend and solve social and productive problems.
3. As a strategy of economic development to promote technological innovation, within a model of economic market.
4. As a strategy of technological innovation in which various actors interact based on government coordination and regulation. The purpose of this study is to determine the challenges in the start-up of companies in Mexico, considering as the main requirement that these companies apply innovative technologies in their processes, and that they are willing to participate in the TTP. Considering this scenario, for the particular case study in Baja California, the dimensions of the participatory sector of medical devices and their expectations of growth were identified, followed by the definition of the needs and the challenges for TTP.

4 Case Study

The analysis of the selected case study can serve as an example, and a similar organization can learn from this experience. Baja California is known as a technological and knowledge frontier, where the geographic characteristics (presented in Fig. 1), the highly specialized and abundant labor force, and government policies to attract foreign investment are just some of the reasons why it is one of the most important industrial cities in Mexico.

In Baja California, several industrial clusters are going through the embryonic and growing phases (considered as emerging industries), as well as other industrial units in stages of maturity and aging (considered industries under risk). For these

Fig. 1 Geographical location of Baja California

industrial sectors, innovation, research, and development are critical factors for emerging industries as well as for mature industries. Figure 2 presents the medical device group standing out in emerging activities as a growing cluster; therefore, it was chosen for analysis.

Baja California has gained international recognition as a leading hub for medical device manufacturing having the largest dedicated workforce in North America with companies such as CareFusion, Medtronic, Covidien, Smiths Medical, NxStage, Nypro Healthcare, Teleflex, DJ Orthopedics, Greatbatch, Welch Allyn, Thermofisher, Fisher and Paykel, Augen, Masimo, Avantti Medi Clear, Õssur, Flextronics Medical, Martech, etc. According to the Tijuana Maquiladora Industry Association, the medical devices cluster is integrated by approximately 67 companies (60% located in Tijuana) and over 45,000 direct jobs (Cluster Medical Devices 2013).

Work starting: The purpose of this study is to determine the challenges in the start-up of companies in Mexico, considering as the main requirement that these companies apply innovative technologies in their processes, and that they are willing to participate in the TTP. Not only is it important to know the challenges for these companies, but the primary objective is also to describe how restrictions impel the innovation process more than obstruct it.

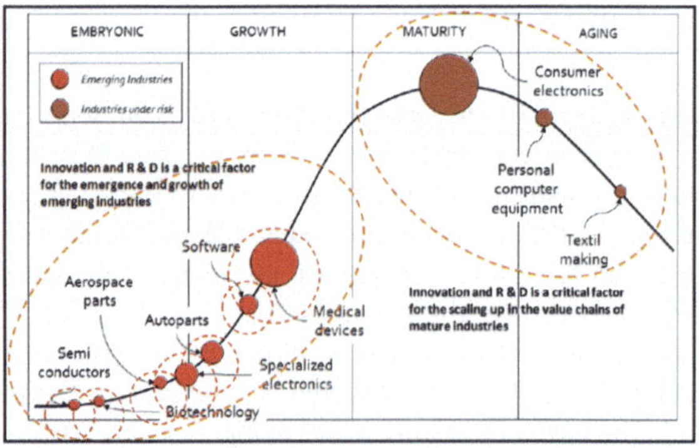

Fig. 2 Innovation and technological development in the policy of industrial clusters in Baja California (Vargas-Leyva 2006)

After analyzing the regional industry, it highlighted the MDC as an emerging industry where innovation, research, and development are critical factors for its growth. The San Diego Union-Tribune affirmed (2017) that the medical sector is bringing much innovation and new technology. They are investing to add value to their products and to reduce their costs. When surveying some representatives of the cluster, the reduction of the times and costs related to the sterilization process was identified as one of the leading areas of opportunity. Among the latest developments has been the establishment of a sterilization plant for medical equipment.

AMC, expected to launch operations early next year, is building a facility in eastern Tijuana. Gerardo Meave, president of AMC, said the high concentration of medical device companies was a key to his company's decision to establish a facility that will use electron-beam technology to sterilize medical products. Establishing a company based on high technology requires a considerable investment; therefore, it is of interest to answer the following questions:

- Why is it necessary to start-up a medical equipment sterilization company in Tijuana?
- What technologies exist for the sterilization process and what is the best option for market conditions?
- What are the challenges in setting up the plant? And, how is it possible to carry out the technology transfer process to innovate processes even with obstacles, based on the experience of AMC?

For the particular case study in Baja California, first, the dimensions of the participatory sector of medical devices and their expectations of growth were identified, highlighting the shortage in the provision of sterilization services in the region. The MDC established in 2012 the commitment to open a sterilizing plant in Tijuana before 2020.

Preparation for the project: Texas A & M University System (R&D) is an HEI expert in sterilization technologies; so AMC started collaborative relations in 2012. As a strategic alliance, the AUBC was included, with the objective of carrying out a transfer of technological knowledge to perpetuate and disseminate the new technology. The collaborative research works began in 2013; a year after in 2014, AMC received support from the IIP of CONACYT in the PROINNOVA modality, where the AUBC was a valuable partner.

In 2015, the construction of the sterilization plant was completed based on the specifications of the Nuclear Safety and Safeguard Commission, and operating licenses were managed. The import of the equipment in the Secretary of Tax Administration concluded at the end of the year. In 2016, MEVEX (the company developing the technology) installed the equipment. In 2017, AMC was in the process of certification in ISO 9000, ISO 13485, and ISO 11137 norms. Finally, the company offers its services to the Medical Device Cluster in 2018.

Realization of the project: The medical devices manufactured locally are sent to other countries to receive the service of sterilization, incurring transportation, and storage costs. In some cases, the sterilized medical devices must be returned to Tijuana before being released. It was estimated if the service was offered in the region, costs related to sterilization could be reduced by up to 40%.

There are three methods of sterilization used in medical devices: EtO (Ethylene Oxide), Gamma irradiation using Cobalt-60, and Electron-Beam (E-Beam) irradiation. Why choose E-Beam over EtO and Gamma? Table 4 presents a comparison between these technologies. E-Beam is considered as the green solution; due to environmental issues such as toxicity of EtO, and handling issues associated with radioactive Cobalt-60, among others, governments around the world are looking at sustainable and green alternatives as E-Beam; it is a "switch on-switch off" technology.

The project lasted five years; in the first instance, it can be considered excessive. However, when planning the development of the project and establishing the critical route, a lot of intermittency in the activities was identified, as well as non-avoidable waiting times. It is of interest to recognize that there are times not considered in the planning phase, so initially, we had an expected time of 3 years to start operations instead of 5.

The strategic collaboration for the development of the project is presented in Fig. 3. Texas A & M University System (R&D) signed a collaboration agreement to train the company and the AUBC in the use of E-Beam technology for a year with the possibility of making updates in the future; the E-Beam technology is being studied for possible future applications. MEVEX provided the E-Beam system and installed it. CONACYT and SEDECO financed part of the project considering it important for the development of regional competitiveness. The AUBC participated with a research group on prototype development activities, operation manuals, work instructions, and operation analysis.

Setting up the study: AMC will guarantee a dose of radiation, not sterility. How is this possible? For the process of granting the sterilization service first, the client performs a microbiological analysis of his product; the count will depend on the

Table 4 Comparison of sterilization techniques (Avantti Medi Clear 2017)

E-Beam	EtO	Gamma
Electricity is the source Majority of the medical devices are compatible with E-Beam Extremely high product throughput (minutes) E-Beam process adapts to costumers needs Proven technology and worldwide applicability Switch on–switch off technology	Toxic gas Flammable, explosive and highly toxic carcinogen Slow process (days) High operation cost Penetration is difficult for some materials Process eaves behind residuals	A radioactive isotope is the source Environmental issues associated with the handling of Cobalt-60 Slow process (usually about 8 h) The process cannot customize for individual packages

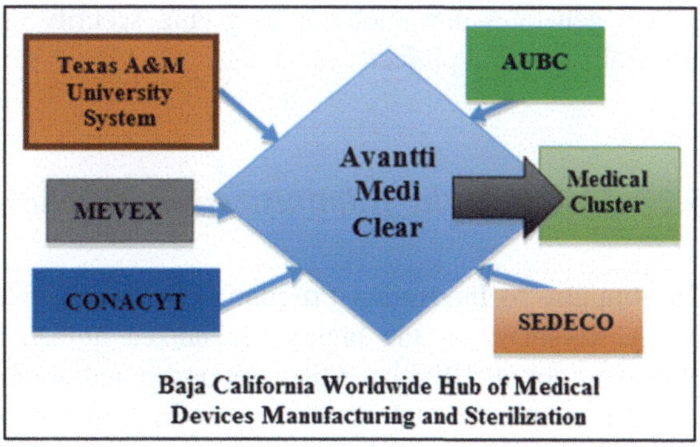

Fig. 3 Strategic collaboration (adapted from Avantti Medi Clear 2017)

parameters of the clean room where the product is manufactured. Then, considering this count, AMC carries out a dosimetry study according to the dimensions of the product, dimensions of the box used in the packaging, arrangement of the individual packages in the box, type of material, and the microbiological count. It is validated that sterility is achieved according to these characteristics when a specific dose is applied. Therefore, a dose is guaranteed, not sterility. The sterility will depend on whether the company maintains or reduces its microbiological count.

5 Learned Lessons

Through the E-Beam technology transfer process, business risk situations were presented. There are described and the way in which they were addressed to overcome the challenges presented.

5.1 Long Start-up Times for the Company

The process of technology transfer tends to be slow, due, among other things, to the size of the investment and the learning curve necessary to carry out the introduction of the process in the national territory. The company workers accomplish this task. The implementation of the new production process suffers several modifications from situations not considered in the critical route planning.

When the times for the realization of the project are lengthened, it is important that investors never lose the conviction in the business, they must constantly think about the incentives, and have full confidence in the project. For investors to have this security, the technology to be transferred must have the characteristic of not becoming obsolete in the short term. Also, there is the expectation of looking for new applications because it is a differential technology. As a consequence, the project generates new motivations giving security to the investor to be the economically feasible project.

5.2 Limits on the Administrative Structure of HEI

The Impulse to Innovation Program is very attractive for companies seeking for taking advantage of the highly specialized human resource generated by HEI. However, not all HEIs have the experience and administrative structure to engage with the industry in high-technology and innovation processes. It is important that companies interested in linking have a general agreement of collaboration with universities before making a specific agreement for a particular project; this generates a greater commitment from both parties.

5.3 The Requirements for Certification in the Standards Are Contradictory

In the experience of the case study, the team realized that the procedure for certification in the mandatory standards seems contradictory. For example, to operate the process, you must be certified, however, to certify you are asked to have validations of the process already operating. The operations manager recommends a validation with alternative products. In this case, the products to be sterilized in the projected service are medical devices, but to achieve the certification, we carried out sterilization runs to other products such as dog food to generate historical data that would validate the process.

5.4 Requirement of Highly Specialized Personnel

Although research centers and universities have a highly specialized and training staff, generally the technology transferred is scarce and commonly unknown. To solve this situation, funds were allocated for the training of undergraduate and graduate students, as well as teachers who participated actively in the project. At the same time, the participants committed to the dissemination of the new technology in the academic and industrial sector, as well as the processes and standards related.

5.5 *Disinformation About the Technology by the Agencies that Grant the Operating Permits*

Finally, one of the most recurring obstacles was related to legal administrative procedures. A large number of people who request different procedures surpass the government institutions granting the permits. Other instances, such as customs, are unaware of the technology and take considerable time to make a decision fearing the technology may be used for purposes not defined in the specifications. In these cases, the administrative process should privilege the permanent communication with the different institutions, sensitizing and educating about the new technology. It is necessary to break paradigms regarding the use of new technology; educating and informing about this will allow the obtaining of licenses to be less complicated. Companies interested in making the transfer must be willing to disseminate the new technology and generate trust and expectation in it.

6 Conclusions

The technology transfer process is a challenge in itself for companies that risk several resources to provide innovative products and services. In fact, Mexican innovations are on a constant basis, which are registered mostly by transnational companies; this accounts for the great creativity of Mexican professionals who, even with the current socioeconomic conditions, considered highly restrictive, becomes a fertile ground for both domestic and foreign private investment.

The company AMC is an example of perseverance to provide an innovative service for the medical device sector with the aim of making it more competitive. It is possible to replicate this case study in a different place in México. The applications of the E-Beam are not limited to the sterilization of medical devices. It is possible to preserve the shelf life of perishable foods such as fruits and meats. Another application is the cross-linking of particles for the modification of properties in plastic materials.

The general methodologies that other companies can learn from this case are as follows: (a) approach the specialized manufacturing sectors and know their needs, (b) identify the strategic alliances to obtain financing and knowledge, (c) make strategic alliances for knowledge management, (d) develop the transfer project for the creation of the technology-based company, and (e) provide the product or service. There are other industrial sectors carrying out projects of high magnitude and impact both social and economic. However, the MDC stands out as a fast-growing emerging sector and is therefore believed to remain of interest for future research.

References

Acuña, P. (1993). Vinculación universidad-sector productivo. *Revista de la educación superior, 22*(87), 1–15. Retrieved July 1, 2017, from http://publicaciones.anuies.mx/pdfs/revista/Revista87_S2A3ES.pdf.

Avantti Medi Clear. (2017). Why choose E-Beam?. Avantti Medi Clear. Retrieved November 2, 2017, from http://avanttimediclear.com/e-beam/why-choose-e-beam/.

Cámara de Diputados del H. Congreso de la Unión. (2015). Ley de ciencia y tecnología. 1–49.

Cluster Medical Devices. (2013). Asociación Industrial de Productos Médicos de las. Retrieved November 2, 2017, from http://www.investinbaja.gob.mx/docs/en/industries/medical-equipment-and-supplies/Cluster_de_Productos_Medicos.pdf.

Consejo Nacional de Ciencia y Tecnología (CONACYT). (2014–2018). Programa Especial de Ciencia, Tecnología e Innovación. (2014–2018). Retrieved July 1, 2017, from https://www.conacyt.gob.mx/images/conacyt/PECiTI_2014-2018.pdf.

CONACYT. (2014). Programa especial de ciencia, tecnología e innovación 2014–2018. 1–108.

de Ciencia, Programa Especial. (2014). Tecnología e Innovación 2014–2018. *Poder Ejecutivo Federal, México*. DOF: 30/07/2014. Recuperado en http://201.147.98.8/dofdia/2014/jul14/pdf/30jul14_conacyt2.pdf.

de Diputados, C. (2002). Ley de Ciencia y Tecnología. *Gaceta Parlamentaria*, (987-II). Última reforma publicada DOF 08-12-2015. Retrieved July 1, 2017, from http://www.diputados.gob.mx/LeyesBiblio/pdf/242_081215.pdf.

de Diputados, C. (2014). Consejo Nacional De Ciencia Y Tecnologia. Plan Nacional de Desarrollo 2013–2018. *D. Of. la Fed., 6,* 6–82.

Dutréinit, G. (2001). *Celso Garrido y Giovana Valenti*. México, UAM: Sistema Nacional de Innovación. Temas para el debate en México.

Elorz, S. K. (2003). *La creación de empresas de base tecnológica: Una experiencia práctica*. España: Proyecto para la promoción de empresas innovadoras de base tecnológica.

Etzkowitz, H. (1993). Technology transfer: The second academic revolution. *Technology Access Report, 6,* 7–9.

Farias-Zuñiga, A. (2012). *El Programa de Estímulos a la Innovación, Restricciones e incentivos a la innovación en México*. Cámara de Diputados/ LXI Legislatura, México, D.F.: Centro de Estudios Sociales y de Opinión Pública H.

Góngora, B., Gabriel, M. G., & Antonia, E. (2017). apoyo a la innovación de la PyME en México. Un estudio exploratorio Investigación y Ciencia. ISSN 1665-4412. Retrieved July 1, , from http://www.redalyc.org/articulo.oa?id=67413393004.

Gutiérrez, N. G. (2004). La vinculación en el ámbito científico-tecnológico de México. Instituciones de Educación Superior en interacción con distintos actores Revista Latinoamericana de Estudios Educativos (México) [en línea] 2004, XXXIV (2° trimestre). ISSN 0185-1284. Retrieved July 18, 2017, from http://www.redalyc.org/articulo.oa?id=27034204.

Mauricio, J. (2017). Abandona Conacyt red de Oficinas de Transferencia de Tecnología. Ciudad de México: *El financiero*. Retrieved July 1, 2017, from http://www.elfinanciero.com.mx/opinion/abandona-conacyt-red-de-oficinas-de-transferencia-de-tecnologia.html.

Jiménez. (2016). Relaciones Universidad- Empresa: hacia una productividad basada en innovación Revista Gestión y Tendencias (gesten) volumen 1 número 2 año 2016. ISSN: 0719-6768. https://doi.org/10.11565/gesten.v2i1.11.

Leal, M. (2012) Factores determinantes de la innovación tecnológica de las PYMES del sector confección Revista de Ciencias Sociales (Ve), vol. XVIII, núm. 3, julio-septiembre, 2012 (pp. 540–552).

Leydesdorff, L., & Etzkowitz, H. (2001). The transformation of university-industry-government relations into a triple helix of innovation. *Electronic Journal of Sociology*, *5*, 4. Retrieved July 1, 2017, from https://www.researchgate.net/profile/Loet_Leydesdorff/publication/248040198_The_transformation_of_university-industrygovernment_relations/links/5663e04008ae15e7463142e8.pdf.

López, E. (2008). Un criterio de eficiencia para la concepción y la evaluación de las políticas públicas. Revista de Economía Institucional, *10*(18), Bogotá, 2008.

Malkin, D. (2012). Evaluación de la OECD del Sector de las Nuevas Empresas basadas en el Conocimiento: México. Organ Para La Coop Y El Desarro Económico OECD 97.

Martínez, L. M., Leyva, M. E. Y., Barraza, A. (2010). Importancia de la vinculación en las Instituciones de Educación Superior. Retrieved July 11, 2017, from www.imef.org.mx/publicaciones/.

Moreno Zagal, M., Maggi Yáñez, R. E. (2011) Estrategias de vinculación de las universidades mexicanas con las empresas. XI Congreso Nacional de Investigación Educativa, Memorias, 1–10.

Morales, M., Ortiz, C. Y., Arias, M. (2012). Factores determinantes de los procesos de innovación: una mirada a la situación en Latinoamérica. Revista Escuela de Administración de Negocios (on line) 2012, (Enero-Junio) ISSN 0120-8160. Retrieved July 1, 2017, from http://www.redalyc.org/articulo.oa?id=20626818015.

Mungaray, A., López, S., & Moctezuma, P. (2013a). La Adicionalidad de los Fondos Públicos en la Innovación Empresarial Mexicana. *El Caso de Baja California, 2001–2010,* 2001–2010.

Mungaray, A., López, S., & Moctezuma, P. (2013b). La adicionalidad de los fondos públicos en la innovación empresarial mexicana: el caso de Baja California, 2001-2010. *Revista de la educación superior, 42*(167), 41–62.

Nelson, R. (Ed.). (1993). *National innovation system: a comparative analysis.* Nueva York: Oxford University Press.

OECD. (2005). *Oslo manual: guidelines for collecting and interpreting innovation data.* Tragsa Group: Third Edit.

OCDE (2012). Evaluación de la OCDE del sector de nuevas empresas basadas en el conocimiento: México. Retrieved July 1, 2017, from http://www.oecd.org/centrodemexico/Evaluaci%C3%B3n_de_la_OCDE_del_sector_de_las_nuevas_empresas_%20IMPRENTA-1.pdf.

Oxford Business Group. (2017). Leading medical devices exporter in Mexico's health sector experiences rapid growth | Mexico 2017 | Oxford Business Group. Retrieved October 5, 2017 from, https://www.oxfordbusinessgroup.com/analysis/latin-america's-leading-exporter-medical-devices-continues-experience-rapid-growth-necessary-devices.

Pérez Hernández, C.C. et al. (2017). Evolución de la capacidad tecnológica en México. Aplicación del análisis estadístico multivariante de cluster. Contaduría y Administración, *62*(2), 505–527.

Petrillo, J. D., Doumecq, J. C., Petrillo, M. I., & Morcela, A. O. (2013). La Gestión de la Innovación tecnológica y el desarrollo territorial. Red EmprendeSur 2013: Fomento y apoyo al emprendimiento y la innovación. Consultado en. https://www.researchgate.net/.

Programa Especial de Ciencia, Tecnología e Innovación. (2015–2019). Retrieved July 1, 2017, from http://www.siicyt.gob.mx/index.php/normatividad/estatales/programas-estatales/1191-baja-california/file.

Programa Nacional de Innovación. (2011). Comité Intersectorial para la Innovación. Retrieved July 19, 2017, from http://www.economia.gob.mx/files/comunidad_negocios/innovacion/Programa_Nacional_de_Innovacion.pdf.

Rivas-Rodríguez, R., Lino, M. (2014). Falta flexibilidad para transferir tecnología. Ciudad de México: El economista. Retrieved July 7, 2017, from http://eleconomista.com.mx/entretenimiento/2014/06/17/falta-flexibilidad-transferir-tecnologia.

Rivera, M., Valladares, V. (2016). Redalyc. Financiamiento de la Ciencia, la Tecnología y la Innovación en las Regiones de México.

Peiro, H. R. (2002). La economía basada en el conocimiento. Las condi-ciones de los estados mexicanos. Information Society Commission) tomado de: http://www.razonypalabra.org.mx/anteriores, (49).

Rojas-Pescio, H. G., & Roa-Petrasic, V. A. (2016). *Tecnología: Resultados De Investigación, 27,* 67–79.

Sánchez, C., Ríos, H. (2011). La economía del conocimiento como base del crecimiento económico en México. *Enl@ce: Revista Venezolana de Información, Tecnología y Conocimiento, 8*(2) 43–60.

Secretaría de Gobernación. (2013). Plan Nacional de Desarrollo 2013–2018. *Gobierno de la República. México. DF*. http://pnd.gob.mx/.
Stott, L., & Ramil, X. (2014). *Metodología para el desarrollo de estudios de caso*. ITD, UPM: Centro de innovación en tecnología para el desarrollo humano.
Solís, A. (2016). Informe de Gobierno: Gasto en ciencia y tecnología no alcanzará 1% del PIB en 2016 • Forbes Mexico. Retrieved October 5, 2017 from https://www.forbes.com.mx/pena-nieto-presume-gastar-mas-en-ciencia-y-tecnologia-que-fox-y-calderon/.
The San Diego Union Tribune. (2017) Medical device sector creates jobs in Tijuana. Retrieved August 11, 2017, from http://www.sandiegouniontribune.com/news/border-baja-california/sdut-medical-device-sector-a-bright-spot-for-tijuana-2012jun15-htmlstory.html.
The World Bank. (2003). Construir Sociedades de Conocimiento: Nuevos Desafíos para la Educación Terciaria.
Vargas-Leyva, M. R. (2006). *Perfiles de Competencias Profesionales Demandados por Cinco Sectores Productivos en la Ciudad de Tijuana*. Tijuana: Instituto Tecnológico de Tijuana Ediciones Del XXXV Aniversario.
Torres, P. A. (2010). La competitividad, calidad y pertinencia de los currículos: una exigencia inaplazable del proceso de globalización económica. *Memorias*, *8*(13), 266–276. Retrieved July 1, 2017, from https://revistas.ucc.edu.co/index.php/me/article/view/203/204.
World Economic Forum. (2017). The Global Competitiveness Report. Inside Report The Global Competitiveness Report 2016–2017. Klaus Schwab.

Applying Action Research to Academic Patent-Licensing Path

Daniel Barrón-Pastor, Norma García-Calderón and Julio Alcantar

Abstract In this chapter, an action research program and two projects that followed the patent-licensing path are presented. The program had the objective of increasing patent-licensing outputs and researching the processes and factors that influence the possibility of success of these activities. The active research methodology proved to be a valuable approach for simultaneously solving both the research goals and the problem-solving goals. The first project presented is related to the development of a new product to human diagnostics. The second one refers to new technology development, with a potential application in electronics. Both projects are the objective of extensive collaboration, particularly in the early research stages, for posteriorly going through the R&D and the patent-licensing activities. The projects also follow an iterative, stage-gated process and were influenced by internal and external factors. The most relevant findings reveal that the experience presented could be useful for developing patent-licensing projects in different contexts and under restrictive conditions.

Keywords R&D management · Patent strategy · Licensing

1 Introduction

The patent-licensing activities were rare events for academic institutions until the 1980s, wherein the Bayh–Dole Act (US Public Law #96-517) allowed them to patent inventions funded with federal money, and to retain the royalties that these patents could generate (Argyres and Liebeskind 1998). Thus, the patent-licensing paths "*constitute immediate, measurable market acceptance for outputs of aca-*

D. Barrón-Pastor (✉) · N. García-Calderón
Instituto Potosino de Investigación Científica y Tecnológica, San Luis Potosí, Mexico
e-mail: daniel.barron@ipicyt.edu.mx

J. Alcantar
Centro de Investigación y Desarrollo Tecnologico en Electroquimica,
Santiago de Querétaro, Mexico

G. Cortes-Robles et al. (eds.), *Managing Innovation in Highly Restrictive Environments*, Management and Industrial Engineering,
https://doi.org/10.1007/978-3-319-93716-8_4

demic research" (Perkmann et al. 2013). OECD members and other countries applied similar rights and diverse incentives for increasing the patent-licensing activities in the academic organizations (Geuna and Nesta 2006; Sampat 2009). In response, academic institutions had been transforming the governance of knowledge transfer activities and implementing diverse support structures, such as the knowledge or technology transfer offices (Markmann et al. 2008; Geuna and Muscio 2009). However, empirical evidence showed that few academic institutions had been making profits from the patent-licensing activities, while most institutions are losing money (Mowery et al. 2001; Sampat 2006). These drove us to question: Why was so difficult for academic institutions to succeed in patent licensing with the purpose to commercialize knowledge? How can be improved the institutional patent-licensing activities without a negative impact on research activities? And not less important, what are the factors that influence the possibility of success in these activities?

This chapter presents action research (AR) program performed between 2008 and 2016 in a public research center in Mexico. The purposes of this program were improving the institutional results related to increasing patent-licensing outputs, while researching the factors that influence the possibility of success of these activities. A detailed analysis of two projects of academic research following the patent-licensing paths is presented and discussed.

In the first section, it is presented the AR methodology and the guidelines proposed by Henfridsson and Lindgren (2007). The second section describes the results in the following order: 1. The context and goals of the AR program. 2. This section describes the processes of formalization of the collaboration between the Intellectual Property Department (IP team) and two R&D teams. 3. The implementation and development of both AR projects, and 4. The analysis of the experience and how the objectives are fulfilled. Section three presents an iterative, stage-gated process and the external factors that influenced the project's development. Finally, the conclusion and remarks are presented.

2 The Action Research Methodology

Action research (AR) has been recognized as a valid research method in applied fields such as organization development, education (Coghlan and Brannick 2014), new product development (Henfridsson and Lindgren 2007), and lately proposed for patent planning (Soranzo et al. 2017). This approach emphasizes the collaboration between researchers and practitioners in solving both the research goals and the problem-solving goals. It is important to note that the "*AR differs from case study research in that the researcher is directly involved in planned organizational change. Unlike the case study researcher, who seeks to study organizational phenomena but not to change them, the action researcher is concerned to create organizational change and simultaneously to study the process*" (Avison et al. 2001). Therefore, AR projects are highly situational and different variants have been proposed for achieving a different set of goals (Baskerville and Wood-Harper 1996). In this diversity, Henfridsson and Lindgren (2007) proposed four general guidelines for the management of AR projects, which were followed during this project. The four guidelines are as follows:

(a) *The identification and specification of a situation where the problems exist (context)*: This guideline refers to the need of defining the context wherein the problem resides. In other words, defining time, geo-economic, organizational issues, and those factors that intervene in the identification and generation of the problem. Also, it is important to show how the problem is discovered by the researcher, the action researcher, or evolves from research–client interactions. As mentioned by Avison et al. (2001) "*each AR project, to some extent at least, is unique*" and highly situational. Thus, defining the initial context is very important to the AR project itself, but it is also important for evaluating the changes made during the AR and communicating the experiences.
(b) *The formalization of the collaboration between the practitioners and the researchers (formalization)*: This guideline arises from issues of control and authority management from the beginning of an AR project and during its development. For instance, Avison et al. (2001) explained that control structures could be formal, informal, and evolved, and should include action warrants and processes for renegotiation and/or cancellation of the project.
(c) *The implementation of full learning cycles of understanding, supporting and improving practice (implementation)*: This guideline refers to the sequence of steps by which AR is conducted. AR projects are iterative processes of a four stages cycle: diagnosis–action planning and taking–evaluation–reflection. In each stage, researchers and practitioners should be involved in an adequate time, with a systematic collection of data and interpretation methods.
(d) *The systematic collection of data and suitable methods of interpretation (analysis)*: This guideline refers to the kind of data used and generated during the AR project, how they are selected, systematically collected, and analyzed.

In the next section, the AR program and the detailed analysis of two projects are presented, following the same order of the four guidelines proposed by Henfridsson and Lindgren (2007).

3 The Active Research Program

The AR program and projects presented in this chapter share some context and goals. (Section 3.1) Both projects were initiated by informal collaboration agreements (Sect. 3.2), while each project has its development path. (Section 3.3) The interpretation of the experience was presented and the analysis of some internal and external factors that influenced the projects was performed. (Section 3.4) An iterative, stage-gated process found, was presented in the discussion section.

3.1 The Context of the AR Program and Projects

The Instituto Potosino de Investigación Científica y Tecnológica (IPICYT) was created on November 24, 2000, as a local government initiative. It was integrated into the National Council of Science and Technology (CONACYT) network of public research centers in 2002, allowing it to receive federal funds. Located in San Luis Potosí (SLP), a city with a population of almost one million in the center of Mexico and not far from the three main cities of the country: Mexico City, Guadalajara, and Monterrey. The most important economic activities in the SLP city are manufacture, logistic services, and commerce. Nevertheless, smaller populations, mostly dedicated to the agriculture and animal husbandry, surround the city but some had large mining operations. Also, there are large natural areas (The Huasteca rainforest and the Altiplano arid zone) that have complex ecosystems with several endemic species that face anthropogenic impacts. Based on these regional characteristics, the IPICYT was designed as a multidisciplinary science and technology research center with capabilities on Molecular Biology, Environmental Sciences, Geosciences, Advanced Materials, and Applied Mathematics, with the missions of impel research, higher education, and get involved in the regional economic development.

For developing the "third mission" of economic development, in 2002 the IPICYT hired an industry–academy relationship officer. Their objective was to increase profit by fostering the contract research and consultancy with the industrial sector and the government. Between 2005 and 2007, IPICYT filed two patent applications per year by outsourcing the complete IP management process. Nevertheless, it proved to be very costly and inefficient.

In 2008, the IP Team was integrated, initiating the implementation of the strategy wherein it was planned as the 'Active research' program, which is presented in this paper. At the beginning of the AR program, the institutional goal was about increasing patent-licensing outputs more sustainably. This goal generated a situation wherein, on the one hand; the researchers were required to produce patentable knowledge that could be licensed to companies (with no prior experience in these topics). On the other hand, the IP team had to develop a sustainable process for increasing the possibility so that basic science findings could be incorporated in the patent-licensing path for generating profits for the academic institution. To reach these goals, the IP team focused on two opportunities: (1) Identifying science findings and research topics that could be developed into patentable technologies or products (for generating a pipeline of potential patentable cases) and (2) To analyze the market opportunity and potential to be licensed (for selecting those cases with licensing potential). The results of this first stage were disclosed in Barron-Pastor and García-Calderon (2014).

3.2 Toward a Formal Collaboration

As the AR program evolved, two cases were selected for developing long-term AR projects. Both cases shared the following criteria: (a) the science findings developed by the R&D teams were in a very early stage, (b) the potential application and market opportunity was detected by the IP team and were not foreseen by R&D teams, and (c) the main researchers of the R&D teams were perceived with openness to collaborate, as they allowed the IP team to participate in the concept definition and R&D planning. Within the planning meetings, the IP team offered to initiate a collaborative AR project, increasing the mutual commitment and support for developing a collaborative R&D project. The AR project also encompassed the patent-licensing activities and allowed tracking the evolution of each case. The agreement to initiate the AR project with each R&D team was informal, hand-shaking accompanied with hopes of mutual benefits from the collaboration.

3.3 Implementation and Development

AR PROJECT 1 (AR1): CANDIDA PROJECT
In collaboration with Irene Castaño and Alejandro de las Peñas.

Project Background: Candida infections are the first source of fungal infection in humans. They cause organ invasion and blood infections associated with up to 33–51% mortality rate. Some factors that increase the risk of these infections are prolonged stays at intensive care units, immune-compromised patients (e.g., AIDS transplant, cancer). Four Candida species (*C. albicans, C. glabrata, C. parapsilosis* and *C. tropicalis*) are responsible for up to 95% of *Candida* infections.

In some cases, this *Candida* develops resistance to common antifungals. A timely diagnosis is critical to give the correct treatment, avoid complications, and death. Currently, diagnosis based on culture and biochemical tests take up to 4–12 days, with an accuracy of 65–85%. During this time, the risk of complications and death increases. There is a global need for a rapid identification test in blood and urine, clearly identified in most developed countries, but less important in developing ones.

Chronology: By performing basic research questions about the evolutionary and regulatory functions of some regulatory DNA sequences of *Candida glabrata*, the R&D team discovered unique DNA repeated sequences in *C. glabrata* that shown to be absent in any other related or non-related species. While the R&D team was making the scientific questions about their origin and functions, the IP team estimated that the DNA sequences could be useful to design an identification assay. The reasons behind this conclusion are (1) the novelty of the DNA sequences against state-of-the-art and bioinformatics information available allows estimating both the potential patentability and the non-cross reaction with other sequenced organisms. (2) The idea was verified based on the unsatisfied need and market opportunity analysis, and (3) the low cost of performing the basic experiments for generating the proof of concept. With this information, a first grant proposal was drafted, and the project was funded with the governmental support.

During 2008–2009, a series of experiments were carried out to achieve a wide variety of DNA fragments for identifying this fungus against reference strains and clinical samples. In 2009 and 2010, the first family of patent applications of the *C. glabrata* identification kit was filed[1] followed by a presentation of relevant results and the *alpha* prototype to the most important hospitals and clinical laboratories in Mexico. The feedback is shown as a design requirement also to identify the three most prevalent Candida species related to morbidity and mortality rates (*C. albicans, C. parapsilosis,* and *C. tropicalis*).

Thus, the new PCR kit should allow the identification of the four most prevalent Candida species that correspond to 95% of Candida clinical cases, which differ in the antibiotic treatment to be suitable for clinical practice. With this feedback, the second round of concept definition was set up. For each species involved, it was required to validate: (1) the need and the market opportunity, (2) the state-of-the-art and bioinformatic DNA sequence's analysis for finding potential targets for identifying the fungus, (3) the technical and economic feasibility of performing the experiments, and (4) the statistical model required for the proof of concept. A second grant proposal was presented in early 2010, finally accepted after several issues, in the middle of 2013.

From 2010 to 2012, no financial aids were available. Thus, inventors returned to their basic research projects abandoning the Candida kit project momentarily. However, the IP team continued the technology transfer efforts with the further presentation of results to potential licensors, investors, and users.

In late 2013, a third grant was earned within a student entrepreneurial contest, allowing to perform the sensibility and specificity tests against a statistically representative number of clinical samples. This grant allowed proving the usefulness of the prototype against the four species DNA sequences included. Although the funding was not enough, the R&D team managed to collaborate with other research institutions and hospitals, for validating the proof of concept. In an exchange strategy, hospitals provided the samples and the R&D team carried out the experiments. In 2014, three patent families were filed for the additional set of sequences.[2] A new round of interviews with diagnosis labs, hospitals, and final users was carried out. A new market feasibility study and financial estimations were made. However, funding was shortened, and only the U.S. and Mexican patents were filed.

At the beginning of 2015, the Health Ministry launched an Innovation Fund (FINNOVA program). In this program, human health was a crucial topic.

---

[1]EP2410052, ES2564416T3, DK2410052T3, MX2009002935, and WO2010107292.

[2]MX2014012841, MX2014012840, MX2014012842, US2015119277, US2015176084, US2015176085, WO2015176085, WO2015060704, and WO2015060705.

Thus, an external advisory committee evaluated health projects close to the market. The goal of this evaluation was to analyze the market opportunity, the possible interactions, and to suggest some recommendations for launching the product's commercialization. The external committee made valuable observations, for instance, changing the Technological Platform from Traditional PCR to Real-Time PCR (q-PCR). The difference between both platforms increases the test sensibility, but also implies the redesign of the molecular tools, R&D strategy, and business plan.

In the middle of 2015, an institutional technology transfer mission took place. A go-to-market plan was designed with an international technology transfer company, and several top multinational companies were interested in the product. However, the technological platform was modified once more, since the top molecular diagnosis companies replaced q-PCR with a new technique TaqMan Probes. The technical barrier implied the redesigning of the prototype and the collection of new evidence about the product performance. Also, a new IP strategy and business plans. Moreover, on May 2016, the USPTO Invention Act suffered some changes regarding judicial exceptions. Jurisprudence related to Natural Law blocked protection on diagnosis methods using the pathogenic genetic material to identify infections in biological samples (USPTO 2016). Therefore, IP protection in the USA was severely affected for three Candida species. *C. glabrata* has strong protection in Europe (where Candida infections are a severe health problem), while the four species are protected and suitable for the Mexican market.

During the evaluation of the feedback and requirements for developing the TaqMan Probes, the foresight analysis indicated that the TaqMan Probe technology could be replaced with an emerging technology: the DNA Point-of-Care (DNA-POC). Currently, new concept selection is under discussion for selecting the technological platform (TaqMan Probe) or DNA-POC technology. The grant proposal including the TaqMan Probe has not been accepted yet by government health-related grants, and we are approaching international groups looking for collaboration to develop DNA-POC tests.

AR PROJECT 2 (AR2): REPROGRAMMABLE LOGIC GATES
In collaboration with Eric Campos Canton.

Project background: Reprogrammable dynamic logic gates based on chaos computing were developed to test Chua's theory. The concept applies to a wide variety of electronic devices such as multivibrators and semiconductors. Therefore, it can be integrated into computers, tablets, smartphones, and so on. The dynamic logic gates perform, in a single gate, all the logic operations (AND, OR, NOT, NOR, XNOR, and XNOT) needed for processing data, instead of requiring a gate for each operation. Dynamic logic gates reduce the processor's size, while simultaneously increase memory and processing capability in the same space. However, the internal architecture of electronic devices requires some adaptations to allow the working conditions within the proposed technology. It is noteworthy that the miniaturization is still a challenge.

Chronology: In 2010, basic research efforts performed by Eric Campos, a researcher of the Applied Mathematics Division, achieved the application of Chua's computing theory. In collaboration with the IP Team, a concept definition was carried out for improving a first rough prototype. The purpose was that a single dynamic logic gate could perform all logical operations. Although the researcher published his findings, the first logic gates were IP protected in the USA.[3] In 2011, a technology assessment study was performed in collaboration with the Texas and Arizona Universities (ATTP program). The study revealed that the technical and market feasibility were congruent with a technology push strategy. Also, the study indicated that the dynamic logic gates platform could have a wide variety of applications that should be explored.

During 2012, a new product was designed: a dynamically reconfigurable multivibrator also based on the chaos reprogrammable computing. This all-in-one multivibrator device could be used in several pulse generators, a rising flank trigger, and a full RS flip-flop device. This technology was protected in 2013–14 in six countries where it was more feasible to be commercialized, according to the technology assessment study.[4] Some interviews were performed in universities, with stakeholders, and consultants for promoting the technology locally. The feedback shows the interest in presenting the technology to a company that participates in some design groups of next-generation microprocessors. Also, the study pointed out that the miniaturization and integration of real-life devices were desirable, but extremely challenging because of the lack of some capabilities at a national level.

In 2014, the research team obtained special funding for developing a prototype of a dynamic reconfigurable FPGA (FDPGA) applying the same mathematical platform. It was used accordingly and allowed to file a US patent application protecting the FDPGA concept in 2015.

In the middle of 2015, an institutional technology transfer mission took place. A go-to-market plan portfolio was designed with an international technology transfer company and was presented to companies that develop microprocessors, particularly those with participation in design groups of next-generation microprocessors in Europe, Israel, and the US. It was difficult to get some responses from most companies, and seeking comments or feedback on the technology was even harder. Just a few and short interviews could be made. In one of them, an executive told us: "*it is quite difficult that a company could embrace a foreign technology in this sector. If we require something, we develop it in-house and sometimes it could be outsourced by research contracts*..."

Meanwhile, technical issues emerged in the miniaturization of the FDPGA prototype. The situation got worse due to an increase of 25% in the US dollar–MX peso exchange rate, which produced a severe budgetary reduction. Both situations

[3]Prov. Appl. No. 61/502, 502, further non-provisional US Patent 8,587,343B2.

[4]WO2013066143, IN3454DEN2014, AU2012331722, CA2853681, EP2775618, JP5847955, US8823464, and MX326790

affected the project to become unfavorable. On the other hand, the technological opportunity moved away: a reduction in manufacturing costs of ASIC microprocessors was replacing rapidly the use of multivibrators and lower cost hybrid ASIC–FPGA microprocessors, which modified the trend of FPGA adoption. Finally, the emergence of research in quantum physics changed the long-term development expectative of the future microprocessors technology. The sum of these issues discouraged the R&D team, and the project got into a standoff.

3.4 Interpretation of the Experience

The projects presented in this chapter were developed by The Instituto Potosino de Investigación Científica y Tecnológica (IPICYT) between 2008 and 2016. Thus, the AR program and both projects shared a similar historical context, institutional conditions, goals, and leadership (Sect. 3.1). In this sense, it might be highlighted that since 2004, the institution began to implement patent-licensing activities, pushed by new national "third mission" policies and incentives for increasing the participation of research on economic development. This major change at national and institutional level pushed the R&D teams toward getting involved in the patent-licensing activities. Within 2004–2008, the institution outsourced the patent activities, but it showed to be very costly and inefficient. In 2008, an internal IP office was set with the aims of increasing the institutional patent activities and outputs. Within the implementation activities, the IP team began to search R&D projects for including them in the pipeline of potential patentable developments, wherein the R&D teams for developing an AR program were found.

In both projects, the initial idea emerged from basic scientific research findings discovered by the R&D teams. In the earliest stage, the researchers had no intention to follow-up a technological development, neither a patent-licensing path; even if the R&D team recognized that the discoveries could be applied to solve a need. However, the IP team convinced them to get involved in these activities with the following arguments: (1) there were little R&D efforts required which could have enough information for filing a patent application; (2) there were an unmet need and a market opportunity; (3) the IP team might help them during the overall activities; and (4) got involved in these activities should increase the potential of attracting more resources for their R&D projects and improving their institutional evaluations. As the collaboration between IP and R&D teams begun, the easygoing and openness to work, led to initiate informal agreements (Sect. 3.2), for developing the AR projects presented in Sect. 3.3.

During the AR projects, among different R&D and business planning tools used, the Koen's Fuzzy Front End framework (2002) was found to be useful to simultaneously identify and analyze the internal and external factors that were influencing the project development.

Internal factors that influenced the project development:

Institutional culture: The institution and R&D personnel were highly involved in scientific research activities that were measured and evaluated with academic parameters. Their culture was based on the traditional missions of teaching and research. However, new "third mission" policies pushed them to face economic development activities that at the beginning of the AR program were unknown to most of the researchers.

Institutional leadership: The purpose of this leadership is to allow the R&D teams to develop their work with freedom and responsibility. Further, both R&D teams were highly committed to the institution and for achieving their goals.

Scientific disciplinary approaches: An important cultural factor that differentiated the teams (and therefore the projects) was the scientific approach for modeling and solving problems.

The first project AR1 (related to Candida's molecular identification kit) was the responsibility of an R&D team of molecular biologists. The team chooses to follow the "New Product Development" (NPD) path. In this project, researchers were reluctant to extrapolate the methodology for identifying different pathogens of clinical interest. Focusing the effort on developing new products around the identification of Candida species that were well known by the team. This NPD strategy focused the AR1–R&D efforts to develop a single product by adding different characteristics as the capacity to identify more *Candida* species; the optimization of the processes, some statistical proves (sensibility and accuracy), among others.

On the other hand, the AR2 (a platform based on chaos mathematics with diverse applications in electronics) was under the responsibility of a mathematician who chose to follow a "New Technology Development" (NTD) path, generating diverse applications that follow parallel NPD paths. These differences were described in the definition of concept stage. In this project, the researcher was very open to extrapolate the theoretical findings to seek a NTD and to develop diverse applications within parallel NPD processes. This NTD strategy moved the goal toward proving the usefulness of the concept in different applications in parallel, rather than optimizing a single product to a real scale operation.

The IP team should balance the focus of the AR1 toward a more open NPD path. On the other hand, the openness of the AR2 should move the NTD path toward a more focused NPD path. Despite the chosen strategy, both R&D teams decide to follow a different development approach (NPD or NTP) with the parallel development of new products.

External factors that influenced the project development: Among the identified external factors that influenced the projects development, the most relevant were: (1) resources required and funding, (2) technological trends, and (3) The accessibility or openness of the sector that promotes research or innovation initiatives and that accept new products.

Resources required and funding: The AR1 project required larger amounts of resources than AR2 in the early stages of R&D. AR1 required: (1) time-consuming bioinformatic assays for developing the in silico target planning; (2) several experimental procedures for identifying the right DNA primers and their PCR conditions; (3) costly statistical experiments for achieving the proof of concept; (4) the optimization of the procedures related to the clinical conditions; and (5) to implement the patent-licensing activities, wherein external feedback was retrieved. Since the earliest feedback, potential licensors set as a requirement, that the proof of concept must include the complete information to face regulatory issues (FDA for instance) to commercialize the product. This process demands a huge amount of money for achieving each cycle. Furthermore, attracting resources from grants was an intermittent effort, and it was obtained after several attempts. On the other hand, AR2 theoretical and in silico planning and rough prototyping were quite cheap; allowing almost a constant funding of these stages. However, the miniaturization toward real scale applications had enormous costs. Therefore, patent-licensing strategies were set for transferring the concepts to those companies that could afford the investments toward real applications.

The financial resources required for each stage of development and the continuity of funding were the most astringent external factor for both projects. The scarcity of resources delayed the activities and outputs of AR1. In the case of AR2, scarcity made it not possible to get involved in miniaturization activities. In both cases, resources availability limited the R&D planning, the implementation of an international patent strategy, and the achievement of the licensing activities.

Technological trends: In the 8 years of development in the AR1 project, the diagnosis technological platforms changed twice during the project, and again, a new platform is emerging. Meanwhile, the AR2 was impacted by an almost punctual, massive change of actual and foresighted emerging technological platforms, moving away from the market opportunities once identified. These rapid technological platforms change and move away from the market opportunity and technical feasibility of the technology, and the prototypes developed in both projects. The potential licensing opportunities continue to exist, but require a considerable effort to adapt the current developments to the new technologies.

The accessibility or openness of a sector (sectorial openness): The AR1 activities were related to the biomedical sector dedicated to human clinical diagnosis. The sector is open and is relatively easy to present and obtain feedback from national and international companies. In the interviews achieved, even with top-level R&D managers, it was not difficult to get valuable feedback or answers to questions like, if they were interested; what they required to be interested? And even, who could be interested in our product proposal? The openness of the people involved in the top R&D management of biomedical companies was an unexpected discovery. On the other hand, the AR2 required to be presented to the microprocessors and microelectronics sectors: it was very hard to obtain even confirmation by e-mail, and few phone call interviews were achieved. A straightforward communication led us to understand an in-house innovation sectorial issue.

AR program and AR projects' objectives fulfillment

The AR program had the institutional objective of increasing the patent-licensing outputs in a more sustainable way. This objective was fulfilled as (1) the patent-licensing activities, and outputs related to both projects increased considerably; and (2) the resources required for developing the international patent-licensing strategies were included in the grant proposals of both projects, reducing the investment.

Also, the problem-solving objectives of the teams involved in the projects were achieved: (1) long-term collaboration within the IP and R&D teams involved in AR1 and AR2 was set; (2) R&D activities were incorporated into the patent-licensing path, increasing the outputs: (a) seven patent families integrated with 32 patents were filed in different countries, most of them have been granted or are still under evaluation; (b) technological and economical portfolios were presented to national and international companies seeking licensing opportunities; (c) top-level interviews were achieved, retrieving relevant feedback; and (d) potential licensing opportunities were kept open.

Finally, the organizational research objectives also were fulfilled as (1) the AR methodology proved to be useful for achieving patent-licensing activities in an academic institution; (2) some of the internal and external factors involved in the patent-licensing path were identified and analyzed, allowing some risk management opportunities; and (3) during the activities performed, it was identified as an iterative, stage-gated process with clear steps that will be further analyzed in the discussion section.

4 Discussion

Research produces valuable knowledge. The challenge is to transform this intangible resource into innovation. This is one of the goals of the academic's third mission policies for participating in the economic development. The patent-licensing paths *"constitute immediate, measurable market acceptance for outputs of academic research"* (Perkmann et al. 2013). However, this process has been a high-risk long-term task that requires diverse resources for its development and for increasing the possibilities of success. Moreover, '*the trip from lab to market'* is highly situational: internal and external factors affect the conditions and the decision-making within the processes. Therefore, just a few organizations are succeeding in these activities (Mowery et al. 2001; Sampat 2009).

The motivation for performing the AR program and the projects shown in this chapter is to promote and move that academic research into innovations. The strategy is to follow the patent-licensing path. In the AR program, two crucial factors identified were: (1) an iterative, stage-gated process and (2) to transform some of the internal and external factors involved that allow a risk, into management opportunities and then to increase the possibilities of success in diverse contexts.

The identified iterative, stage-gated process

During the AR program and projects performed, it was identified an iterative, stage-gated process with clear steps: (a) defining a concept and R&D planning; (b) attracting the necessary resources; (c) developing the activities and evaluating the outputs; (d) when required, performing the patent-licensing activities; and (e) retrieving the feedback and redefining the concept. The iterative process that we found is depicted in Fig. 1 and later explained in the following pages.

Defining a concept: Both projects began with an external, internal, and trend analysis process (FFE) for defining a concept (Koen and Fountoulakis 2002). In this FFE-like process, the idea selection and enrichment are nurtured with the technical, economic feasibility, and capabilities analyses. The market opportunity selection and enrichment are supported by the state of the art, the technological trends, and the market information analyses. Another relevant source of information is the feedback of potential licensors, investors, and users.

In each concept definition stage, several enrichment rounds were performed on different days. At the end of each meeting, some specific tasks were identified. Also, some feasible schedules to perform each task by the participants are proposed. It is necessary to emphasize that the personal commitment was the only source of motivation among the project's participants. It was required to set a follow-up process for tracking the task development. Usually, the requirement to present a grant proposal accelerated the task development, the number of planning actions, and thus the meetings performed increased.

One of the most important issues was related to the team's configuration. R&D researchers had a strong scientific profile, but a limited experience in the intellectual property processes and business backgrounds; while the IP team had scientific,

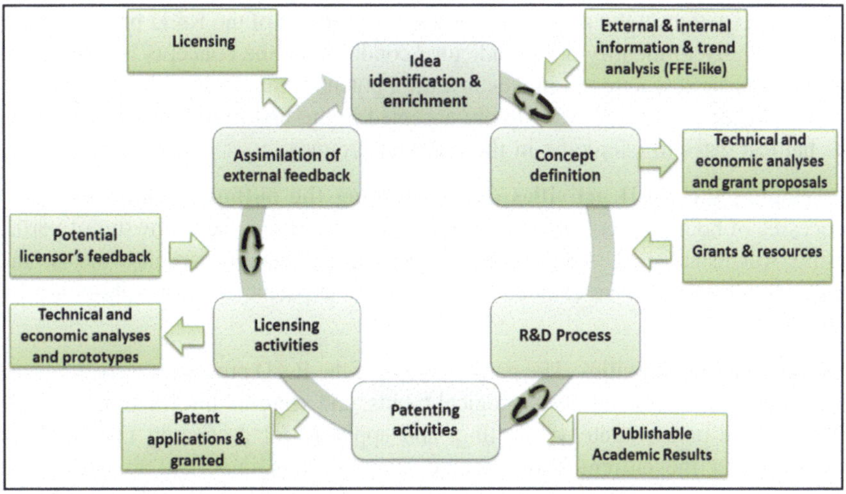

Fig. 1 Patent-licensing path process

legal, and business backgrounds. The R&D and IP teams jointly contributed to identify and enrich the concept for developing potential novel applications for a specific market opportunity. This configuration is consistent with Fisher and Oberholzer (2013), which propose that the multidisciplinary interaction within R&D and IP teams produced a more efficient concept generation.

Also, one of the most important factors during the concept generation and enrichment stages was retrieved by external feedback. Top-level communications had a stronger influence on the decision-making of each project than middle-level personnel (even if the advice was made toward solving a similar problem). While this phenomenon has its roots in cultural behavior, it is required to listen and retrieve the feedback based on the argument, not its origin. The final user feedback was also important and had surprising effects on the project development. When the R&D and IP teams got closer to the final user's needs of the AR1 project, the final users were real people that had suffered or were suffering a chronic health condition. The communication of their needs increases the team's commitment to generating a potential solution.

R&D plan: During the concept definition stage, emerged diverse sets of things to be done. In our experience, it was required to define and prioritize them for achieving a modular action plan, and adapt them to the activities that could be financed by the grants available. In our experience, the R&D activities and the resources attracted to support these initiatives became more efficient when performed in parallel with patent-licensing activities.

Attracting the resources required: This stage is highlighted because it was the main gate that enables the R&D and scaling-up processes as well as the patent-licensing strategies of both projects exposed. In this stage, the concepts defined and the R&D planning to be developed might be translated and adapted into grant proposals, according to the terms and conditions of each program.

In our experience, the acceptance of a grant proposal of the R&D projects offers an opportunity to identify a contradictory condition as the concepts became more distant to basic research and moves toward a potential innovation, the funding was scarcer. When a proposal was rapidly accepted in the first round, it became one of the hardest gates to overcome in the following rounds.

Developing the R&D activities and evaluating the outputs: During the R&D processes of both projects, diverse meetings for performing the follow-up of partial results was established. Also, academic publishing, scientific articles, thesis, and congress summaries were evaluated for keeping the patent strategy without a prior publication.

Patent-licensing activities: The partial results of the R&D processes were analyzed against the state of the art, technological trends, and sectorial information available to define the patent strategy, drafting, and applying for protecting inventions in countries that the current budget could support. It is important to notice that implementing an international patent strategy could be very costly. Thus, R&D and patent-licensing activities compete for the available budget, creating a second resource stage-gated process.

When the intellectual property strategy was set, and the patent applications were filed; also a "technological package" starts its preparation. The technological package includes various presentations about the new technology and related products, prototypes, and their economic potential within the identified market opportunities, without forgetting the risk management issues. Presentations in congresses and specific forums were achieved, and specific information was distributed to potential licensors, investors, and users for retrieving their feedback. As a result of these promotional activities, some interviews with top-level R&D personnel of national and international companies, potential licensors, and investors were achieved, and licensing options kept open. We thought that by accomplishing the required tasks of potential licensors, the product or the technology could get into the licensing-investing negotiations, which was the hardest gate to open in the complete cycle.

The final stage was retrieving the feedback into a new FFE-like process, to enrich the concept definition for the next R&D -funding round. The feedback was crucial for taking another decision. The feedback received from the AR1 project allows us to be flexible about the concept definition of the project. We thought that if the funding stage-gate had not been so restrictive, the project AR1 would have reached the market earlier. On the other hand, although funding was not an obstacle, insufficient user's feedback on the project AR2 prevented the team from reducing the technological risk.

External factors and risk management

Among the identified external factors that influenced the project's development, the most important one was the availability of financial resources and the intermittent funding. The initial stages of concept definition and R&D planning demand few resources in both projects, but the costs increase exponentially to develop the laboratory proof of concept and move toward a real scale validation. In our experience, funding shortage limited the R&D development, the IP strategy, and the licensing efforts. Also, scarcity and intermittent funding produces a delay in the project's development and, therefore in their outputs. In both projects, the technological trends moved away from the market opportunities that were identified within the concept definition stages.

Managing the risk of scarcity and intermittent funding seems to be not so hard during planning stages, but is very hard to be tackled for performing R&D or patent-licensing activities. It is important to notice that the amount of resources define the activities that could be performed. While waiting for better times could be an option, seeking alliances with higher risk investors (when available) could be a solution for attracting the resources required to approach a potential licensing deal or establishing a spin-off company. In these last options, it is required to have realistic technological and financial estimations.

On the other hand, keeping constant surveillance of the evolution of different competing technologies and products could mitigate the risk that involved the technological change. Hence, it is necessary to constantly observe the platforms and products in the patent documents and the markets. While the information available always will be imperfect, surveillance could capture some signals of a major technological change event.

Next steps and future research
State-of-the-art, a technology landscaping, and patent activities have been scaled to give attention to more than 20 projects in parallel within the IPICYT. Simultaneously, some services were open to other academic institutions and companies. Also, more than 20 workshops and short courses have been developed in the past 5 years. The training was oriented to IP managers, researchers, and graduate students, as well as R&D, and project management personnel of companies.

Moreover, a new Active Research project is in progress, consisting of scaling-up the AR capabilities to collaborate within a multi-institutional and multidisciplinary project about bioenergetics.

In the patent-licensing path topic, a further action research is required in collaboration with academic institutions and companies. The purpose will be to identify and analyze the influence of internal and external factors in different organizational and historical contexts.

5 Conclusions

Applying the action research (AR) methodology to academic patent-licensing path proved to be a valuable approach for simultaneously fulfilling institutional and problem-solving objectives, without losing the focus on the organizational processes and their influencing factors.

In this chapter, an AR program that included two projects was presented. The projects were chosen from several others because (1) the IP and the R&D team's interaction were initiated with early basic scientific research findings. (2) The projects developed through a 'new product development' (AR1), and a 'new technology development' plus parallel 'new product development' (AR2). Finally, (3) both projects achieved to go through national and international patent-licensing activities.

From the analysis of the projects, we found (1) an iterative, stage-gated process with clear steps, and (2) some internal and external factors that influenced the projects. Both the process and the factors were discussed, seeking to share the experience and learning process.

References

Argyres, N., & Liebeskind, J. P. (1998). Privatizing the intellectual commons: Universities and the commercialization of biotechnology. *Journal of Economic Behavior & Organization, 35* (1998), 427–454.

Avison, D., Bakersville, R., & Myers, M. (2001). Controlling action research projects. *Information Technology & People, 14*(1), 28–45. https://doi.org/10.1108/09593840110384762.

Barron-Pastor, D., & García-Calderon, N. (2014). *Conceptos básicos de Gestión de Tecnología y Propiedad Intelectual para investigadores y emprendedores científico tecnológicos, ID4BIZ.* CICY, INECOL Y CINVESTAV: IPICYT. ISBN 978-607-7823-26-1.

Baskerville, R. L., & Wood-Harper, A. T. (1996). A critical perspective on action research as a method for information systems research. *Journal of information technology, 11,* 235–246.

Coghlan, D., & Brannick, T. (2014). *Doing action research in your own organization* (4th ed.). London: Sage. ISBN 978-1-4462-7257-2.

Fisher III, W. W., & Oberhozler-Gee, F. (2013). Strategic management of intellectual property—An integrated approach. *California Management Review, 55*(4), 157–183. ISSN 0008-1256.

Geuna, A., & Nesta, L. (2006). University patenting and its effects on academic research: The emerging European evidence. *Research Policy, 35*(6), 790–807.

Geuna, A., & Muscio, A. (2009). The governance of university knowledge transfer: A critical review of the literature. *Minerva 47*, 93–114. https://doi.org/10.1007/s11024-009-9118-2

Henfridsson, O., & Lindgren, R. 2007. Action research in new product development. In N. Kock (Ed.) *Information systems action research: an applied of emerging concepts and methods* (p. 193–216). Springer Science + Business Media LLC.

Koen, P., Ajamian G., Burkart R., Clamen A., Fisher E., Fountoulakis S., et al. (2002) Fuzzy front end: effective methods, tools, and techniques. In P. Belliveau., A. Griffin., & S. Somermeyer (Eds.) *The PDMA ToolBook for new product development* (p. 5–35) John Wiley & Sons, Inc.

Markmann, G. D., Siegel, D. S., & Wright, M. (2008). Research and technology commercialization. *Journal of Management Studies, 45,* 8.

Mowery, D. C., Nelson, R. R., Sampat, B. N., & Ziedonis, A. (2001). The growth of patenting and licensing by US universities: an assessment of the effects of the Bayh-Dole Act of 1980. *Research Policy, 30*(1), 99–119.

Perkmann, M., Tartari, V., McKelvey, M., Autio, E., Broström, A., D'Este, P., et al. (2013). Academic engagement and commercialisation: A review of the literature on university-industry relations. *Research Policy, 42,* 423–442.

Sampat, B. (2006). Patenting and US academic research in the 20th century: The world before and after Bayh-Dole. *Research Policy, 35,* 772–789.

Sampat B. (2009). The Bayh-Dole model in developing countries: Reflections on the Indian Bill on Publicly funded Intellectual Property. UNCTAD-ICTSD Project on IPRs and sustainable Development. Policy Brief number 5, October 2009. Retrieved February 18, 2017, from http://unctad.org/en/Docs/iprs_pb20095_en.pdf.

Soranzo, B., Nosella, A., & Filippini, R. (2017). Redesigning patent management process: an action research study. *Management Decision, 55*(6). https://doi.org/10.1108/md-04-2016-0226

USPTO United States Patent and Trademark Office. Retrieved February 14, 2017, from https://www.uspto.gov/sites/default/files/documents/ieg-may-2016-ex.pdf.

Mass Customization Process in Companies from the Housing Sector in Brazil

D. C. Fettermann, G. L. Tortorella and Carlos M. Taboada

Abstract This paper aims at assessing how Mass Customization (MC) is being implemented in housing companies throughout and a comparison with the recommended approaches existent in the literature. Based on this comparison, the work identifies opportunities for improvement in the way the construction and housing industry employs the MC strategy. The used method comprises of a multi-case study with housing companies located in the northeast of Brazil. All companies are undergoing the implementation of MC strategy. This study allows managers from housing companies to address their product strategy to benefit from the mass production while properly customizing their products to customers' needs. As a result, the research provides an argument to comprehend better the methodological differences between the recommended customization process and the actual strategy from housing companies from developing economy countries, such as Brazil.

Keywords Customization · Construction · Buildings · Developing countries

1 Introduction

Companies tend to increase the variety of the product portfolio available on the market to meet the growing diversity of customer needs (Pine II 1993). However, the simple adoption of this strategy results in a reduction of company's operational performance, since a larger portfolio can result in higher manufacturing and overhead costs, inventory levels, and production lead times (Salvador et al. 2002). This cost impact is also associated with the level of available customization. In general, higher levels of customization are related to higher operating costs; while

D. C. Fettermann (✉) · G. L. Tortorella · C. M. Taboada
Department of Industrial Engineering, Federal University of Santa Catarina, Universidade Federal de Santa Catarina, Campus Universitário Trindade, Postal Box 476. 88040900, Florianopolis, Brazil
e-mail: d.fettermann@ufsc.br

G. Cortes-Robles et al. (eds.), *Managing Innovation in Highly Restrictive Environments*, Management and Industrial Engineering,
https://doi.org/10.1007/978-3-319-93716-8_5

offering a wider range of products and services cover a greater number of customers and their needs (Fogliatto et al. 2012).

The adoption of MC (Mass Customization) consists on the ability to provide customized products through flexible production processes with high volumes and low costs (Da Silveira et al. 2001; Tu et al. 2007). With the purpose to achieve the customization benefits, customers should be seen as a unique and indispensable resource for integrating the customization process (Peppers and Rogers 1997; Gilmore and Pine II 2000). Sievänen (2002) describes the process of MC as a product differentiation that seeks to meet the individual needs of the largest number of customers possible. Thus, MC could be understood as the challenge to meet the individual needs of customers while maintaining the manufacturing efficiency at similar levels to mass production (Jiao and Tseng 2004; Kumar and Phrommathed 2005). Although customers may desire a variety of product attributes, it is recommended to limit the set of customization options to achieve economies of scale in the manufacturing processes (Kotha 1996; Massyn et al. 2015). Therefore, it is important for the company to identify which requirements are feasible to be customized and align its portfolio with what customers would like most to customize (Åhlström and Westbrook 1999; Bentzien et al. 2012; Chakraborty and Das 2014).

In this sense, the application of MC strategy has been acknowledged as an efficient alternative to manufacturing companies (Pine II 1993; Gilmore and Pine 1997; Da Silveira et al. 2001; Fogliatto et al. 2012). The literature reports the implementation of this strategy, such as the manufacture of bicycles (Kotha 1996), Hewlett-Packard (Feitzinger and Lee 1997), Dell Computers (Peppers and Rogers 1997), modular furniture, (Fettermann and Echeveste 2011), or automobile industry (Bardacki and Whitelock 2004; Ro et al. 2007; Liao et al. 2013; Fettermann and Freitas 2017). Further, this strategy has also been widespread in other sectors beyond manufacturing, as the development of services (Voss and Hsuan 2009), tourism (Cheng and Han 2013), and software development (Krueger 2001).

Additionally to these applications, some researchers emphasize that an opportunity is still latent for developing the MC strategy concerning the housing sector, in which the goal is to provide quality housing at an affordable cost (Duarte 2005; Hofman et al. 2006; Shin et al. 2008; Formoso et al. 2011; Rocha et al. 2012; Kendall 2013). Studies applying MC in the housing sector are found in developed economy countries, such as USA (Kendall and Teicher 2000; Willetts et al. 2014), Great Britain (Barlow 1998), Taiwan (Tu and Wei 2013), Australia (Sahin et al. 2016), and Japan (Barlow et al. 2003; Noguchi 2013). Further researches investigate the implementation of MC in the production of inputs for the housing industry (e.g., Hofman et al. 2006; Halman et al. 2008; Shin et al. 2008; Nahmens and Bindroo 2011). Despite this evidence, there is a wide variety of available recommendations that hinder the development of this strategy in housing companies (Rocha et al. 2012; Fettermann and Echeveste 2014).

Thus, this chapter aims at assessing how MC is being implemented in housing companies throughout a comparison with the recommended approaches existent in the literature. Based on this comparison, the work identifies opportunities for improvement in the way the construction and housing industry employs the MC strategy. The proposed method comprises of a multi-case study with housing companies that are undergoing the implementation of MC strategy and are located in the northeast of Brazil. The comparison among the companies is qualitative-based and provides a descriptive analysis of the opportunities and similarities in the literature. As a result, the research provides an argument to comprehend better the methodological differences between the recommended customization process and the actual strategy from housing companies from developing economy countries, such as Brazil. Besides the theoretical contribution, this study allows managers from housing companies to address their product strategy to benefit from the mass production, while properly customizing their products to customers' needs.

This study unfolds as follows. Section 2 provides a literature review on MC, both in general and specifically within the housing sector. The third section presents the description of the proposed research method, whose results are described and discussed in Sect. 4. Finally, Sect. 5 presents the final remarks and further research opportunities.

2 Mass Customization

MC objective is to attend a large number of customers and, at the same time, deal with them individually (Davis 1987; Ulrich 1995; Blecker et al. 2003). With the goal to competitively offer this higher variety of products, companies must develop the capability of designing, producing, and marketing quality products in a short time, and at an affordable cost to customers (Tseng and Jiao 2007). Therefore, companies must invest in the standardization of products, while promoting the variety of features valued by customers (Pine II 1993). According to the portfolio strategy, companies can be classified into three different categories: (i) production, (ii) semi-customized, and (iii) customized. Companies categorized into production are directed to the advantages of production scale, producing standardized units in less time and cost. The semi-customized companies analyze the feasibility of customization unit attributes and provide a set of viable options to be customized. These companies are the ones that usually have a higher possibility of implementing the MC strategy. Finally, the customized companies offer customers complete freedom for creating original and unique designs (Noguchi and Hernández-Velasco 2005; Massyn et al. 2015).

Despite the increasing value perception from customer's perspective, to add product options can also create chaos in the production processes (denoted as 'mass confusion'), reducing the competitive advantage of MC (Huffmann and Kahn 1998). Thus, companies must identify which product attributes are associated with customers' needs, and select the customizations that will attend such attributes (Squire et al. 2004). Further, customers have increased their standard requirements over time, and are less willing to limit their choice to standard products (Tseng and Jiao 2007); entailing a value addition and greater customer willingness to pay for this customized option (Franke et al. 2010; Ulrich 2011). Such willingness tends to allow higher profit margins for companies (Robert and Meyer 1991), and increased customer satisfaction level (Jiao and Tseng 2004). However, excessive varieties of options also result in higher costs in design, production, inventory, and logistics (Da Silveira et al. 2001). To define what can be customized, companies must plan their processes to efficiently meet the customization requirements (Rocha et al. 2012), taking into consideration the aimed market scope and the most valued features by customers.

The 'customer focus' is a term deemed as the ability of an organization to meet customers' needs with a high level of satisfaction (Zeithaml 1988; Barlow and Ozaki 2003). The attributes are considered as variables that affect the buying decision of a customer, e.g., product functionality, quality, brand, design, etc. (Lihra et al. 2012). However, the price difference that customers can accept is limited and depends on the perceived value by customization (Hart 1995; Huffmann and Kahn 1998; Naderi et al. 2012). Additionally, the delivery time of customized products cannot significantly differ from the standardized products, since customers may assign considerable importance to the delivery time when deciding on purchasing (Squire et al. 2004; Lihra et al. 2012). Pine II (1993) and Wang et al. (2015) mention that customer value is the result of a balance between quality, cost, and delivery time. From the company's perspective, the customization process also allows the interaction with customers, enabling the collection of market information and establish customer loyalty (Piller 2004; Fogliatto et al. 2012).

To customize products, the establishment of the proper flow of information and activities is necessary (Kumar and Phrommathed 2005). This flow incorporates knowledge about the market behavior and translates it into specifications of design and production. Several authors have proposed complementary methods for addressing such flow more effectively (e.g., Gilmore and Pine II 1997; Duray et al. 2000; Da Silveira et al. 2001; Fettermann and Echeveste 2011). In general, these methods comprise five steps to guide the customization process. The first step determines the options that will be offered for customization. At this stage, it is recommended to verify the need for customization of the product, the possibility to customize it, and the availability of technology and logistics for production activities (Da Silveira et al. 2001; Fettermann et al. 2012; Fogliatto et al. 2012.).

In the second step, it is considered the way that customers define their choices according to customization options. One of the main critical success factors for this step is mapping the customers' needs and translating them into custom product design (Duray and Milligan 1999; Ogawa and Piller 2006). This process can be carried out with the aid of a designer capable of supporting the customer during the execution of the customized product design. Also, the customer project experience during the customization process offers to customers a valuable experience (Franke et al. 2010; Merle et al. 2010; Moreau 2011). The use of an Internet-based product configurator has also been used for this purpose (Hellander and Jiao 2002; Fettermann et al. 2012; Trentin et al. 2014). Step three aims at transferring the customization options from the point of sale to the production processes. An example to undertake this transfer is the modular furniture alternative (Fettermann and Echeveste 2014). The companies from this sector usually apply the CAD software (PROMOB®) to identify the modules and forward the list of products for the production through the use of EDI (Electronic Data Interchange) (Fettermann and Echeveste 2011). The fourth step comprises of the transformation of the production expectations, obtained from the previous step, into manufacturing instructions (Da Silveira et al. 2001). There is a variety of tools to support this step, such as templates for manufacturing cost optimization (Du et al. 2003), the use of CAD and CAM technologies (Kotha 1996), and other manufacturing technologies such as FMS—Flexible Manufacturing System (Fettermann and Echeveste 2011). Moreover, more flexible production arrangements, such as manufacturing cells, may also contribute to the production of custom items (Tseng and Jiao 1996). Finally, the last step consists of the delivery and use of the product by the customer (Lizarralde 2011).

Specifically, in the case of construction and housing companies, the application of MC strategy has been recommended as a way to meet customer requirements, who usually present higher purchasing power and wish to customize the room according to their individual needs (Barlow et al. 2003; Karadimitriou 2005). In this sense, construction and housing companies have been offering customization options since the beginning of the construction (Elliman and Orange 2003). Thus, the customization of housing has been an important business strategy to promote the real estate market (Barlow and Ozaki 2003; Barlow et al. 2003; Noguchi 2003; Noguchi and Hernández-Velasco 2005). There are several determining factors in the purchase decision of the customer, and most of them depend on the project itself, such as location, environmental features, price, etc. It is still possible to sell the internal space of the building, being it possible to be customized by the client according to their needs (Roy and Roy 2016). Hence, the housing developments provide a wide range of options for internal customization, such as finishing improvements, automation tools, post-occupancy services, among others (Barlow et al. 2003; Bentzien et al. 2012). Further, customization in housing industry should consider both products and services. The products may be considered as attributes determined by components and material specifications; while the design and construction may be considered as part of services (Noguchi and Hernández-Velasco 2005; Naderi et al. 2012).

3 Method

This research investigates the customization process of housing companies for housing development. This research can be characterized as exploratory, as it aims to understand the phenomenon and provide more explicit knowledge (Denzin and Lincoln 1994). A qualitative approach was used to understand the phenomenon from the perspective of the participants that studied the situation. Thus, in-depth interviews were conducted using semi-structured questionnaire (Denzin and

Lincoln 1994) with representatives of the involved companies undergoing the implementation of customization processes.

The questionnaire was elaborated to identify how housing companies carry out each step of the MC process. Before the application of the questionnaire (see Appendix A), a pre-test was performed to validate the proposed questions. The questionnaire pre-test was conducted with two experts from the housing sector, whose suggestions were collected and, later, incorporated into the questionnaire.

The studied housing companies are all located in the northeast of Brazil and develop projects of semi-customized or customized categories of buildings. Twelve medium-sized housing companies were identified in this region, whose projects fit the aimed categories aforementioned. These companies are developing at least two residential buildings simultaneously. Among these companies, we selected three of them according to their willingness to participate in the research. The senior managers of these companies were invited for a 90-minute interview, whose contents were later transcript.

The obtained results were consolidated and descriptively presented to identify the most frequent procedures used by companies to implement the MC. Subsequently, a qualitative comparison is undertaken to raise improvement opportunities within the studied companies.

4 Results

The characteristics of the studied companies are presented in Table 1. The portfolio strategy of Companies A and C is mainly semi-customized, while Company B is fully customized. Although all the participating companies are at least 10years operating in this sector, two of them are usually involved with higher buildings (from 18 to 20 floors) and one focuses on smaller residential buildings.

In the customization process, there is the release point (RP) in which the customers' specifications begin to be considered in the product design (Duray et al. 2000). Table 2 shows the RP according to the steps of the housing project in the three studied companies. Companies A and C begin to include customers' specifications in the masonry demarcation stage, enabling the customization of the subsequent activities according to customers' needs. However, company B includes

Table 1 Description of the characteristics of the studied companies

	Company A	Company B	Company C
Number of employees	230	300	178
Operating time (years)	10	40	28
Number of floors limit of residential buildings	18	2	20
Customization type	Semi-customized	Customized	Semi-customized

customers' specifications since the stage of 'acquisition of the land', giving them complete freedom to develop their project.

Previously, the Company A has given total freedom to the customer change the floor finish, walls, provision of sanitary appliances and masonry through the submission of a project carried out by a qualified professional; however, this procedure has been abandoned by the company. In turn, the company allowed customers to select the arrangement of masonry according to a predefined set of options, which were based on previous experiences. Company B is responsible for land sales and offers complete freedom to customers for hiring engineers and architects to customize their units. The only limitation for customizing concern the urban requirements established for the location, such as setbacks, the height of the building, occupancy, and the utilization rate. Further, Company C sets the customization options catalog according to the characteristics of the project and previous experiences. The customer can select from a set of options the interior of the building, without interfering with the building structure. Among the three studied companies,

Table 2 Comparison of the release point (RP) of the studied companies

A	B	C	Company
	X		1. Land acquisition
			2. Development of standardized projects
			3. Leasing foundation
			4. Implementation of infrastructure (shoe)
			5. Implementation of the structure (columns and slabs)
X		X	6. Demarcation of masonry (layout plant)
			7. Pipe installations
			8. Sanitary pipe
			9. Drain pipe of splits
			10. Wiring apartments
			11. Gas installation
			12. Settlement coating
			13. Plaster lining
			14. Floor regularization cemented
			15. Settlement coating (floor and baseboard)
			16. Bench application
			17. Application of granite filets on the floor of the box
			18. Painting (wall and ceiling)
			19. Application of the tanks in services
			20. Settlements of the doors of the apartments
			21. Sanitary ware installation
			22. Installation of metal fittings
			23. Showers installation
			24. Breakers installation and mirrors

no structured method or technique was identified to defining the building attributes that will be available for customization.

According to the Company A, historically, changing the location of the masonry is the customization most requested by customers, followed by changes in electrical points, hydraulic points, and floor finish. Therefore, any customization request includes the analysis of its technical feasibility and its effect on the cost of the company's product. At Company B, the project units are offered so the customer can use the same or completely different project, as long as it respects the established urban requirements. While providing freedom for customization, a qualified professional should do the housing project and the company should carry out its implementation. Company C sets a deadline for customizing the unit, which is performed based on an available set of attributes options. Overall, the completion of units' customization is carried out by professionals from outside the companies.

Regarding the information transfer, all companies manage it manually, using paper for this communication. This communication is accompanied by the design of the building and performed by an engineer or architect. In the case of Companies A and C, the data transfer is mainly performed in the digital form. In Company B, due to the higher level of customization available, the customization project is transferred directly to the professional responsible for the project, mainly occurring in digitized form. Further, in Company A, any customization request must go through a technical feasibility analysis. This analysis, especially regarding the architectural, electrical, and hydraulic design, is performed by the company's projects department; subsequently, the customized project is approved by the customer. Company C undergoes a situation similar to company A, which is primarily performed a compatibility analysis of custom design with the original project design. Then, the custom design is available for the implementation of production. In the case of company B, since it is fully customized, there is a predefined template project, entailing that the cost of labor and productivity will be defined only by the customer's choice. The client designs according to his needs with the professional of his choice, while the company engineering carries out the analysis and translation of specifications for production instructions.

Further, although company A does not provide any post-employment service, customer demand was mentioned by this type of service and business opportunity for the company. Company B, despite offering a higher level of customization also does not provide any post-employment service. Company C offers the possibility of customizing some attributes after the delivery of the unit after evaluating the request.

Table 3 synthesizes the characteristics of the customization strategy of the companies and compares them with the literature recommendations at each step of the MC process. From this comparison, it was possible to identify that several of the recommendations mentioned in the literature are not applied in the studied companies. Thus, few development opportunities have been indicated for the companies according to each step of the customization process. None of the mentioned opportunities had its financial assessment carried out.

Table 3 Comparison of companies' MC strategy and the recommendations in the literature

Recommendations	Authors	Cases	Opportunity
Step 1—Definition of options			
Identify customizable items of greater value-added capabilities	Da Silveira et al. (2001); Tseng et al. (2010); Jiao et al. (2003), Gemser and Perks (2015)	Evaluation of customization requests from previous ventures	The development of a product family, with customizable items the building most valued by the client allows the company greater standardization of options keeping production scale advantages
Study of market segmentation	Drejer and Gudmundsson (2002);Jiao et al. (2003); Grenciand Watts (2007)	It is not carried out by companies	The market segmentation study may contribute to the companies to provide a set of customizable items with greater ability to add different value for each market segment/enterprise
Integrating the sales process to the design of new projects	Da Silveira et al. (2001); Jiao et al. (2003); Helander and Jiao (2002)	It is not carried out systematically by companies	Direct contact with the customer during the sales process is an important source of market behavior of the information and can be used in defining customizable items the next developments
Set a portfolio of more optimized customizable attributes	Meyer and Lehnard (1997); Bare and Cox (2008)	Evaluation of customization requests from previous ventures	Identify customization items that can be standardized between the different projects of the company, obtaining greater benefits of scale of production
Using standardization methods for (communalization)	Simpson (2004); Jiao and Tseng (2000); Robertson and Ulrich (1998)	It is not carried out by companies	The use of formal methods of standardization of components contributes to reducing the number of items (part numbers and SKUs) to be managed by the company
Step 2—Data transfer from customers' requests to production			
Online configurator system to perform the customization unit	Franke and Piller (2003); Fogliatto and Da Silveira (2008); Franke et al. (2010); Fettermann et al. (2017)	It is not carried out by companies	The possibility of the customer himself perform the customization project is mentioned as a value added to the product factor and can be supported by the product configurator
Develop a support configuration process (Decision Support System)	Jiao et al. (2003); Tseng and Jiao (2007)	It is not carried out by companies	The provision of auxiliary projects for the customization of dwellings such as architectural design services, auxiliary projects, furniture design, and decoration can contribute to add value to the product

(continued)

Table 3 (continued)

Recommendations	Authors	Cases	Opportunity
Develop a support configuration process (Decision Support System)	Tseng and Jiao (1996); Hellander and Jiao (2002); Salvador and Forza (2004); Grenci and Watts (2007); Barco et al. (2017)	It is not carried out by companies	Show the customer a procedure to perform the property customization may contribute to the customization process, supporting the client during the selection of options
Three-dimensional functionality in the product module	Dai et al. (2003)	It is not carried out by companies	Helps customers have a broader view of its customization options; see the product before assembly or manufacture
Step 3—Data transfer from customers' requests to production			
Use of EDI systems for communication between the point of sale and production	Da Silveira et al. (2001)	It is not carried out by companies	An integrated system between the project sector and production can provide greater flexibility in the production process of the company
Step 4—Translation of customers' requests for design and production instructions			
Manufacturing Cells	Tseng and Jiao (1996)	It is not carried out by companies	The manufacturing concept cell may be transferred to the custom designed area, wherein each cell is responsible for the customized design of a specific environment of the housing
Use systems for customization of cost optimization	Du et al. (2003)	It is not carried out by companies	
Generic BOM matrix (Bill of Material) (GBOM)	Zhang et al. (2005)	It is not carried out by companies	The knowledge about the variety of products is organized and there is a master process (MP) knowing the variety of the process. It is generated a production planning and corresponding matrices to manage relations between GBOM and MP. It also contributes to the analysis of the customization of feasibility
Step 5—Delivery of the customized product and its use by the customer			
Providing customization post-occupancy services	Lizarralde (2011), Enshassi et al. (2014), Feng et al. (2015)	Held partially by the company C	The availability of customization services such as floor changes, masonry, electrical and finishing services. Also, customization design service after the product occupation and during their use stage is also a possibility for companies

5 Contributions to the Theory

Mass customization in the building industry is a relatively new concept (Khalili-Araghi and Kolarevic 2016). The literature review indicated several options for actions during the various stages of the housing customization process. Mass customization in the building industry is a relatively new concept (Khalili-Araghi and Kolarevic 2016). The literature review indicated several options for actions during the various stages of the housing customization process. Among the practices identified in the literature, the greatest number of practices was verified in the first step of the customization process, named as 'definition of options'. In this phase, many opportunities were identified for increasing the value of customizable items (Da Silveira et al. 2001; Tseng et al. 2010), market segmentation (Grenci and Watts 2007), use of sales information for new projects (Jiao et al. 2003; Helander and Jiao 2002), optimization of customizable option's portfolio (Bare and Cox 2008) and communalization methods (Simpson 2004; Jiao and Tseng 2000).

In the case of industrialized products, such as modulated furniture, automobiles, among others, the MC strategy is mainly supported by the application of modularity in the product (Wang et al. 2014; Zhang et al. 2014). In the Mass Customization of buildings, it was verified that this strategy is little supported by the modularity of the products, particularly for the cases studied. These cases present traditional constructive techniques, based on reinforced concrete and without the use of industrialized structural elements. In these cases, the characteristics of construction do not allow the same advantage of scale with the use of modularity as other industrial products. Despite that, the advantages with the application of MC in buildings are immense. In this sense, the identification of the customizable items that most add value to the customer is presented as an important factor for MC (Tseng et al. 2010). Another opportunity is the possibility of customization of the product in post-sale time (occupation) (Lizarralde 2011; Enshassi et al. 2014; Feng et al. 2015). The long life cycle of the product (housing) allows the application of several cycles of customization, unlike other industrial products. This possibility allows added value to the customer through the housing customization during the guarantee time of building.

6 Contributions to the Practice

This research shows some opportunities to develop the MC strategy in the building sector (Table 3). The literature recommendations for MC in housing are presented according to the five stages of the customization process (Fig. 1). This review allows the selection of the most appropriate practices for the implementation of MC

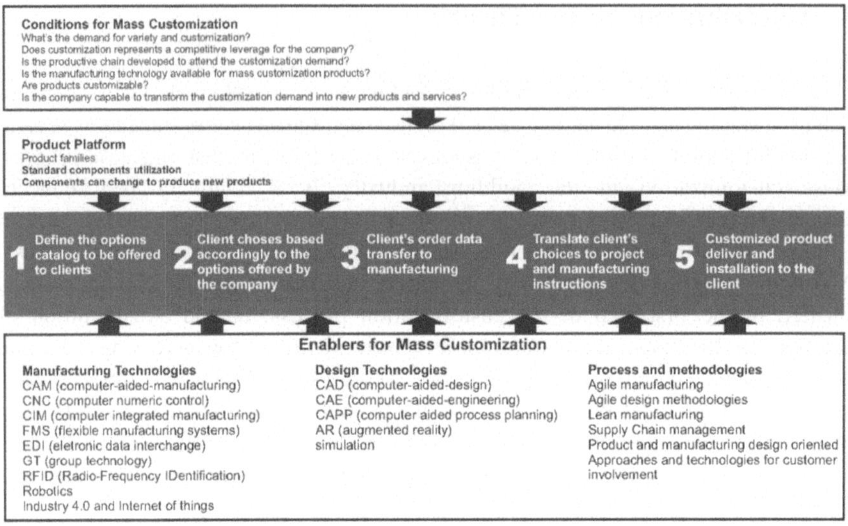

Fig. 1 Interface mapping between client and production for a customizable product

strategy in the company. In general, companies analyzed in Brazil do not apply the MC practices recommended in the literature. However, all studied companies pointed the use of MC strategy as beneficial to their business. As in other developing countries, Brazil has a high housing deficit (D'Ottaviano 2014). The lack of housing in the country still does not require companies to use strategies, such as MC, to add value to their products (buildings). Thus, the use of policies for reducing Brazilian housing lack, as in other developing countries, may contribute to the greater use of MC by building's companies.

7 Final Considerations

This study aimed to identify how the mass customization process is undertaken in housing companies and compare it with the recommended process in the literature. A multi-case study was conducted in three different housing companies located in the northeast of Brazil to identify differences between the methods and techniques applied by companies and recommended by the literature as a way to identify improvement opportunities for companies in the sector. Due to the scope of the MC, it was possible to identify several practices employed in other sectors that can be adapted and have satisfactory results for MC in housing. The results identified several opportunities for improvement, especially during the first two steps of the customization process. These steps aim at identifying the customization possibilities and present them to the customer. Most of the improvement opportunities identified for these steps do not require significant investment for their implementation. Also, they tend to use information already available in the company or easily accessible from their customer database. From this, the information collected from the market, obtained mainly during the sales process, provides a systematic manner in the project design process to contribute to the development of MC strategy by the studied companies.

Among the three housing companies where the study was undertaken, it was identified that the quantity of methods and techniques used is still low. Still, few methods and techniques are formally applied in a structured and systematic way, reducing the benefits of its application. Further, it was identified several development opportunities for companies in the sector, such as market segmentation studies, online configurator systems, and providing post-employment services. The MC in housing can be further studied and improved, bringing new technologies, especially for the treatment and subsequent use of the information received by customers in the MC process in the housing units. As a recommendation for future work, we suggest the application of this diagnosis in housing companies from other regions, including developed economy countries, so that results can be widely compared.

Appendix A

Questionnaire for construction companies that apply the Mass Customization strategy.

IDENTIFICATION	
Identification of the applicator:_____________date: ____ / __ / ___ questionnaire number: ___	
Construction company: _________________________	**Interviewee's position: __________________**

	Give your opinion on these issues:
	What is the number of employees in the construction process? And in the area of management?
	What is the construction company's time of operation?
	At present, how many projects does the construction company have?
	Does the construction company operate in any other region? And city?
	What is the number of floors of the building?
	How many units are per floor?
	In your opinion, how would you classify the customization offered by the company? Semi-Customized or Customized?
	The release point (RP) is the point in the production process where the customer's specifications are considered and where product customization is effectively started. In what stage is the construction company's DP? Land Acquisition Elaboration of the Standardized Projects Foundation Lease Infrastructure Execution Execution of the Structure (columns and slabs) Demarcation of Masonry (Layout of Plants) Facilities Piping Pipes and fittings Split's Drainage Piping Wiring Apartments Gas installation Coating Settlement Plaster Liner Cement regularization floor Laying Coating (Floor and Footer) Bench Top Application Application of granite fillets on the floor of the box Painting (Wall and Ceiling) Application of tanks in the service areas Apartment Doors Settlements Sanitary ware installation Installation of sanitary fittings Shower Installation Installation of Breakers and Mirrors
1	What and how many options are given to the customer to choose from in each unit? (Finishes, Walls, Floors, Number of Rooms)
1	How are these options mentioned above defined? Is there any tool or method for this?
1	In your opinion, is there a need for more product options to meet customer demands? What types of options should be used?
1	In relation to services, the clients request some type of service that the constructor can not attend? What types of services should be incorporated?
1	Would the company sell more units if it could or offer more alternatives for customization? Do you think it is good to include these options in the construction?
1	With regard to the services and products that may occur in the post-occupation of housing? Such as, changing masonry, finishes, lighting? Does the company offer support and options for unit customization?
1	What requests or type of customer requests are not usually answered by the company? For what reasons?

2	Which product choice choices are most often made by customers?
2	How is the choice made by the customer? Is there any difficulty or problem during this process? What is the source of this contact? Salesman Computer interface Another
2	What is the orientation that is given to the client to choose the alternatives of customization?
2	In your opinion, would the use of a program that allows the customer to simulate how the customization alternatives would be useful for the company and for the customers?
3	What is the procedure for transferring customer request data to the building?
3	What are the means of communication used for this transfer?
3	Are there problems / errors related to the transfer of these requests?
3	When the customer chooses some alternative that is not provided in the options catalogue, how does this request take place?
4	What are the main difficulties encountered in meeting the variety of customer customization needs?
4	Which functional sectors of the company are most involved in this process?
4	Is there any manufacturing technology to support the customization process?
4	For customized homes, how is production organized and programmed? Are there any differences from standardized ones?
4	Does the company apply any of these processes or methodologies? Could these processes or methodologies help the company meet customized demand? Agile manufacturing Lean production Customer-oriented design and manufacturing Supply chain management Product development process Product Platform Modularization Cite others
4	Is there a difference in deadlines for the delivery of a customized and a standardized unit? Justify.
5	How does the construction and delivery of the customized unit to the customer take place? Is there any follow-up to this process?
5	Does the company monitor the use, reuse or alteration of the housing unit after delivery?
5	Is there any product / service provided by the company after the sale and delivery process?

About Customization

Comments

References

Åhlström, P., & Westbrook, R. (1999). Implications of mass customization for operations management: an exploratory survey. *International Journal of Operations & Production Management, 19*(3), 262–275. https://doi.org/10.1108/01443579910249705.

Barco, A. F., Vareilles, É., Gaborit, P., & Aldanondo, M. (2017). Building renovation adopts mass customization. *Journal of Intelligent Information Systems 49*(1), 119–146. https://doi.org/10.1007/s10844-016-0431-6.

Bardacki, A., & Whitelock, J. (2004). How "ready" are customers for mass customization? An exploratory investigation. *European Journal of Marketing, 38*(11–12), 1396–1416. https://doi.org/10.1108/03090560410560164.

Bare, M., & Cox, J. (2008). Applying principles of mass customization to improve the empirical product development process. *Journal of Intelligent Manufacturing, 19*(5), 565–576. https://doi.org/10.1007/s10845-008-0134-0.

Barlow, J. (1998). From craft production to mass customisation? Customer-focused approaches to housebuilding. In *Proceedings of IGLC 6th Annual Conference*, Sao Paulo.

Barlow, J., Childerhouse, P., Gann, D., Hong-Minh, S., Naim, M., & Ozaki, R. (2003). Choice and delivery in housebuilding: lessons from Japan for UK housebuilders. *Building Research & Information, 31*(2), 134–145. https://doi.org/10.1080/09613210302003.

Barlow, J., & Ozaki, R. (2003). Achieving customer focus in private housebuilding: current practice and lessons from other industries. *Housing Studies, 18*(1), 87–101. https://doi.org/10.1080/0267303032000076858.

Bentzien, V., Rottke, N., & Zietz, J. (2012). Affordability and Germany's low homeownership rate. *International Journal of Housing Markets and Analysis, 5*(3), 289–312. https://doi.org/10.1108/17538271211243616.

Blecker, T., Abdelkafi, N., Kaluza, B., & Friedrich, G. (2003). Variety steering concept for mass customization. Discussion Papers of the University of Klagenfurt, No. 2003/04. Retrieved June 17, 2013, from http://mpra.ub.uni-muenchen.de/5251/1/MPRA_paper_5251.pdf.

Chakraborty, B., & Das, S. (2014). MGNREGA and Water Management: Sustainability Issues of Built Forms in Rural India. *Journal of Construction in Developing Countries, 19*(2), 33–50.

Cheng, G., & Han, Z. (2013). Mass customization in the tourism industry of China. *Advanced Materials Research, 683,* 917–920. https://doi.org/10.4028/www.scientific.net/AMR.683.917.

Da Silveira, G., Borenstein, D., & Fogliatto, F. (2001). Mass customization: Literature review and research directions. *International Journal of Production Economics, 72*(1), 1–13. https://doi.org/10.1016/S0925-5273(00)00079-7.

Dai, K., Li, Y., Han, J., Lu, X., & Zhang, S. (2003). An interactive web system for integrated three-dimensional customization. *Computers in Industry, 57,* 827–837. https://doi.org/10.1016/j.compind.2006.04.017.

Davis, S. (1987). *Future perfect.* MA: Addison-Wesley Publishing.

Denzin, N. K., & Lincoln, Y. S. (1994). *Handbook of qualitative research.* Sage publications, Inc.

D'Ottaviano, C. (2014). Housing policy in Brazil and the social rent program in São Paulo. *Caderno CRH, 27*(71), 255–266. https://doi.org/10.1590/S0103-49792014000200003.

Drejer, A., & Gudmundsson, A. (2002). Towards multiple product development. *Technovation, 22* (12), 733–745. https://doi.org/10.1016/s0166-4972(02)00024-x.

Du, X., Jiao, J., & Tseng, M. (2003). Modelling platform-based product configuration using programmed attributed graph grammars. *Journal of Engineering Design, 14*(2), 145–167. https://doi.org/10.1080/0954482031000091482.

Duarte, J. (2005). A discursive grammar for customizing mass housing: the case of Siza's houses at Malagueira. *Automation in Construction, 14*(2), 265–275. https://doi.org/10.1016/j.autcon.2004.07.013.

Duray, R., & Milligan, G. (1999). Improving customer satisfaction through mass customization. *Quality Progress, 32*(8), 60–66.

Duray, R., Ward, P., Milligan, G., & Berry, W. (2000). Approaches to mass customization: configurations and empirical validation. *Journal of Operations Management, 18*(6), 605–625. https://doi.org/10.1016/S0272-6963(00)00043-7.

Elliman, T., & Orange, G. (2003). Developing distributed design capabilities in the construction supply chain. *Construction Innovation, 3*(1), 15–26. https://doi.org/10.1108/14714170310814828.

Enshassi, A. A., Arain, F., & El-Rayyes, Y. (2014). Post-evaluation System in Construction Projects in Gaza Strip-Palestine. *Journal of Construction in Developing Countries, 19*(2), 51–73.

Feitzinger, E., & Lee, H. (1997). Mass customization at Hewlett-Packard: the power of postponement. *Harvard Business Review, 75*(1), 116–123.

Feng, X., Yan, D., & Hong, T. (2015). Simulation of occupancy in buildings. *Energy and Buildings, 87,* 348–359. https://doi.org/10.1016/j.enbuild.2014.11.067.

Fettermann, D. C., Echeveste, M. E. S., & Tortorella, G. L. (2017). The benchmarking of the use of toolkit for mass customization in the automobile industry. *Benchmarking: An International Journal, 24*(6), 1767–1783. https://doi.org/10.1108/BIJ-01-2016-0002.

Fettermann, D. C., & Freitas, F. C. (2017). Automobile variety in emerging countries: A comparative study between Brazil and USA. *Engineering Journal, 21*(4), 325–338. https://doi.org/10.4186/ej.2017.21.4.325.

Fettermann, D., & Echeveste, M. (2011). Desenvolvimento de produto para customização em massa: alternativas para o setor de móveis modulados. *Espacios, 32*(4), 12.

Fettermann, D., & Echeveste, M. (2014). New product development for mass customization: a systematic review. *Production and Manufacturing Research, 2*(1), 266–290. https://doi.org/10.1080/21693277.2014.910715.

Fettermann, D., Echeveste, M., & Ten Caten, C. (2012). When and how to use the online configurator in the automobile industry. *IEEE Latin America Transactions, 10*(6), 2331–2341. https://doi.org/10.1109/TLA.2012.6418140.

Fogliatto, F., & Da Silveira, G. (2008). Mass customization: a method for market segmentation and choice menu design. *International Journal of Production Economics, 111*(2), 606–622. https://doi.org/10.1016/j.ijpe.2007.02.034.

Fogliatto, F., Da Silveira, G., & Borenstein, D. (2012). The mass customization decade: an updated review of the literature. *International Journal of Production Economics, 138*(1), 14–25. https://doi.org/10.1016/j.ijpe.2012.03.002.

Formoso, C. T., Leite, F., & Miron, L. I. G. (2011). Clients requirement management in social housing: A case study on the residential leasing program in Brazil. *Journal of Construction in Developing Countries, 16*(2), 47–67.

Franke, N., & Piller, F. (2003). Key research issues in user interaction with configuration toolkits in mass customization. *International Journal of Technology Management, 26,* 578–599. https://doi.org/10.1504/IJTM.2003.003424.

Franke, N., Schreier, M., & Kaiser, U. (2010). The "I designed it myself" effect in mass customization. *Management Science, 56*(1), 125–140. https://doi.org/10.1287/mnsc.1090.1077.

Gemser, G., & Perks, H. (2015). Co-Creation with Customers: An Evolving Innovation Research Field. *Journal of Product Innovation Management, 32*(5), 660–665. https://doi.org/10.1111/jpim.12279.

Gilmore, J., & Pine, B., II. (1997). The four faces of mass customization. *Harvard Business Review, 75*(1), 91.

Gilmore, J., & Pine, J. (2000). *Markets of one: Creating customer-unique value through mass customization*. Harvard Business Press.

Grenci, R., & Watts, C. (2007). Maximizing customer value via mass customized e-consumer services. *Business Horizons, 50*(2), 123–132. https://doi.org/10.1016/j.bushor.2006.08.006.

Halman, J., Voordijk, J., & Reymen, I. (2008). Modular approaches in Dutch house building: an exploratory survey. *Housing Studies, 23*(5), 781–799. https://doi.org/10.1080/02673030802293208.

Hart, C. (1995). Mass customization: conceptual underpinnings, opportunities and limits. *International Journal of Service Industry Management, 6*(2), 36–45. https://doi.org/10.1108/09564239510084932.

Hellander, M., & Jiao, J. (2002). Research on E-product development (ePD) for mass customization. *Technovation, 22*(11), 717–724. https://doi.org/10.1016/S0166-4972(01)00074-8.

Hofman, E., Halman, J., & Ion, R. (2006). Variation in housing design: identifying customer preferences. *Housing Studies, 21*(6), 929–943. https://doi.org/10.1080/02673030600917842.

Huffmann, C., & Kahn, B. (1998). Variety for sale: mass customization or mass confusion? *Journal of Retailing, 74*(4), 491–513. https://doi.org/10.1016/S0022-4359(99)80105-5.

Jiao, J., & Tseng, M. (2000). Fundamentals of product family architecture. *Integrated Manufacturing Systems, 11*(7), 469–483. https://doi.org/10.1108/09576060010349776.

Jiao, J., & Tseng, M. (2004). Customizability analysis in design for mass customization. *Computer-Aided Design, 36*(8), 745–757. https://doi.org/10.1016/j.cad.2003.09.012.

Jiao, J., Ma, Q., & Tseng, M. (2003). Toward high value-added products and services: mass customization and beyond. *Technovation, 23*(10), 809–821. https://doi.org/10.1016/S0166-4972(02)00023-8.

Karadimitriou, N. (2005). Changing the way UK cities are built: The shifting urban policy and the adaptation of London's housebuilders. *Journal of Housing and the Built Environment, 20*(3), 271–286. https://doi.org/10.1007/s10901-005-9010-9.

Kendall, S., & Teicher, J. (2000). *Residential Open Building*. London and New York: E & FN Spon.

Kendall, S. H. (2013). The next wave in housing personalization: customized residential fit-out. In P. A. E. Piroozfar., & F. T. Piller (Eds.), *Mass customization and personalization in architecture and construction*. Routedge.

Khalili-Araghi, S., & Kolarevic, B. (2016). Development of a framework for dimensional customization system: A novel method for customer participation. *Journal of Building Engineering, 5*, 231–238. https://doi.org/10.1016/j.jobe.2016.01.001.

Kotha, S. (1996). From mass production to mass customization: the case of the national industrial bicycle company of Japan. *European Management Journal, 14*(5), 442–450. https://doi.org/10.1016/0263-2373(96)00037-0.

Krueger, C. (2001). Easing the transition to software mass customization. *Software Product-Family Engineering, Springer, Berlin Heidelberg, 2002*, 282–293. https://doi.org/10.1007/3-540-47833-7_25.

Kumar, S., & Phrommathed, P. (2005). *New product development: an empirical study of the effects of innovation strategy, organizational learning, and market conditions*. New York: Springer.

Liao, K., Deng, X., & Marsillac, E. (2013). Factors that influence Chinese automotive suppliers' mass customization capabilities. *International Journal of Production Economics, 146*(1), 25–36. https://doi.org/10.1016/j.ijpe.2013.01.014.

Lihra, T., Buehlmann, U., & Graf, R. (2012). Customer preferences for customized household furniture. *Journal of Forest Economics, 18*(2), 94–112. https://doi.org/10.1016/j.jfe.2011.11.001.

Lizarralde, G. (2011). Stakeholder participation and incremental housing in subsidized housing projects in Colombia and South Africa. *Habitat International, 35*(2), 175–187. https://doi.org/10.1016/j.habitatint.2010.08.001.

Massyn, M., McGaffin, R., Viruly, F., & Hopkins, N. (2015). The challenge of developing higher density, affordable housing in the inner city of Cape Town. *International Journal of Housing Markets and Analysis, 8*(3), 412–428. https://doi.org/10.1108/IJHMA-11-2014-0049.

Merle, A., Chandon, J., Roux, E., & Alizon, F. (2010). Perceived value of the mass-customized product and mass customization experience for individual consumers. *Production and Operations Management, 19*(5), 503–514. https://doi.org/10.1111/j.1937-5956.2010.01131.x.

Meyer, M., & Lehnerd, A. (1997). *The power of product platforms: Building value and cost leadership*. New York: Free Press.
Moreau, C. (2011). Inviting the amateurs into the studio: understanding how consumer engagement in product design creates value. *Journal of Product Innovation Management, 28*(3), 409–410. https://doi.org/10.1111/j.1540-5885.2011.00813.x.
Naderi, I., Sharbatoghlie, A., & Vafaeimehr, A. (2012). Housing valuation model: an investigation of residential properties in Tehran. *International Journal of Housing Markets and Analysis, 5* (1), 20–40. https://doi.org/10.1108/17538271211206644.
Nahmens, I., & Bindroo, V. (2011). Is customization fruitful in industrialized homebuilding industry? *Journal of Construction Engineeringand Management, 137*(12), 1027–1035.
Noguchi, M. (2003). The effect of the quality-oriented production approach on the delivery of prefabricated homes in Japan. *Journal of Housing and the Built Environment, 18*(4), 353–364. https://doi.org/10.1023/B:JOHO.0000005759.07212.00.
Noguchi, M. (2013). Commercialisation principles for low carbon mass customised housing delivery in Japan. In P. A. E. Piroozfar., & F. T. Piller (Eds.), *Mass customization and personalization in architecture and construction*. Routedge.
Noguchi, M., & Hernández-Velasco, C. (2005). A 'mass custom design' approach to upgrading conventional housing development in Mexico. *Habitat International, 29*(2), 325–336. https://doi.org/10.1016/j.habitatint.2003.11.005.
Ogawa, S., & Piller, F. (2006). Reducing the risks of new product development. *MIT Sloan Management Review, 47*(2), 65.
Peppers, D., & Rogers, M. (1997). *Enterprise one to one: Tools for competing in the interactive age*. New York: Bantam Doubleday Dell Publishing Group.
Piller, F. (2004). Mass customization: reflections on the state of the concept. *International Journal of Flexible Manufacturing Systems, 16*(4), 313–334. https://doi.org/10.1007/s10696-005-5170-x.
Pine, J., II. (1993). *Mass customization: the new frontier in business competition*. Boston: Harvard Business School Press.
Ro, Y., Liker, J., & Fixson, S. (2007). Modularity as a strategy for supply chain coordination: the case of U.S. auto. *IEEE Transactions on Engineering Management, 54*(1), 172–189. https://doi.org/10.1109/TEM.2006.889075.
Robert, E., & Meyer, M. (1991). Product strategy and corporate success. Sloan School of Management, Massachusetts Institute of Technology (MIT). https://dspace.mit.edu/bitstream/handle/1721.1/2334/SWP-3239-23273127.pdf.
Robertson, D., & Ulrich, K. (1998). Planning for product platforms. *Sloan Management Review, 39*(4), 19–31.
Rocha, C., Formoso, C., & Santos, A. (2012). An overview of the customisation strategies developed by four organisations of the house-building sector. In *Proceedings of 20th Conference of the International Group for Lean Construction*, San Diego, USA.
Roy, U., & Roy, M. (2016). Space standardisation of low-income housing units in India. *International Journal of Housing Markets and Analysis, 9*(1), 88-107. https://doi.org/10.1108/ijhma-12-2014-0057.
Sahin, O., Miller, D., & Mohamed, S. (2016). Value-based modelling: an Australian case of off-site manufactured buildings. *International Journal of Construction Management*, 1–19. https://doi.org/10.1080/15623599.2016.1247774.
Salvador, F., & Forza, C. (2004). Configuring products to address the customization responsiveness squeeze: A survey of management issues and opportunities. *International Journal of Production Economics, 91*(3), 273–291. https://doi.org/10.1016/j.ijpe.2003.09.003.
Salvador, F., Forza, C., & Rungtusanatham, M. (2002). Modularity, product variety, production volume, and component sourcing: theorizing beyond generic prescriptions. *Journal of Operations Management, 20*(5), 549–575. https://doi.org/10.1016/S0272-6963(02)00027-X.
Shin, Y., An, S., Cho, H., Kim, G., & Kang, K. (2008). Application of information technology for mass customization in the housing construction industry in Korea. *Automation in Construction, 17*(7), 831–838. https://doi.org/10.1016/j.autcon.2008.02.010.

Sievänen, M. (2002). What is customization? In *Proceedings of 9th International Annual Conference of European Operations Management Association*, Copenhagen, Denmark. https://pdfs.semanticscholar.org/b988/6946f76dfdfb0f4d8411a5500ffdda7a1a88.pdf.

Simpson, T. (2004). Product platform design and customization: status and promise. *AI EDAM: Artificial Intelligence for Engineering Design, Analysis and Manufacturing, 18*(1), 3–20. https://doi.org/10.1017/S0890060404040028.

Squire, B., Readman, J., Brown, S., & Bessant, J. (2004). Mass customization: the key to customer value? *Production Planning and Control: The Management of Operations, 15*(4), 459–471. https://doi.org/10.1080/0953728042000238755.

Trentin, A., Perin, E., & Forza, C. (2014). Increasing the consumer-perceived benefits of a mass-customization experience through sales-configurator capabilities. *Computers in Industry, 65*(4), 693–705. https://doi.org/10.1016/j.compind.2014.02.004.

Tseng, M., & Jiao, J. (1996). Design for mass customization. *CIRP Annals-Manufacturing Technology, 45*(1), 153–156. https://doi.org/10.1016/S0007-8506(07)63036-4.

Tseng, M., & Jiao, J. (2007). Mass Customization. In Salvendy, G (Ed.), *Handbook of industrial engineering: Technology and operations Management* (3rd ed.) New York: John Wiley & Sons.

Tseng, M., Jiao, R., & Wang, C. (2010). Design for mass personalization. *CIRP Annals-Manufacturing Technology, 59*(1), 175–178. https://doi.org/10.1016/j.cirp.2010.03.097.

Tu, K.J., Wei, H.Y. (2013). Emerging trends and concepts of mass customisation in Taiwan's housing industry. In: P. A. E. Piroozfar., & F. T. Piller (Eds) *Mass customization and personalization in architecture and construction*. Routedge.

Tu, Y., Xie, S., & Fung, R. (2007). Product development cost estimation in mass customization. *IEEE Transactions on Engineering Management, 54*(1), 29–40. https://doi.org/10.1109/TEM.2006.889065.

Ulrich, K. (1995). The role of product architecture in the manufacturing firm. *Research Policy, 24* (3), 419–440. https://doi.org/10.1016/0048-7333(94)00775-3.

Ulrich, K. (2011). Design, creation of artifacts in society: Utility and complexity. University of Pennsylvania.

Voss, C., & Hsuan, J. (2009). Service architecture and modularity. *Decision Sciences, 40*(3), 541–569. https://doi.org/10.1111/j.1540-5915.2009.00241.x.

Wang, Z., Chen, L., Zhao, X., & Zhou, W. (2014). Modularity in building mass customization capability: The mediating effects of customizationknowledge utilization and business process improvement. *Technovation, 34*(11), 678–687. https://doi.org/10.1016/j.technovation.2014.09.002.

Wang, Q., Wang, Z., & Zhao, X. (2015). Strategic orientations and mass customisation capability: the moderating effect of product life cycle. *International Journal of Production Research, 53* (17), 5278–5295. https://doi.org/10.1080/00207543.2015.1027012.

Zeithaml, V. (1988). Consumer perceptions of price, quality, and value: a means-end model and synthesis of evidence. *Journal of Marketing, 52,* 2–22. https://doi.org/10.2307/1251446.

Zhang, M., Chen, Y., & Tseng, M. (2005). Distributed knowledge management for product and process variety in mass customization. *International Journal of Computer Applications in Technology, 23*(1), 13–30. https://doi.org/10.1504/IJCAT.2005.006419.

Zhang, M., Zhao, X., & Qi, Y. (2014). The effects of organizational flatness, coordination, and product modularity on mass customization capability. *International Journal of Production Economics, 158,* 145–155. https://doi.org/10.1016/j.ijpe.2014.07.032.

A Talent Management Model for Innovation and Competitiveness in Complex Domains: A Case Study in the Latin American Energy Sector

Yasmín Hernández, Ramón Zatarain and Lucía Barrón

Abstract Companies require highly qualified electricians because they have to be competitive in a globalized world and a changing environment. Additionally, they have to face the generational change, since knowledgeable personnel is in the retirement process, and new human resources have to be trained to get the required competencies: knowledge, skills, and attitudes. Talent management is concerned with employee recruitment, development, and administration, and suggests the integration of learning as a fundament for talent management strategies. We have developed a training model, which provides online and personalized instruction within the talent management framework based on three axes: knowledge, skills, and attitudes. The training model aims to complement traditional training with computer-based training. An important aspect of the model is an intelligent training system. This intelligent system adapts the instruction relying on a trainee model, which represents the state of the trainee. The trainee model includes knowledge, skills, attitudes, affect, and learning styles, and it is built as the trainee interacts with the training system. Emotions are recognized using facial expressions and training situation. An animated agent who uses the trainee model to show emotions and empathy presents the instruction. Here, we present the training model, and we discuss the current results in the implantation of the training model.

Y. Hernández (✉)
Instituto Nacional de Electricidad y Energías Limpias, Cuernavaca, Mexico
e-mail: yasmin.hernandez@ineel.mx

R. Zatarain · L. Barrón
Instituto Tecnológico de Culiacán, Culiacán, Mexico
e-mail: rzatarain@itculiacan.edu.mx

L. Barrón
e-mail: lbarron@itculiacan.edu.mx

G. Cortes-Robles et al. (eds.), *Managing Innovation in Highly Restrictive Environments*, Management and Industrial Engineering,
https://doi.org/10.1007/978-3-319-93716-8_6

1 Introduction

Talent management is increasingly an essential process in modern companies as they compete in a globalized world. Talent management is concerned with to identify, attract, recruit, retain, and produce the most talented employees available in the job market and to achieve the objectives of the organization (Cunningham 2007). Talent management involves employee recruitment, development, and management, where learning becomes an important driver of these processes. Companies have to look beyond traditional learning activities, because learning happens in several ways within an organization and across generations (Ruiz-Hau 2017).

Talent management systems comprise of several processes related to training and development of personnel. More than 45% of learning executives of high-impact learning organizations value the incorporation of talent management as one of their key training policies (Bersin 2009). In this way, training is a strategic tool to provide knowledge, develop skills, and change attitudes for performance improvement aligned with talent management and strategic planning.

In the electrical field, as in many other domains, training is imperative because its procedures involve a high risk of accidents resulting in arc flash and electric shock, among other risks, which can injure workers. Similarly, accidents would break electrical equipment and damage installations. Another aspect is that experienced personnel are in retirement process and the knowledge has to be transferred to novice workers. And, most of the times, there is a gap between the required competencies and the competencies of new employees.

Traditional training has been successful; however, it is expensive, takes a lot of time, and it is not available in a continuous way. Electricians need ongoing training on normal and abnormal situations. However, there are several restrictions, for example, electrical installations are not available for training because they are in production on a regular basis. Moreover, the personnel have to take productive days to travel to training centers, which become training into an expensive process.

On the other hand, although training is carefully planned and executed, work accidents still happen. This situation could be due, to a large degree, to human factors such as attitudes and emotions, which are involved in behavior and learning.

There are several situations in the context of work and training where attitudes are key to the result since attitudes act as a filter where only the important things for the individual go through the filter. Attitude is a psychological concept; it is a mental and emotional state, which characterizes the person. It also is called a hypothetical construct, a concept that cannot be observed directly but can be only inferred from the actions of the people (Perloff 2017). In this way, attitudes are involved in the behavior of the personnel and, therefore, in their reactions to certain circumstances. For example, if workers have a positive attitude toward learning, then training takes place in a shorter period and with successful results. Likewise, it is necessary that people have a positive attitude toward safety aspects to avoid accidents at work. Attitudes are the result of individual life, education, rising, and are influenced by the surrounding reality; they change continuously thorough experiences.

On the other hand, emotions are central to motivation and consequently for learning. Several studies state the importance of emotions in almost every aspect of human life (Damasio 1996). Some of the most important aspects are creativity, decision-making, and communication. Communication is an important piece of learning, so it is needed to understand instruction not only from conventional learning but also as a communication process where, as we already mentioned, emotions play an influential role (Reeves and Nass 1996). Therefore, in a training environment, it is necessary to understand which emotional states are relevant for learning and motivation, such a way they are recognized and shown to the trainees, and to act in consequence to support trainees in the learning process and also to try to motivate them according to particular training states. Furthermore, it is imperative to understand which pedagogical actions promote a positive emotional state and which are pedagogically suitable considering the specific trainee (Hernández et al. 2015).

Another key aspect of training is the learning styles since it is recognized that every person prefers different techniques, methods, and modes of learning. Learning styles classify general approaches in which people learn. Every person has a combination of learning styles; however, some individuals have a preferred learning style over the other learning styles. Other people benefit from different styles in different situations. Dominant learning styles drive the people learning progression. Also, learning styles change how people represent experiences internally, how to recall memory, and how people choose words (Willingham et al. 2015).

These human factors can improve learning if they are considered in the planning and design of the training. Therefore, they can be included in the development of intelligent training systems to complement the traditional training and to allow distance training and self-training. Including attitudes, affect, and learning styles provides adaptive training. That is to say, the training can be personalized according to particular needs of each person.

For several years, we have worked in a framework for talent management as a fundamental base for competitiveness and innovation with encouraging results. In this chapter, we describe the overall framework and how its components are connected. We have work in companies related to electric industry to probe our research. However, the models we have proposed are applied to almost any domain since all companies require well-trained personnel.

We have worked on a training model for talent management. An important component of the model is composed of intelligent training systems for several electrical fields. The intelligent system adapts the instruction relying on a trainee model, which represents the state of the trainee. The trainee model includes knowledge, skills, attitudes, affect, and learning styles. The student model relies on Bayesian networks (Sucar 2015) and it is built as the trainee interacts with the training system. The trainee emotion is recognized using facial expressions and training situation. The facial expression recognizer relies on neural networks and the theory of Ekman and Friesen (1978). Training situation is used to recognize emotions as stated in the OCC model (Ortony et al. 1988).

An animated agent who uses the trainee model to show emotion and empathy presents the instruction with the purpose to have a more natural interaction. We incorporated these constructs in the training model to foster social interaction and to obtain believability. In this way, trainees can be engaged, and hence their learning could be improved.

With the training model and intelligent systems complementing the training we have composed a blended training framework, aiming to allow distance and self-training. This training model allows training more people, in less time and with fewer costs. Also, it responds to the needs of energy sectors, which demand several thousands of workers to be skilled in the diverse topics of the electrical field (SENER, SEP and CONACYT 2014).

In this paper, we present the training model, we describe its components, and we discuss the current results in the implantation of the training model. The rest of the chapter is organized as follows. In next section, the training model is described. Then, the following sections depict the intelligent training system and its components. Finally, conclusions and future work are presented.

2 Training Model for Talent Management

It is well known that human resources constitute the most valuable active resources of organizations, since they have to do all the processes, and consequently they have an important impact on conclusive outcomes. For that reason, companies pay attention in the design of mechanisms to administer their human resources, specifically strategies for talent management. In corporations, the combination of talent management and learning represents an important tendency in corporate development and learning (Bersin 2009).

A talent management model has been implemented, as a way to achieve the strategic objectives and it is aimed to have the maximum value creation for the company, customers, employees, and society.

The main actions to achieve these strategic objectives have been a training program, to maintain and to enhance human capital, but also to promote learning and innovation. Throughout the years, a significant effort has been deployed to train and to develop human capital in all the processes of electricity. These efforts are primarily focused on training and the achievement of academic diplomas or degrees, but also, as an incentive for value input through the proposal of innovative developments as fundamentals for competitiveness.

Training programs are then established to ensure the acquisition of knowledge in the personnel, skills, and attitudes to improve performance in their functions and tasks on their current appointment and the following superior appointment.

Recently, some research on emergent methodologies and technologies supports training for competencies development and traditional training. The study underlines e-learning systems, blended models, and intelligent systems for personalized learning (Rodríguez et al. 2011).

Consequently, to develop strategies to manage human resources and, more specific, strategies for talent management is crucial to any company. We have designed and implemented a Talent Management Framework based on four axes: involvement in training, experience at work, a contribution of added value, and academic degrees. In Fig. 1, the Talent Management Framework is shown.

Training is a fundamental axis, and it is a foundation for the other axis due to its importance and impact on the human resources performance. Companies have recognized its importance and have considered training as a strategic activity. Also, companies have realized that their productivity depends, in no small extent, on the human resources, which, in turn, require an efficient training program. Training strategies are more important when the integrity of people is threatened by potential accidents. The electrical domain involves the threat of arc flash, and electric shock, among other risks for workers. Additionally, accidents could harm electrical systems, equipment, and installations.

In light of talent management, we propose a training model, which consists of complementing traditional training with intelligent systems that provide efficient and personalized training. The customized training is achieved by considering several states of the trainees. In this way, we have a training model based on fundamentals of blended learning integrating three elements: (i) classes with a certified instructor in the classroom, (ii) an intelligent training system, and (iii) field practice with certified electricians. Figure 2 shows the training model.

The blended training model aims to provide a practical, quick, secure, and inexpensive training. The blended training model is based on the traditional training programs. Classical training in electrical domains is composed of theoretic and classroom courses, and field practice in actual electrical installations.

With the first constituent, the classroom courses, trainees have interaction with a certified instructor, which instructs them about electrical concepts and procedures in

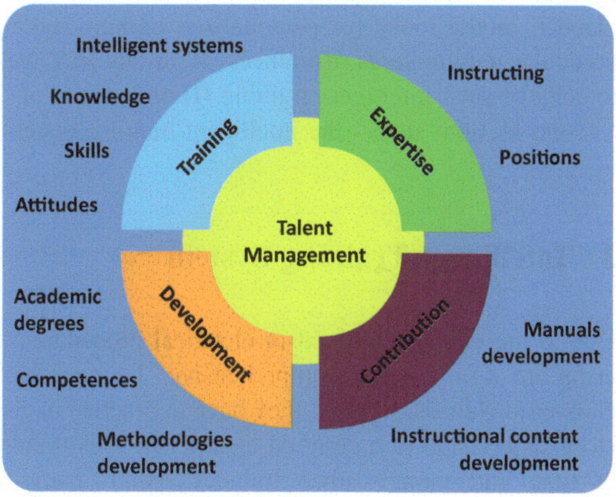

Fig. 1 A talent management model. The model is based on four axes: training, working experience, value contribution, and academic development (Hernández et al. 2012)

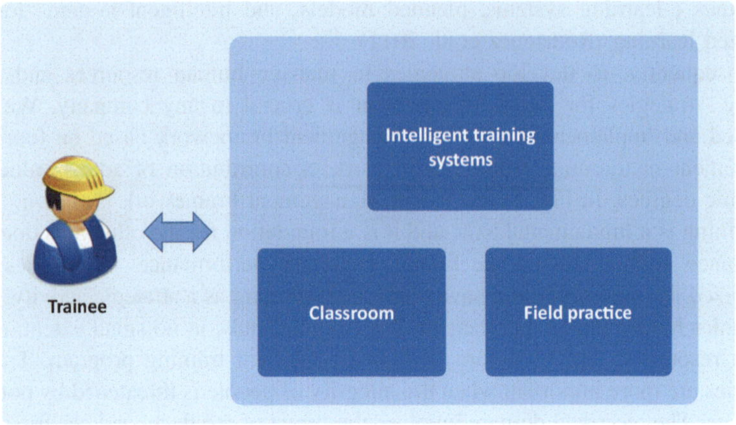

Fig. 2 Training model. The model consists of (i) classroom classes, (ii) intelligent training systems, and (iii) field practice, in this way a blended learning framework is composed

the classroom. The second constituent of the blended model consists of intelligent training system, which supports the classroom courses. With these systems, trainees strengthen electrical concepts and execute procedures in a graphical environment. The third constituent in the field of practice is a real scenario. When workers have attended and approved the required classroom courses, they are ready to perform an auxiliary position, helping certified electricians to conduct electrical procedures; this activity represents a field practice in real environments.

The instructor designs the classroom course. He selects topics to be taught in the course, and also the instructor plans the course in the intelligent training system. However, the teaching and testing materials are already designed in the system. The big challenge for instructors is to communicate his working experience besides to teach the theoretic concepts.

In a distance learning modality, in the intelligent training system, trainees practice the electrical procedures, and also they can evaluate their progress. The blended model looks for personalized training, taking into account personal and particular requirements of trainees. With this aim, we have developed a route map to achieve such intelligent training (Hernández et al. 2016a, b).

Next sections depict the intelligent training system and their components.

3 Intelligent Training System

Due to the danger involved in electrical procedures, the curricula to train electricians are rigorous and comprehensive. Trainees have to attend and to approve classroom courses, but also they have to practice in electrical installations under the close supervision of a certified electrician. This training methodology is expensive in time and money since trainees spend many days attending courses and developing field practice. Several training systems have been developed to support the traditional training in different electrical fields. For example, functioning and maintenance of high-voltage overhead power lines (Ayala et al. 2016), electrical tests to substation primary equipment (Hernández and Pérez 2016a), for maintenance tests to protections (Hernández and Pérez 2016b), and for electricity distribution substations operation (Hernández et al. 2016a, b). We are working on the integration of the intelligent training systems in these developments.

The intelligent training system helps to have shorter periods of training since trainees can practice electrical procedures before to have field practice, and the training is adapted to particular needs of trainees. Thus, trainees learn the electrical concepts and practice the procedures included in the course by interacting with an animated pedagogical agent in the intelligent training system, besides attending the course and having field practice. The training system includes virtual environments, which represent accurately electrical scenarios, and the procedures are performed following standard procedures using animations. The instructor experience stated how environment and animations are built and presented. For instance, the procedure states that the electrician has to wear safety gear, but the instructor expands the instruction stating that all the elements of the procedure have to be identified. Thus, the trainees can observe, analyze, and think about safety aspects.

We designed the architecture of the intelligent training system considering the components of the architecture of classical intelligent tutoring systems (Woolf 2008). These systems adapt the instruction to particular needs using a student model (Sottilare et al. 2013), which represents the student state. Figure 3 shows a diagram of the intelligent training system and its interactions with trainee and instructors.

The system is composed of trainee model, trainer module, electrical knowledge module, and the interface module. The trainee model is an important constituent of the intelligent training system; it is built as the trainee interacts with the intelligent training and it records results of practices. Additionally, the instructor updates the

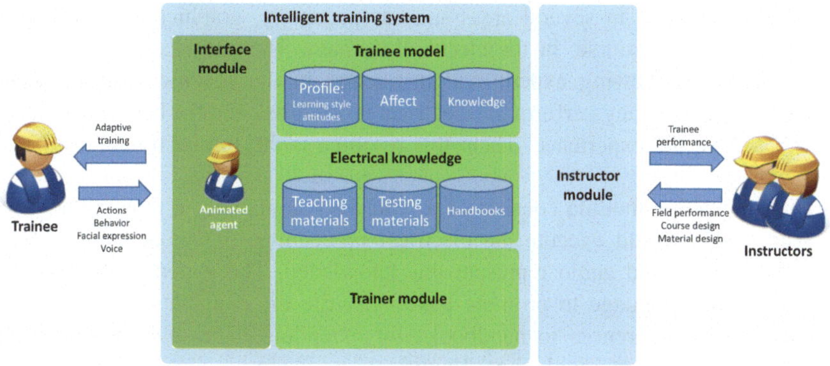

Fig. 3 Intelligent training system architecture. The system is composed of four main components: training model, electrical knowledge module, trainer model, and an interface module. This architecture is based on the intelligent tutoring systems classical architecture

trainee model taking into account the performance of the trainee in class and the practice field. The instructor uses the instructor module to update the trainee model.

The trainee model is a representation of the knowledge and emotional states of trainees. This representation includes the profile of the trainee, and it comprises attitudes and learning styles. With the base on the particular electrical topic, the intelligent training system adapts the instruction and the practice to the performance and knowledge of the trainee. Also, the system can suggest reviewing specific procedures steps or to studying specific topics. The trainee model is also helpful for instructors since they can modify the instruction in the classroom, to adjust the field practice, to recommend trainees to attend other training courses, and, finally, to concede a certification with the base on the information in the model. Moreover, the trainee model is helpful to design new courses, new teaching materials, and new testing materials and to redesign training courses and material.

The intelligent training system includes a virtual environment, which is a graphical representation of the electrical environment, which enables the practice in realistic environments before to practice in real electrical installations. The training system provides a safe environment to practice and learn without the risks of damaging people, expensive equipment, or facilities.

The virtual environment also includes an animated pedagogical agent, which deploy emotions and empathy to instruct accordingly. The agent shows empathy based on the emotions of the trainee stored in the trainee model. Our proposal for the agent includes the characteristics of electricians; for example, the agent fulfills the safety and institutional requirements such as wearing the safety helmet and uniform, among other features. This research has been guided by the assumption where we think that the characterization of the animated agent as an electrician help to trainees welcome the training environment. Also, empathy is a design attribute for the agent, which can understand other people and their emotions. We have integrated empathy to obtain credibility for the agent and trainees' commitment (Hone 2006).

The electrical knowledge database includes training, and testing material designed by experts in several areas such as pedagogy, graphics designers, programmers, and, of course, in the electrical domain. The knowledge domain was obtained by interviewing experts and instructors in the specific electrical field. Experts stated how they perform the electrical procedures, and the best way to teach them based on the experience. Then, this resource is the formal training to be an instructor.

The knowledge module consists of a repository containing a representation of how an expert would execute the electrical procedures. 3D models added with animations, text, and audio represent this knowledge. The experts, who used an understandable language to promote learning, write the explanations. The explanations contain references to handbooks for extending them and for promoting knowledge and use of regulations. Empirical knowledge is also included in this component in the form of additional information; this knowledge is the result of experience rather than standard procedures. The explanations are also recorded in audio for reinforcing learning.

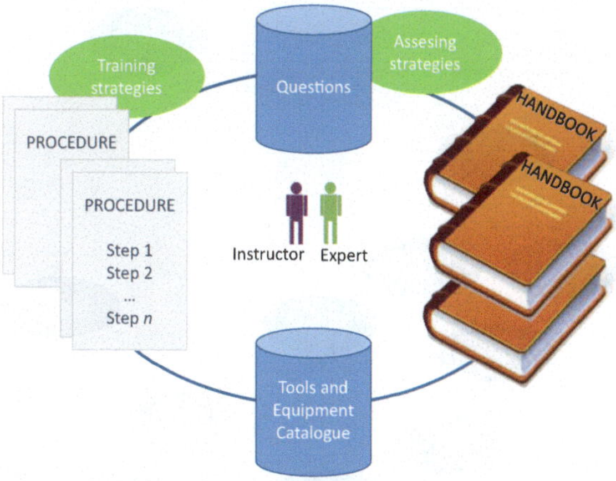

Fig. 4 Electrical knowledge model: electrical procedures descriptions, questions databases, equipment catalogs, and reference books (Hernández and Pérez 2016a)

A repository of questions about tests is also included. The questions are written and structured taking into account pedagogical strategies for evaluation developed by experts. Figure 4 presents a diagram showing the different elements involved in the construction of the knowledge domain.

The module of the instructor permits instructors to design training courses considering the structure of the domain knowledge. Thus, an instructor selects which and how many electrical topics and procedures to be included in a course. The design is based on the knowledge of the trainees, represented in the trainee model. The instructor also designs the theoretic exam through the questions database and choosing appropriate items to be integrated into the assessment.

In next section, we describe our proposal for the trainee model based on affect, attitudes, and learning styles; then we describe how we model them.

4 Trainee Knowledge Model

The trainee model is a structure, which represents the current state of trainees. This model is updated as trainees attend courses, attend lessons, practice, and set exams. For the knowledge side, it considers electrical topics learned, correct and incorrect clicks in practice activities, outcomes of the theoretic review, and right and inaccurate clicks in the practical evaluation.

The trainee model is dynamic; it is built according to the trainee progress. This structure is constructed while the electrical knowledge model for the course is passed through. Thus, we rely on an overlay student model (Woolf 2008), where the knowledge of the trainee is a subset of the knowledge of the course, which, in turn, is a subset of the knowledge of the expert. Figure 5 presents the structure of trainee knowledge model.

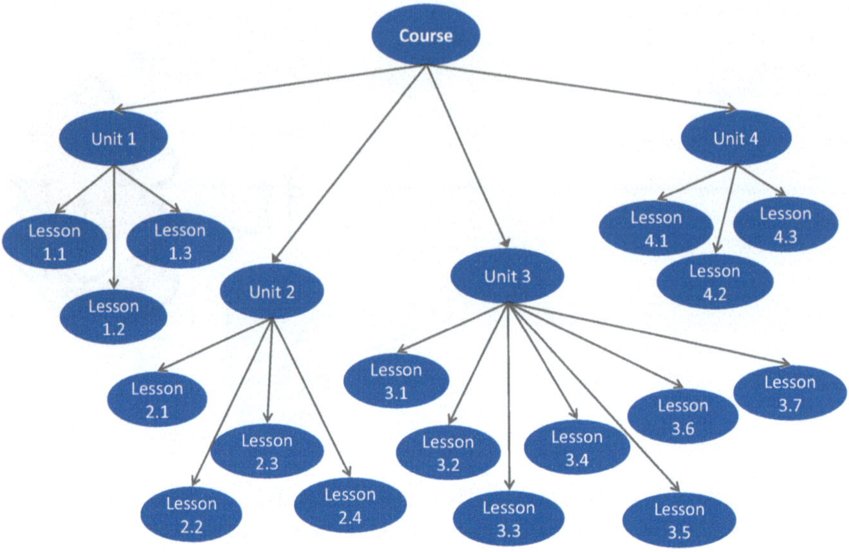

Fig. 5 Trainee knowledge model. The trainee model is composed of the electrical procedures included in the course

The trainee model is used in the practice activities. When a trainee is learning a method, he must visit all the steps sequentially; he cannot move toward a step not yet learned previously. However, when he is practicing, he can review any step he wants, traversing the tree in the order he wishes, since the tree represents his knowledge.

Electrical procedures are arranged into a sequence of steps to promote learning of safety aspects since omissions are dangerous in real work; safety regulations must not be neglected. It is worth to mention that while an electrical procedure is not fully learned, trainees cannot start learning another procedure. This restriction reinforces organized learning of electrical procedures and also avoids possible misunderstandings.

Besides modeling knowledge of the trainee, personal traits such as personality, affect, attitudes, learning styles, and so on, could be modeled to provide a more adaptive instruction. In parallel, we already have developed an affective trainee model based on the contextual information; basically, it considers the results of the interaction between trainees and the training system. We are still working on integrating this model into the intelligent training system.

5 Emotion Recognition

This chapter proposes to incorporate information from several sources to obtain a more precise affective state. We include knowledge about sources of emotions and their observable outcomes. The sources of emotions include training setting and trainees features, while observable outcomes of emotions include facial expressions and voice. We are working on integrating electroencephalogram signals. The trainee model relies on a dynamic Bayesian network (DBN), which associates available evidence with emotions in a probabilistic way. The DBN representing the affective trainee model is shown in Fig. 6, This is a high-level representation since where the vertex is composed by a group of vertices in the comprehensive model.

5.1 *Emotions Through Contextual Information*

To model the reasons causing emotions, the affective trainee model relies on the OCC Cognitive Model of Emotions (Ortony et al.1988). This model provides a causal appraisal of emotions taking into account information about context. In this way, emotions emerge as the consequence of the cognitive assessment of the present situation considering the goals of the individual. The affective state of the trainee represented by a DBN is shown in Fig. 7. This DBN is an abstract representation of the model since each vertex in the DBN is composed by a group of nodes in the comprehensive DBN. This proposed model is based on the research by Conati and Maclaren (2009), and on our earlier research (Hernández et al. 2012).

Emotions are dynamic as they are changing constantly as a consequence of the changing environment. This dynamic behavior of emotions is supported by the DBN. At time t_{n+1}, the emotions are inferred considering the trainee's knowledge, personality, and the training setting at that time, as well as the emotions of the trainee in the previous time. The training settings are explained considering the results of the trainee's actions.

The trainee's cognitive assessment of the current situation-given relevant goals is represented by the edge between the goals and the training setting vertex through the satisfied goals vertex. The edge connecting the satisfied goals vertex and the affective state vertex represents the impact of the cognitive process on the trainee's emotions.

The OCC model states that individual's goals are central to determine emotions, however requesting trainees to express their goals during a training session would be disturbing since they are concentrated in the learning process. Subsequently, in

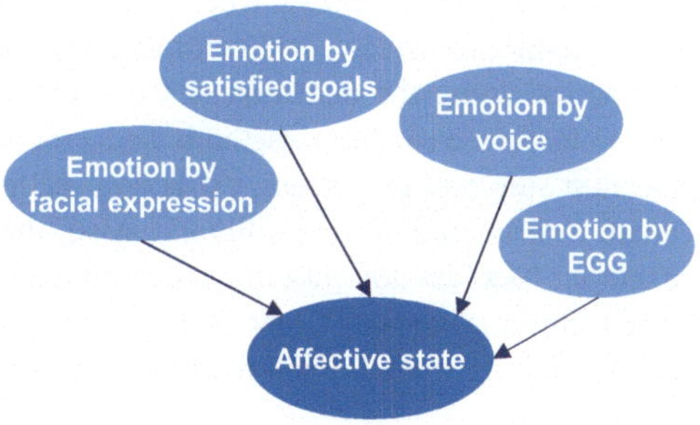

Fig. 6 Affective trainee model. The trainee model is composed of the electrical procedures included in the course

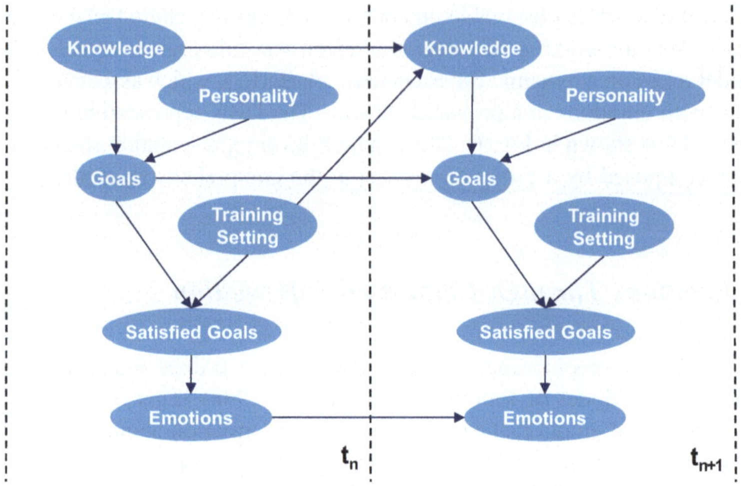

Fig. 7 Dynamic Bayesian network representing the affective student model with base on contextual information (Hernández et al. 2016a, b)

our model goals are deduced considering information from indirect evidence from several sources. We rely on personality and trainee knowledge to predict the trainee's goals.

We model the students' affect based on the causes of the emotions. However, sometimes this representation could have not enough information to provide an accurate emotional state. Therefore, we are working on integrating on observable parameters such as voice, facial expressions, and electroencephalogram (EEG) signals.

5.2 *Emotions Through Facial Expressions*

Considering consequences of emotions, expressions of the face are one means to identify emotions. Emotion detection is made by extracting special face features and by identifying facial expressions. Our approach is based on Ekman's theory (Ekman and Friesen 1978), where human emotions are anger, disgust, fear, happiness, sadness, surprise, and neutral. They are also known as basic emotions. The method for doing the detection of facial expressions has three steps: face identification, feature extraction method, and emotion classification. To build the face detection step and the feature extraction method we worked with open source like OpenCV, a computer vision library. Also, the Neuroph library was used for classification. Next section describes the procedure to recognize affect in human faces.

A technique named "Haar-like feature cascade" (Viola and Jones 2001), was used to identify three human features: face, eyes, and mouth. After the human face is identified, eyes and mouth are then recognized. We produced regions of interests (ROI) around the features we are searching for. After the ROIs and features are located, we applied several transformations to the human face. The transformations are necessary to improve the location of components of the face like edges and distances between face parts (e.g., distance between the left eye and right eye). The modifications of the human image are also important because we eliminate aspects like image colors that are not necessary for the classification.

The method of feature extraction produces a set of 10 values needed to feed the artificial neural network. The network was first trained with a part of the corpus RaFD (Langner et al. 2010) that contains more than eight thousands of human faces. The human faces are from men and women with different ages and acting or posing in different emotions. Figure 8 shows the complete process of feature extraction and classification using a neural network produced with Neuroph. The output of the network (output layer) is the current emotion.

After training the neural network, we tested the network with another part of the RaFD corpus. After evaluating the neural network, we obtained a success rate of 85%. We can say that the network is now ready to be integrated into the training model.

5.3 Emotions Through Voice

The human voice has also been used to define emotions as stated by Han et al. (2014). In our proposal, an audio file is divided into small segments to train a neural network (Zatarain-Cabada et al. 2016). The first step is the segmentation of audio file, which consists of finding the appropriate acoustic segments units for the classification process. The next step is the extraction of characteristics to find acoustic signals that best describe an emotion where each segmented acoustic unit is represented as feature vectors. In this method, there are two very critical

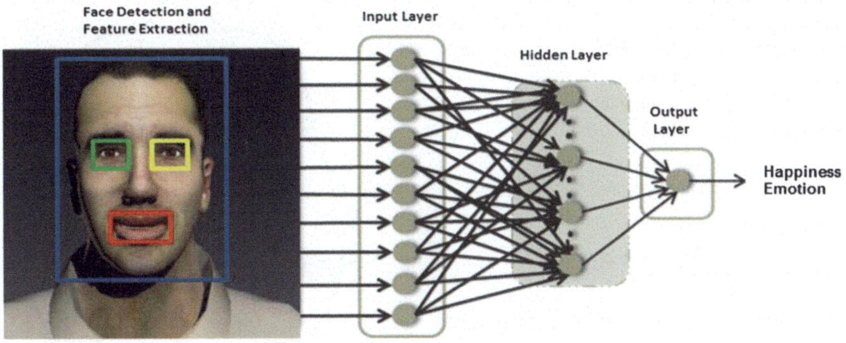

Fig. 8 Feature extraction and emotion classification for facial expressions using a neural network (Hernández et al. 2016a, b)

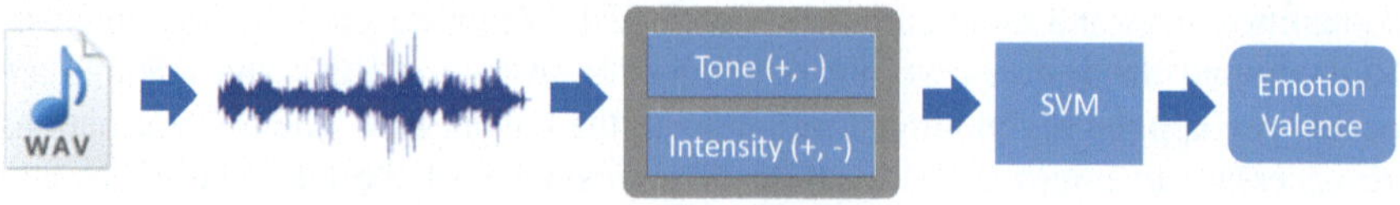

Fig. 9 Method for recognizing emotions in voice via audio files processing

variables: intensity and valence. The model identifies the valence of the audio, positive, or negative.

A total of 45 audio files from 9 different subjects were used, where each audio file is classified as positive or negative. From the audio files, the characteristics are extracted using its spectrogram with the musicg library (Emotiv 2016) to obtain the parameters to train a support vector machine (SVM) of the LibSVM library (Chang and Lin 2011). Our proposal for emotion recognition in voice is depicted in Fig. 9.

5.4 Emotions Through EEG Signals

Recognition of emotions can be done using a headband that emits electroencephalographic (EEG) signals. This type of devices can be very effective for finding signals in real time.

In our proposal, we use an Emotiv Epoc headset (Emotiv 2016). In this model, the valence of emotion is identified through EEG signals obtained from an edf file, which is generated after a session using the headband and the Human–Computer Interface of Emotiv Epoc. The Emotiv Epoc Headband offers 16 information channels from different parts of the brain as shown in Fig. 10.

To properly select the channels of the headset, we considered the work of Mahajan et al. (2014) and the work of Liu et al. (2011). After several experiments,

Fig. 10 Information channels of Emotiv headset. The information of AF3, F3, F4, FC5, and FC6 channels are used to recognize relevant emotions (Emotiv 2016)

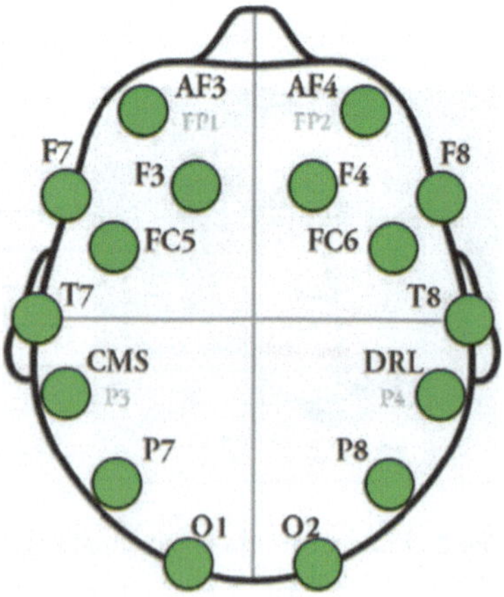

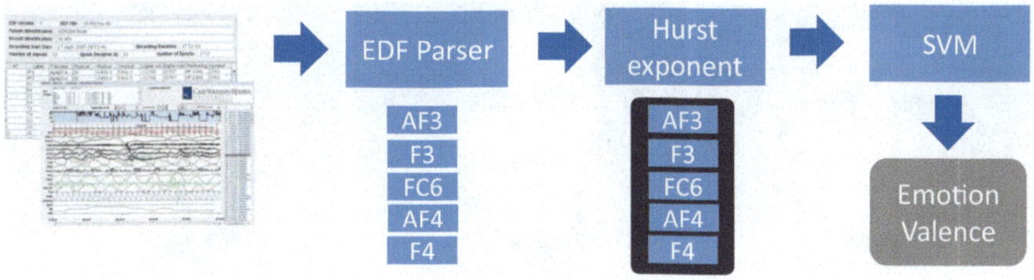

Fig. 11 Emotion recognition model using EEG signals

we selected AF3, F3, F4, FC5, and FC6 channels because they are the closest to human emotions (Zatarain-Cabada et al. 2016). The main feature extracted from EEG signals is the Hurst Exponent, which is used in time series analysis to identify the unstable behavior of EEG signals to identify trends in the data. The Hurst Exponent was chosen for the good results, 71.38%, shown in Wang et al. (2014).

The training dataset was obtained from experiments carried out with an affective Tutoring System to learn Java programming and with a group of 10 students. All test students were shown small videos to induce emotions as they resumed their brain information (Barrón-Estrada et al. 2015).

We trained a SVM with the information obtained from the 10 subjects, which were classified in positive, negative, and neutral. Our proposal for emotion recognition via EEG signals is depicted in Fig. 11.

6 Animated Agent

We have developed an animated pedagogical agent, whose actions will be based on the inferred emotional state of the trainees to exhibit an empathic behavior.

Activities in the training system are presented to trainees via an animated agent. The animated agents are an important movement in the achievement of a more natural human–computer interaction (Breese and Ball 2008). These agents interact with the students using a wide range of nonverbal behavior such as emotions, gaze, deictic gestures, and facial expressions directly. At the same time, animated agents accompany students cohabiting learning environments. Animated pedagogical agents impact significantly the learning environments, because users have the impression that they are interacting with a human and not with a computer (Sagae et al. 2012); as a result, trainees observe a different behavior from a traditional system, and similar to human behavior. Animated pedagogical agents present some behaviors that are exhibited by intelligent tutoring systems as well. However, animated pedagogical agents also exhibit some behaviors which are particular of these characters, such as to demonstrate complicated tasks, to observe and help the trainee in the performance of the procedure, and to guide trainees in virtual spaces (Wang et al. 2008).

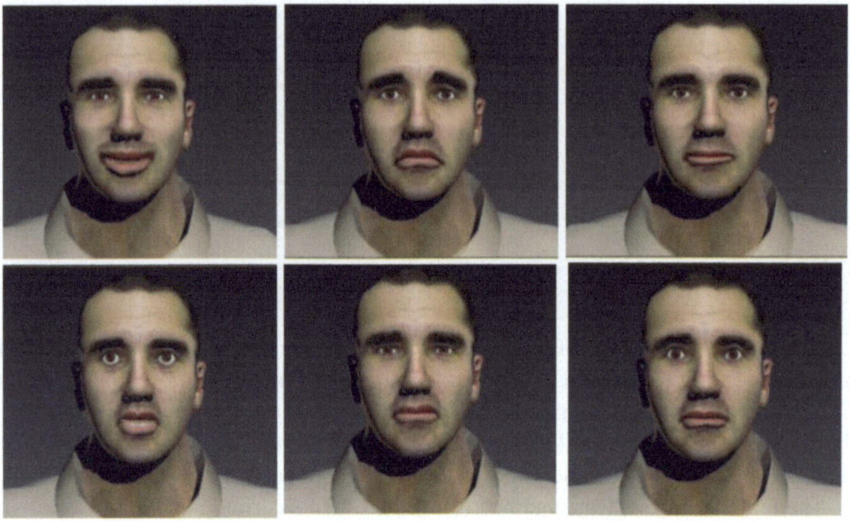

Fig. 12 An animated agent is depicting different facial expressions based on the basic emotions proposed by Ekman and Friesen (1978). Upper, left to right: *happiness*, *sadness*, and *anger*. Bottom, left to right: *surprise*, *disgust*, and *fear* (Hernández et al. 2016a, b)

To develop the agent, we are using the characteristics of operators, such as wearing uniform and safety helmet, among other features. We believe that by representing the tutor as an electrician, operators will accept the training environment, as it is more natural for electricians and hence useful for learning. Figure 12 shows the animated agent who is depicting different facial expressions based on the basic emotions proposed by Ekman and Friesen (1978).

Our initial approach to foster positive emotional states in users is focusing on the development of an emphatic agent. Empathy is the capacity and a gift to realize, comprehend, and feel the emotions of other people. This concept is integrated into animated agents in order to achieve credibility, social contact, collaboration, and commitment of the user (Hone 2006). Initially, the animated agent exhibits the identified emotions in trainees according to the emotion model already described.

To assess the present design of the animated agent, we surveyed to collect information to refine the agent in his physical characteristics and his capability to express emotions (Hernández et al. 2016a, b). This investigation provided us with knowledge, which we are using to enhance the agent aspect, for instance, they stated the convenience of having several agents with different physical characteristics and genres such a way users can choose it. All participating people coincided with the benefits of an animated pedagogical agent in the learning process. Concerning the three-dimensional representation of the agent, who is representing a certified instructor wearing the obligatory uniform, the majority of participants stated that it helps inexperienced workers to become familiar with the training and the working environment. However, there were some opposing views on the physical performance of the agent, but these opinions have to do with faults on the agent, which can be worked out by upgrading the agent. Accordingly, the majority of participating individuals considered animated pedagogical agents as a strong element of the training environment. Participants also stated some recommendations to advance in the design of the empathic agent; they proposed to customize the agent according to particular learning contexts, this is according to the specific training situation where different capabilities of the instructors are required. In the development of the agent, we assume that the physical aspect of the agent, similar to a real instructor dressed in uniform, would benefit his acceptance. The results of the study verified this hypothesis even when it was not explicitly included in the survey.

The results of the study were useful indicating the effectiveness of emotions physical realization. Results indicated that *the participants clearly and easy identified sadness, happiness, and surprise*, while some discrepancies arise on the recognition of *anger* and *disgust*.

7 Attitude Model

Human resources constitute the most valuable, active resources of any organization since they have the knowledge and experience of all activities and, consequently, they have a solid impact on the concluding outcomes. Therefore, it is required to design methods for the training and retention of personnel. These methods must allow personnel to achieve the required profiles for the good performance of the company.

Attitudes have a strong impact on the formation of competencies and values, as they have to do with affect, beliefs, and behavior. Affect involves feelings that people experience and may or may not refer to a particular object or event. Beliefs are knowledge about the probability that an object or event is associated with a given attribute. Behaviors are defined as the actions of an individual. Each of these individual phenomena is fundamental to create attitudes and transform existing attitudes. Likewise, attitudes have a reciprocal impact on affect, belief, and behavior (Bohner and Wänke 2002).

In the literature, we can find diverse descriptions of attitude, and these definitions, or most of them, coincide on their characteristics: they are shaped throughout individual life, change continuously, and they are guided by the encompassing environment and reality. Human attitudes are composed of three elements: feeling, thought, and behavior. Attitudes emerge as the outcome of the standards, principles, and values that precede them. Attitudes are a tendency to evaluate objects, events, and people, either positively or negatively (Bohner and Wänke 2002). Human attitudes reflect the way in which people feel about someone or something or in certain circumstances, and they foresee the trends of people to behave in a particular manner. They are related to the mental structures built in childhood and which are strengthened continuously in everyday life.

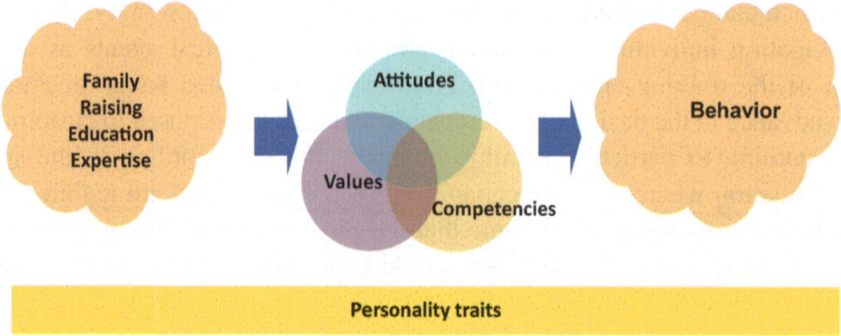

Fig. 13 The configuration of attitudes in the model of attitudes (Hernández et al. 2012)

We have implemented a model of attitudes to identify and relate attitudes and behavior in work with the training needs of the staff. The configuration of attitudes in the attitudes model is presented in Fig. 13.

The implementation of a model for attitudes management requires the identification of relevant attitudes for the work environment and the tasks, responsibilities, and roles of the personnel. Among the relevant attitudes for the work are *learning*, *tolerance*, *commitment*, *loyalty*, *service*, *responsibility*, *safety*, *and teamwork*, among others.

There are several approaches and theories of measurement of attitudes, including numerical and axiomatic. The numerical approach consists of scales of attitudes. While the axiomatic approach evaluates some probing hypotheses on the relationship between traits (Morales 2006).

In this work, we take the numerical approach, and particularly, we used Likert scales (Likert 1932), since they are relatively simple to construct and to answer. We wanted a straightforward method for attitudes measurement in such a way people could feel comfortable and could respond freely.

The Likert scales take advantage of the classic measurement of personality traits in the measurement of attitudes: the total sum of the responses to a series of items that measure, or express the selfsame characteristic positions of the individual in the attitude (characteristic) being measured. This measurement method only assumes that the answer for each item is a function of the position of individuals in the attitude (continuous variable) being measured: the greater the agreement, or disagreement, the more the individual has the trait being measured (Morales, Urosa and Blanco 2003).

In Table 1, an extract of a Likert scale to measure attitude toward learning attitudes is shown. This example contains three items and five possible answers. The responses of the individuals express the extent of agreement with the items, which express opinions. Commonly, a Likert scale contains about 20 items to be considered am acceptable scale. However, some scales could contain any number of items.

Apparently, a measurement scale of attitudes looks similar to a questionnaire, but they have some significant differentiation. The main distinction strives in that a questionnaire consists of questions that are analyzed independently, whereas an

Table 1 Examples of Likert scale. Measurement of attitude toward learning

Please select the answer indicating how much you agree or disagree with each statement		Strongly agree	Agree	Do not know	Disagree	Strongly disagree
1	I invest time and money in training because I consider it is a good investment		X			
2	I attend training courses to get out of the routine					X
3	I study only when I have an exam				X	

attitude measurement scale consists of items that convey the selfsame characteristic or attitude, there is a relationship between the items, and whose answers are summed in an amount indicating in which extent the person has that attitude.

The correlations between the items must be positive, and the shared variance among the items is related to the attitude. The test theories in personality tests and academic performance apply the Likert scales (Morales 2006; Morales et al. 2003).

To construct a scale, we formed teams of experts that know people intended to answer the scale and also they have to know the attitude being measured. The process to construct a scale of attitudes consists of the stages: (1) to describe the attitude to be measured, (2) to build the instrument for measurement, mainly to write down the items, (3) to gather data from a representative sample of the population conducting a pilot study, and (4) perform a statistical analysis of the study results (Morales 2006; Morales et al. 2003).

The procedure for measurement scale construction starts when a mutual definition of the attitude is established. After that, the items of the pilot scale have to be written. The pilot scale is applied in a pilot study with a representative sample of the population. In the pilot survey, data are obtained to carry out statistical tests and to establish which items will compose the definitive scale. Statistical tests consist of Student's t-test (Zimmerman 1987) and coefficient Cronbach's α (Cronbach 1951, Tavakol and Dennick 2011).

In the literature, there are many examples of measurement scales for different purposes and attitudes, which could be used, in diverse studies. However, we decided to construct the Likert scales for our model because they must be planned and designed specifically for the people who are intended to respond, and considering the mission, vision, and objectives of the company.

To decide which attitudes include in the model, we review the literature to find which attitudes are relevant to work. Also, a team of experts, who know the energy process, the company and personal define the attitudes in this work context and decide which attitudes include in the model. The current model of attitudes consists of nine attitudes: *learning*, *safety*, *work behavior* (*work satisfaction*, *commitment to work*, *organizational commitment*), *responsibility*, *service*, *perseverance*, *loyalty*, and *leadership*.

Currently, we are working on a computational model for attitudes, for the time being, we measure them by Likert scales.

8 Learning Styles Component

In the literature different learning theories, which present approaches about how individuals acquire new knowledge, conceptions, and skills can be found. These approaches state diverse and even opposing perspectives about learning; for instance, there are theories center on learners, and there are some theories centered on educators. Learning styles theory states that every single person possesses personal and specific approaches and preferences to acquire and discover knowledge, and also emphasizes that people comprehend and process informative material in very dissimilar manners. Therefore, this theory affirms that the learning process is more related to how the individual learning style is approached than with the intelligence of the people. Researches have proposed various learning styles models including the Felder–Silverman learning Styles Model (Graf and Kinshuk 2007; Blouin 2010; Felder and Silverman 1988). This renowned and largely exploited model is the base for our proposal.

The Felder–Silverman model proposes four categories of learning styles: sensing/intuitive, visual/verbal, active/reflective, and sequential/global.

The active individual acquires knowledge as a result of applying the available information and enjoys teamwork. The reflective individual acquires knowledge as a result of thinking in silence and favors to work on his own. The sensitive person appreciates to study and decipher problems with well-established methods and does not like difficulties. The intuitive person favors prefers and relationships discovery, appreciates innovation and dislikes repetition. The visual individuals retain what they look: photographs, figures, diagrams, timelines, pictures, and presentations. The verbal person grasps knowledge from words, that is, from written and spoken explanations. The sequential individual understands concepts in progressive steps and achieves solutions following logical routes step by step. The global individual understands concepts in big leaps and resolves complex problems once he has captured the general view.

We base the learning style identification on the Soloman and Felder questionnaire, which comprises of 44 items (Felder and Soloman 1993).

In the implementation of our learning styles proposal, we have applied a rules set based on the approach of Savic and Konjovic (2009). Rules consist of a set of teaching instructions for each learning style (Hernández and Rodríguez 2011).

These instruction rules can be applied in intelligent training systems since they are relatively easy to implement. However, to apply these rules means that every explanation, lesson, question, and exercise has to be developed eight times, that is, we need eight different versions of teaching material according to the teaching instructions. In our study case, this effort is acceptable as there are many potential trainees for every learning style who can take advantage of the individualized learning (Hernández and Rodríguez 2011).

9 Conclusions

Latin American utility companies have a complex situation since they have to be competitive in a globalized world and a changing environment. Therefore, utility companies require highly qualified electricians with positive attitudes. However, they face a challenging situation: the generational change. The knowledgeable personnel are in the retirement process, and there is a gap between the present and required competencies of new human resources. Therefore, they have to be trained to get the required competencies in knowledge, skills, and attitudes.

Additionally, facilities demand trained personnel because the electrical procedures are dangerous, and accidents can occur damaging personnel, equipment, or both. Therefore, they have to be trained in normal and abnormal situations on a continuous basis.

Aiming to have practical training, we propose a training model for talent management. Talent management is a key factor in the operation of any organization. Training is an important element of talent management. Our training model is based on blended learning basics as it is composed of classroom classes, intelligent training systems and field practice, such a way that traditional training is complemented with intelligent training systems. Consequently, the system supports instructors when teaching the electrical procedures.

The intelligent training systems provide adaptive and personalized instruction considering several traits and states of employees: knowledge, affective state, learning styles, and attitudes.

Emotions and attitudes are central to have successful training and to reduce work-related accidents. We proposed to integrate several sources to recognize the affective state of the trainee. We have modeled emotions by taking into account contextual information, facial expressions, voice, and EEG signals. We have separated models from these sources, and we are in the process of integrating these models in a multimodal affective framework.

We have constructed an attitude model consisting of relevant attitudes for work and attitude measurement scales. We take a numerical approach, by using Likert scales. Currently, we are working on measuring the attitudes of the employees toward new technologies. These attitudes have a big impact on the acceptation and success of the intelligent training systems; therefore, it has to be considered in the training programs. Also, we are working on conceptualizing a computational model to measure attitudes automatically.

In our proposal, we considered the learning styles of trainees as proposed on the Felder–Silverman learning styles model. We are in the process of building the different versions of the lessons as proposed by the learning styles theory. Since there is a big amount of prospective trainees for different learning styles, we believe that this effort is reasonable.

An animated agent presents personalized instruction, which deploys empathy considering the current emotion in the trainee. We have evaluated the agent design and behavior. The results are encouraging since participating individuals stated positive opinions about how animated agents could improve learning and motivation in trainees.

With these training models, we are trying to focus traits, needs, and aspirations of people in such a way we have elements which allow the company a smooth shifting from talent management to people management. People management is focused on culture, engagement, environment, leadership, empowerment, and fit (Bersin 2015).

In this chapter, we have described our work and experiences with electrical domains. However, the models we have proposed are applied to almost any domain since all companies require well-trained personnel and with positive attitudes.

In conclusion, we propose a blended training model based on these three axes: knowledge, skills, and attitudes, to provide adaptive and distance training in a talent management framework for the energy sector. For time being, we are working into integrating them into a unified platform. However, we have encouraging results in the development and evaluation studies. The aim is to have efficient training in less time and fewer costs.

References

Ayala, A., Galvan, I., Arroyo, G., Perez, M., Muñoz, J., & Rodríguez, E. (2016). Virtual reality training system for maintenance and operation of high voltage overhead power lines. *Virtual Reality, 20*(1), 27–40.

Barrón-Estrada, M. L., Zatarain-Cabada R., Aispuro-Gallegos, C. L., Sosa-Ochoa, C. G., & Lindor-Valdez, M. (2015). Affective environment for java programming using facial and EEG recognition. *Research in Computer Science, 106*, 39–47.

Bersin, J. (2009). Talent management: a training strategy, E-learning!, February 17th. Retrieved April 2017, fromhttp://www.2elearning.com/top-stories/item/2404-Talent-Management-A-Training-Strategy.

Bersin, J. (2015). Why People Management is Replacing Talent Management, Insights on Corporate Talent, Learning, Leadership, and HR Technology, January 25th. Retrieved April 2017, from http://joshbersin.com/2015/01/why-people-management-is-replacing-talent-management/.

Blouin, T. (2010). Felder-Silverman Learning Styles Model. Retrieved 2015, from http://chat.carleton.ca/~tblouin/Felder/felder.html.

Bohner, G., & Wänke, M. (2002). *Attitudes and attitude change*. Sussex: Psychology Press.

Breese, J. S., & Ball, G. (2008). *Modeling emotional state and personality for conversational agents*. Technical report. MSR-TR-98–41.

Chang, C., & Lin, C. (2011). LIBSVM: A library for support vector machines. *ACM Transactions on Intelligent Systems and Technology, 2,* 1–39.

Conati, C., & Mclaren, H. (2009). Empirically building and evaluating a probabilistic model of user affect. *User Modeling and User-Adapted Interaction, 19,* 267–303.

Cronbach, L. J. (1951). Coefficient alpha and the internal structure of tests. *Psychometrika, 16*(3), 297–334.

Cunningham, I. (2007). Talent management: making it real. *Development and Learning in Organizations, 21*(2), 4–6.

Damasio, A. R. (1996). El error de Descartes, Andrés Bello, Santiago, Chile.

Ekman, P., & Friesen, W. (1978). *Facial action coding system: A technique for the measurement of facial movement*. Palo Alto: Consulting Psychologists Press.

Emotiv (2016). Epoc headset. Retrieved 2016, from https://emotiv.com/epoc.php.

Felder, R. M., & Silverman, L. K. (1988). Learning and teaching styles in engineering education. *Journal of Engineering Education, 78*(7), 674–681.

Felder, R. M., & Soloman, B. A. (1993). Learning styles and strategies, NCSU.

Graf, S., & Kinshuk (2007). Providing adaptive courses in learning management systems with respect to learning styles. In *Proceedings of World Conference on E-Learning in Corporate, Government, Healthcare, and Higher Education* 2007 (pp. 2576–2583).

Han, K., Yu, D., & Tashev, I. (2014). Speech emotion recognition using deep neural network and extreme learning machine. In *Proceedings of Interspeech* 2014, (pp. 223–227).

Hernández, Y., & Pérez, M. (2016a). Virtual reality systems for training improvement in electrical distribution substations. In *Proceedings of 16th International Conference on Advanced Learning Technologies and Technology-enhanced Learning, ICALT 2016*. IEEE Press.

Hernández, Y., & Pérez, M. (2016b). Adaptive and blended learning for electrical operators training with virtual reality systems. In *Proceedings of the 8th International Conference on Computer Supported Education* (CSEDU 2016) (Vol. 1, pp. 519–524).

Hernández, Y., & Rodríguez, G. (2011). Learning styles theory for intelligent learning environments. In A. Verbraeck, et al. (Eds.), *memorias de CSEDU 2011* (Vol. 1, pp. 456–459). Netherlands: SciTePress.

Hernández, Y., Sucar, L. E., & Arroyo-Figueroa, G. (2012). Affective modeling for an intelligent educational environment. In A. Peña-Ayala (Ed.), *Intelligent and adaptive educational-learning systems: achievements and trends. Smart innovation, systems and technologies Springer book series* (pp. 3–24). Berlin, Heidelberg: Springer.

Hernández, Y., Arroyo-Figueroa, G., & Sucar, E. (2015). A model of affect and learning for intelligent tutors. *Journal of Universal Computer Science, 21*(7), 912–934.

Hernández, Y., Pérez-Ramírez, M., Zatarain-Cabada, R., Barrón-Estrada, L., & Alor-Hernández, G. (2016a). Designing empathetic animated agents for a B-learning training environment within the electrical domain. Educational Technology & Society, 19(2), 116–131.

Hernández, Y., Pérez, M., Ingram, W., Nava, E., & Ontiveros, N. (2016b). Architecture of an intelligent training system based on virtual environments for electricity distribution substations. Research in Computing science, 129(1), 63–70.

Hone, K. (2006). Empathic agents to reduce user frustration: The effects of varying agent characteristics. *Interacting with Computers, 18*(2), 227–245.

Langner, O., Dotsch, R., Bijlstra, G., Wigboldus, D., Hawk, S., & van Knippenberg, A.(2010). Presentation and validation of the Radboud Faces Database. *Cognition & Emotion*, 1377—1388.

Likert, R. (1932). A Technique for the Measurement of Attitudes. *Archives of Psychology, 140,* 44–53.

Liu, Y., Sourina, O., & Nguyen, M. K. (2011). Real-time EEG-based human emotion recognition and visualization. *Transactions on Computer Sciences* XII, *6670*, (256–277) (Cyberworlds).

Mahajan, R., Bansal, D., & Singh, S. (2014). A real time set up for retrieval of emotional states from human neural responses. *World Academy of Science, Engineering and Technology, 8*(3), 144–149.

Morales, P. (2006). *Medición de actitudes en psicología y educación*. Madrid: UPCO.

Morales, P., Urosa, B., & Blanco, A. (2003). *Construcción de escalas de actitudes tipo Likert: una guía práctica*. Madrid: La Muralla.

Ortony, A., Clore, G. L., & Collins, A. (1988). *The cognitive structure of emotions*. Cambridge University Press.

Perloff, R. M. (2017) The Dynamics of Persuasion: Communication and Attitudes in the Twenty-First Century, Routledge.

Reeves, B., & Nass, C. (1996). *The Media Equation: How people treat computers, television and new media like real people and places*. Cambridge University Press.

Rodríguez, G., Hernández, Yasmín, & Paredes, I. (2011). Emerging technologies for electric power generation personnel training. In A. Magadan (Ed.), *memorias de CERMA 2011* (pp. 445–449). Los Alamitos: IEEE Press.

Ruiz-Hau, J. (2017). What is learning's role in talent management?, Chief Learning Officer, January 17th. Retrieved April 2017, fromhttp://www.clomedia.com/2017/01/17/learnings-role-talent-management/.

Sagae, A., Hobbs, J. R., Wertheim, S., Agar, M., Ho, E. & Johnson, W. L. (2012). Efficient cultural models of verbal behavior for communicative agents. In *Proceedings of International Conference on Intelligent Virtual Agents* (pp. 523–525).

Savic, G., & Konjovic, Z. (2009). Learning style based personalization of SCORM e-learning courses, In *Proceedings of 7th International Symposium on Intelligent Systems and Informatics (SISY 2009)* (pp. 349–53). IEEE.

SENER, SEP & CONACYT. (2014). Programa Estratégico de Formación de Recursos Humanos en Materia Energética Secretaría de Energía, Secretaría de Educación Pública Consejo Nacional de Ciencia y Tecnología. Retrieved April 2017, from https://www.gob.mx/sener/acciones-y-programas/programa-estrategico-de-formacion-de-recursos-humanos-en-materia-energetica-pefrhme-7732.

Skymind. (2016). Musicg. Deep Learning, Spectrogram. Skymind. Retrieved 2016, from http://deeplearning4j.org/canovadoc/musicg/extension/Spectrogram.html.

Sottilare, R., Graesser, A., Hu, X., & Holden, H. (2013). *Design recommendations for intelligent tutoring systems: Learner modeling* (Vol. 1). Orlando, FL: Army Research Laboratory.

Sucar, E. (2015). *Probabilistic graphical models: Principles and applications*. Springer.

Tavakol, M., & Dennick, R. (2011). Making sense of Cronbach's alpha. *International Journal of Medical Education, 2*(2011), 53–55.

Viola, P., & Jones, M. (2001). Rapid object detection using a boosted cascade of simple features. In *Proceedings of the IEEE Computer Society Conference on Computer Vision and Pattern Recognition*, Kauai: IEEE Computer Society.

Wang, N., Johnson, W. L., Mayer, R. E., Rizzo, P., Shaw, E., & Collins, H. (2008). The politeness effect: Pedagogical agents and learning outcomes. *International Journal on Human-Computer Studies, 66*(2), 98–112.

Wang, X. W., Nie, D., & Lu, B. L. (2014). Emotional state classification from EEG data using machine learning approach. *Neurocomputing, 129*, 94–106.

Willingham, D. T., Hughes, E. M., & Dobolyi, D. G. (2015). The scientific status of learning styles theories. *Teaching of Psychology, 42*(1), 266–271.

Woolf, B. P. (2008). *Building intelligent interactive tutors: Student-centered strategies for revolutionizing e-learning*. Morgan Kaufmann.

Zatarain-Cabada, R., Barrón-Estrada, M. L., & Muñoz-Sandoval, G. (2016). PREMOC: Plataforma de reconocimiento multimodal de emociones. *Research in Computing Science, 111*, 97–110.

Zimmerman, D. (1987). Comparative power of student T test and mann-whitney U test for unequal sample sizes and variances. *The Journal of Experimental Education, 55*(3), 171–174.

The Role of ICT in Educational Innovation

Pedro García-Alcaraz, Valeria Martínez-Loya, Jorge Luis García-Alcaraz and Cuauhtémoc Sánchez-Ramírez

Abstract In this chapter, the importance of the Information and Communication Technologies (ICT) in education is emphasized, and a case study describes their integration into a higher education institution, specifically, the Instituto Tecnologico de Colima. Since ICT are considered a tool to innovate teaching practices, as well as providing the possibility of including new didactic strategies that arouse the interest and motivation of students to improve the quality of teaching-learning processes inside and outside of the classroom. The case study reports a survey integrated by 43 items related to ICT, which was administered to 469 students of the Instituto Tecnologico de Colima. Using the WarpPLS Software v5.0, the causal relationships that exist between the uses of ICT (independent variables) and the benefits obtained when using ICT (dependent variables) were analyzed. After the analysis, it is concluded that there is sufficient statistical evidence to ensure that academic uses have a direct and positive impact on students' performance, improving the grades and their psychological state.

Keywords Educational innovation · Benefits of ICT

P. García-Alcaraz (✉)
Centro de Bachillerato Tecnológico Agropecuario
No. 148 e Instituto Tecnologico de Colima, Colima, Mexico
e-mail: alcarazgarciapedro@yahoo.com.mx

V. Martínez-Loya · J. L. García-Alcaraz
Department of Industrial Engineering and Manufacturing,
Universidad Autónoma de Ciudad Juárez, Chihuahua, Mexico
e-mail: al160648@alumnos.uacj.mx

J. L. García-Alcaraz
e-mail: jorge.garcia@uacj.mx

C. Sánchez-Ramírez
Division of Research and Graduate Studies, Instituto Tecnologico de Orizaba,
Veracruz, Mexico
e-mail: csanchez@itorizaba.edu.mx

G. Cortes-Robles et al. (eds.), *Managing Innovation in Highly Restrictive Environments*, Management and Industrial Engineering,
https://doi.org/10.1007/978-3-319-93716-8_7

1 Introduction

Nowadays, using Information and Communication Technologies (ICT) is essential in the teaching-learning process, since many systems are used to disseminate knowledge, store, and distribute information between students and teachers. In that sense, the following concepts provide a better understanding of innovations in the education system.

1.1 Educational Innovation

Education plays a significant role in the future of human beings. In this sense, Ausubel and Colbs (1990; cited by Santamaría, Quintana, Milazzo, y Rodríguez 2005) define education as the set of knowledge, methods, and orders by which the individual is assisted to develop and improve the intellectual, moral, and physical faculties, that is, he or she is taught to survive in the real world. In this regard, León (2007) considers that education seeks the perfection and human security. Hence, the most intricate education's autonomy: education seeks to secure freedom for man, but education demands discipline, submission, conduction, and is guided by signs of compulsion, even authoritarianism, firmness, and directionality.

In this sense, in every educational process, the teaching-learning process must be innovated. Understanding innovation as the introduction of changes that produce improvement, changes that respond to a planned, deliberate, systematized, and intentional process (Salinas 2004). Likewise, for Fullan y Stiegelbauer (1991; cited by Salinas 2008) the processes of innovation attributed to improvements in the teaching-learning processes involve changes related to the incorporation of new material, new behaviors, different teaching practices, new beliefs, and conceptions.

According to Barraza (2005), etymologically the word innovation derives from the Latin ***innovare***, which is derived from ***novus*** (new). Meanwhile, Arias (1996) considers that the word innovation derives from the Latin noun ***innovatio***. Its etymology is novus, which forms the basis of an extensive lexical field: novo, novitas, novius, renovo, renovatio, renovator, innovo, and innovatio. But, what is educational innovation? Several authors define these concepts as a group, for example, Cañal del León (2007, cited by Arias 1996), considers that educational innovation is

> A set of ideas, processes, and strategies, systematized, through which it is intended to introduce and cause changes in the current educational practices. Innovation is not a punctual activity but is a process, a long journey that stops to contemplate life in the classrooms, the organization of the centers, the dynamics of the educational community and the teachers' professional culture. Its purpose is to alter the current reality, modifying conceptions and attitudes, altering methods and interventions, thus improving, or transforming, as appropriate, teaching and learning processes. Innovation, therefore, is associated to change and has a component (explicit or occult) that could be ideological, cognitive, ethical and affective, since innovation appeals to the subjectivity of the individual and to the

> development of its individuality, as well as to the inherent relationships between theory and practice in the educational act.

Consequently, in all processes of educational innovation, the use of Information and Communication Technologies (ICT) is essential; these are defined according to García-Valcárcel (2003; cited by Pérez and Fernández 2005) as

> The set of technologies that allow the acquisition, production, storage, processing, transmission, recording and presentation of information, that may be in diverse ways such as voice, images, and data which is contained in acoustic, optical or electromagnetic signals. The most significant fact about the modern technologies, and what has guided to the real communicative revolution, is the creation of global communication networks.

Finally, ICT are those computational and informatics tools that process, store, synthesize, retrieve, and present information in the most varied forms (Castillo et al. 2010).

1.2 The Importance of ICT in Education

The definition of ICT was explained before, following the importance of ICT in educational innovation is detailed. According to Cox (2004), ICT contribute to improving educational environments providing teaching-learning resources, using motivating artifacts for students, including communication tools and machines to process information. Also, young people must acquire the necessary knowledge and skills about the use of ICT for new demands for labor and academics, so they can improve effectiveness (resource optimization) and efficiency (do things right) of the educational system management.

Belloch (2012) states that the use of ICT presents several advantages in comparison with the resources used in traditional education. Most of these advantages are directly related to the characteristics of ICT, among these can be mentioned the following:

- **Diverse information**: It is possible to access a large amount of information about different topics, which allows students to carry out the analysis of information and assess its quality and credibility.
- **Instructional flexibility**: The pace of learning and the path to follow during the process may differ among students, which demonstrate the necessity for adapting available resources to the diverse needs in the classroom.
- **Complementarity of codes**: Multimedia applications that use various communication codes allow students with different capabilities and cognitive abilities, taking full advantage of the lessons.
- **Increased motivation**: Several studies show that students are more motivated when using ICT. This interest may be an effect of novelty, although it may be due to the increase in motivation since multimedia presentations are more attractive than traditional ones.

- **Collaborative activities B** The proper use of ICT in working groups can enhance collaborative and cooperative activities among students, but also increase collaboration between other centers or institutions.
- **Enhancing educational innovation B** The actual society uses modern technologies that favor new methodologies. Although it is not a cause-and-effect relationship, teachers who are aware of new technologies tend to look for new ways of teaching and new didactic methodologies, more appropriate to current society according to the knowledge and skills that students must develop to adapt to the adult world.

In this regard, Mikre (2011) declares that students who use ICT mostly obtain better academic results than those who do not. Also, Tinio (2002, cited by Mikre 2011) considers a series of implications when using ICT in the education:

- **Active learning**: The learning promoted by the specific uses of ICT encourages greater student participation.
- **Collaborative learning**: The use of ICT in education increases the interaction and cooperation between students, teachers, and experts, regardless of where they are located.
- **Creative learning**: The learning supported by ICT promotes manipulating existing information and creating products.
- **Interactive learning**: ICT are used in learning by promoting the integration of the teaching-learning process.
- **Evaluative learning**: Using ICT in education allows the creation of different learning scenarios when exploring and discovering instead of just listening and memorizing.

According to Ferro and Martínez y Otero (2009), there are some advantages of using ICT in education, such as

- ***Breaking time and space barriers in teaching and learning activities***: In this way, educational institutions can offer courses or programs online, increasing the number of students enrolled and school coverage.
- ***Open and flexible training processes***: This point is related to the previous one because the students can take courses or programs from places located far away from the educational center.
- ***Improving communication between participants of the teaching-learning process***: ICT improve synchronous and asynchronous communication between teachers and students, as a result, they can send homework and ask questions to their teachers from any place and at any time.
- ***Personalized teaching***: The teaching-learning process that uses ICT enables the possibility of adapting the information to the needs and characteristics of the students.
- ***Quick access to information***: Both, students and teachers can use ICT to have faster and more efficient access to information (not only textual and auditory,

but also dynamic), thus reducing the degree of obsolescence, obtaining relevant formation, as well as increasing its degree of authenticity.

- ***The possibility of interacting with information***: When ICT is incorporated into the teaching-learning process; the student ceases to be only an active information processor (reception and memorization), turning into a significant constructor.
- ***Increasing the interest and the motivation of the students***: The ICT motivates the students and captures their attention so that it incites the activity and thought.
- ***Improvement of educational effectiveness***: When having new tools for the information and communication process, more interactive educational resources, and more information, it is easier to develop new didactic methodologies of formative effectiveness.
- ***Making the best use of teachers' time***: On the one hand, ICT facilitate the professional updating of teachers in an easy and faster way. On the other hand, they facilitate systematic practice through self-corrective exercises with immediate feedback about instrumental techniques or general knowledge.
- ***Supplementary activities to support learning***: The teacher can provide supplementary or extra work for students to use ICT, so that students can search for information on the Internet.

Therefore, ICT constitute new supports and channels for shaping, recording, storing, and disseminating informational content. Also, Fernández (2010) considers that the use of computer must be established in the school to favor the following:

- Stimulation of creativity.
- Experimentation and manipulation.
- Respect the learning pace of the students.
- Work in groups to socialize.
- The curiosity and spirit of investigation.

Also, Fernández et al. (2001) consider that ICT are a tool used to arouse interest, maintain motivation, and active participation in the teaching-learning process, so their use should be intensive and extensive in all levels of education.

1.3 ICT in Educational Innovation

Currently, ICT are considered as a tool to innovate teaching practices, as well as they provide the possibility of including new didactic strategies that arouse the interest and motivation of students to improve teaching-learning processes inside and outside the classroom. According to Tesouro y Puiggalí (2006; cited by Domingo y Fuentes 2010), ICT have opened new expectations in the field of education that generate changes in the student's relationship between space and time.

However, the innovation plans must be made in the short term due to the changing and complex world in which we live, where knowledge, educational models and computer processes expire quickly. According to UNESCO (2005: 62; cited by Barriga 2007), innovation is not only the production of new knowledge, but "innovation requires creating new needs in society," since society must understand that the advantages, which can obtain, are greater than the cognitive costs generated in the period of transition between the old and the new situation.

Therefore, Knight (2005) considers that the knowledge society is more demanding, and increasingly asking for electronic resources and technological tools for training in skills. However, the need to have more specialized teachers shows obstacles increasing bureaucracy for centers' accreditation, titles, and hierarchal positions, which shows more levels.

In this sense, Muñoz-Repiso (2007) considers that ICT seek the effectiveness of education because

- They offer great possibilities for functioning of schools and training systems on criteria of efficiency, cost reduction, efficiency, and productivity in the achievement of school learning
- They are an essential element for training people, especially in thinking skills, creativity, problem-solving, and operations.
- They provoke an immense transformation in the system of formal schooling (spaces and times, working conditions for teachers and students).
- ICT offer the possibility to rationalize, individualize, and control the training path of each student.
- ICT allow the incorporation of motivating elements linked to their audiovisual and interactive nature.
- The teacher acquires a new role as guide, adviser, and facilitator of suitable choices for students.

2 ICT Tools in Educational Innovation

There are many ICT tools that can be used in the classroom, which are described below:

- **Simple digital whiteboard**: According to Gallego and Gámiz (2011), this technology system is frequently composed of a computer and a video projector, which allows the projection of digital content in a format suitable for group viewing. It is possible to interact with the projected images by using the computer peripherals: mouse, keyboard, among others. In this sense, it is considered that a simple digital board contributes to the teaching-learning process due to:
  - The development of the learning processes is constructive and creative. Also, it can stimulate students' creativity and imagination.

- It facilitates and supports sensory learning especially in the early ages. Multimedia information comes through different sensory channels; thus, learning is better when is possible to see, hear, and do.
- Increases motivation, since students feel very comfortable in an environment where they can use this type of means.

For Gómez (2005), the whiteboard has the following functionalities:

- Projecting any information from the computer, the Internet or any other device connected to the system (television antenna, video projector, video camera, etc.) onto a screen located in a strategic place in the classroom.
- Easy interaction with the computer from the relevant position, where the screen is placed, including a format large enough. Using them, it is possible to control the software applications, presentations, and reproductions.
- Writing down as in the traditional blackboard. A teacher can write unlimitedly on the screen without deleting since a simple touch opens a new blank document.

In that context, SMART, 2006; BECTA, 2003, 2004a, 2004b; Gallego y Dulac, 2006; Marqués, 2005; cited by García et al. (2008), point out the great benefits in the teaching-learning process when using Digital Whiteboard, highlighting the following:

- It increases the opportunity for interaction between teachers, students, topics, and technology.
- It encourages active participation of students in the teaching-learning process.
- Using different resources (web pages, educational applications, drawings, graphics, videos, etc.) makes classes more attractive and closer to the multimedia environment in which children develop.
- Optimization of teaching time.
- The teacher can record, reuse, and prepare valuable information.
- Technology can adapt to different teaching models, then learning is easier.
- The creativity of the teacher, students, as well as the use of new pedagogical strategies is increased since the teacher needs to design the classes to improve the students' critical thinking.
- It allows showing more visual content to facilitate the understanding of concepts.
- It helps to the collaborative learning.
- It increases the motivation.
- It strengthens the understanding of complex concepts and helps to students' concentration.
- By removing the keyboard, it is especially attractive for young and the disabled children.
- It allows developing social skills.

The mobile phone as a pedagogical tool. At present, the use of mobile devices has increased exponentially in the society, since they are considered as resources that improve education; they have been incorporated into schools. Therefore, there are generic, disciplinary, and teacher competencies that refer to the use of ICT in the teaching-learning process.

In the classroom, it is common to hear: "Turn off your cell phones" However, the use of the cell phone in the classroom is a beginning point to change and capturing students' interest in the educational field. Therefore, the cell phone can be used as a pedagogical tool improving the academic performance of students.

Pinto (2011) considers that mobiles should not be banned in the classroom, but incorporate them into teaching. Besides, Bongarra (2013) states that for students it is great to use the cell phone, and they take the idea with enthusiasm. However, teachers should impose authority, establishing rules about the use of the mobile devices in the classroom, but not prohibit them, as this would encourage their use. Also, it is fundamental to establish norms when using a smartphone in the classroom, creating committees, integrated by students and teachers to promote the appropriate use of these devices.

On the other hand, the use of the cell phone in the classroom generates novel situations or concepts to students express their attitudes and feelings, that is, the collaborative learning is strengthened, thus motivating the discovery of new knowledge.

- **Moodle platform.** At present, it is possible to offer education as mixed and distance modality, to achieve the objectives set out in the institutional program which is possible by a tool called Moodle that allows the virtual teaching the 24 h of the day, the 365 days of the year.

In that sense, Pérez et al. (2008) describe Moodle platform, *Modular Object-Oriented Dynamic Learning Environment*, as an educational content management system (CMS) that enables the organization of courses based on the creation and combination of educational resources managed by the platform itself.

Consequently, the following characterizes Moodle courses:

- They enrich the training/learning process.
- They provide greater motivation for learning.
- They improve communication between educational agents.
- They monitor the teaching-learning process.

Likewise, Moodle courses facilitate group work activities and collaboration in projects (Correa 2005; cited by Marín and Maldonado 2010).

3 The Research Problem and Hypotheses

The Instituto Tecnologico de Colima (ITC) is a university, which is integrating ICT in its study plans to meet generic and specific competencies. This institution offers the following bachelor's degrees: Architecture, Accounting, Management,

Environmental Engineering, Biochemical Engineering, Engineering in Business Management, Industrial Engineering, Informatics Engineering, Mechatronics Engineering, Computer Systems Engineering. As well as a master's degree in Computer Systems and another in Sustainable Architecture and Urban Management.

Today, the number of students enrolled is 2992 in the face-to-face modality. In this institution, some programs include the use of ICT by default. However, others are obliged to use them to facilitate the teaching-learning process.

3.1 The Research Problem

The university has a good ICT infrastructure with the aim of improving teaching-learning processes, among which are: wireless and wired Internet (80 Mbps bandwidth), computer laboratories, all classrooms have wireless Internet, multimedia projectors, and some interactive whiteboards. Therefore, this research seeks to determine a quantitative measure of the effect of ICT uses on the benefits that students obtain in the teaching-learning process.

As mentioned before, the ICT are widely used in the teaching-learning process; for instance in classrooms, their use has been gradually implemented. Although saying that there are so many ways of using the technological tools and knowing their constructivist approach, it does not mean that they are well used inside of the educational centers, since many of these centers and institutions still teaching by using the traditional methods. Hence, the significant learning does not take place, and the technology is not properly used. In this sense, in the Instituto Tecnologico de Colima, the uses that give students to ICT have been identified. However, it is important to analyze the causal relationships between the independent variables (uses) and the dependent variables (benefits) that students obtain when using ICT in the classroom.

3.2 Research Objective

The objective of this research is to analyze the impact that has the use of the ICT in the Instituto Tecnologico de Colima on the benefits that students obtain when using them in the classroom.

3.3 *The Development of Research Questions and Hypotheses*

The research questions considered in this research project are as follows:

- What is the impact that the academic uses of ICT have on the recreational uses that the undergraduate students of the Instituto Tecnologico de Colima give them?
- What is the impact that the academic uses of ICT have on the psychological benefits of the undergraduate students of the Instituto Tecnologico de Colima?
- What is the impact that academic uses of ICT have on the academic benefits of the undergraduate students of the Instituto Tecnologico de Colima?
- What is the impact that the recreational uses of ICT have on the social benefits of the undergraduate students of the Instituto Tecnologico de Colima?
- What is the impact that the recreational uses of ICT have on the psychological benefits of the undergraduate students of the Instituto Tecnologico de Colima?
- What is the impact that the psychological uses of ICT have on the social benefits of the undergraduate students of the Instituto Tecnologico de Colima?
- What is the impact that the psychological benefits have on the academic benefits of the undergraduate students of the Instituto Tecnologico de Colima?
- What is the impact that social benefits have on the academic benefits of the undergraduate students of the Instituto Tecnologico de Colima?

The following hypotheses, which are illustrated in Fig. 1 are proposed to answer the previous research questions

H_1 The Academic Uses that undergraduate students of the Instituto Tecnologico de Colima give to ICT have a direct and positive impact on the Recreational Uses.

H_2 The Academic Uses that undergraduate students of the Instituto Tecnologico de Colima give to ICT have a direct and positive impact on the Psychological Benefits of students.

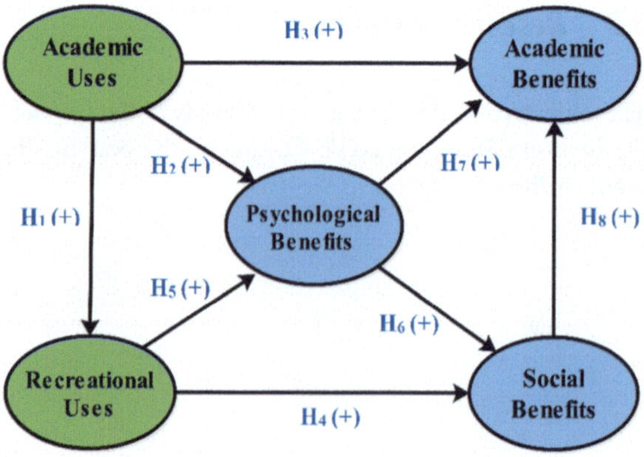

Fig. 1 Proposed hypotheses

H_3 The Academic Uses that undergraduate students of the Instituto Tecnologico de Colima give to ICT have a direct and positive impact on the Academic Benefits.

H_4 The Recreational Uses that undergraduate students of the Instituto Tecnologico de Colima give to ICT have a direct and positive impact on the Social Benefits.

H_5 The recreational uses that the undergraduate students of the Instituto Tecnologico de Colima give to ICT have a direct and positive impact on the Psychological Benefits

H_6 The Psychological Benefits obtained by the undergraduate students of the Instituto Tecnologico de Colima have a direct and positive impact on their Social Benefits.

H_7 The Psychological Benefits obtained by the undergraduate students of the Instituto Tecnologico de Colima through the appropriate use of ICT have a direct and positive impact on the Academic Benefits obtained.

H_8 The Social Benefits obtained by an undergraduate student of the Instituto Tecnologico de Colima have a direct and positive impact on the Academic Benefits obtained.

4 Materials and Methods

To achieve the objective stated above it is necessary to follow the methodology described in the following sections.

4.1 Literature Review

The literature review is performed to identify information related to the problem and determine the different attributes considered to study the integration of ICT and their impact on academic efficiency. Several databases are useful to accomplish this purpose, such as ScienceDirect, Springer, EBSCOhost, among others.

4.2 Questionnaire Development

According to the information obtained, a questionnaire is developed to collect data (Bisquerra et al. 2009; Aravena et al. 2006). The final questionnaire comprises 43 items, and it was administered to 469 students of the Instituto Tecnologico de Colima to validate it.

Table 1 Rating scale used

Value	Meaning
1	ICT are never used/The benefit is never obtained when using ICT
2	ICT are almost never used/The benefit is slightly obtained when using ICT
3	ICT are regularly used/A benefit is regularly obtained by using ICT
4	ICT are used almost always/The benefit is usually obtained when using ICT
5	ICT are always used/The benefit is always obtained when using ICT

The 43 items are divided into three sections. The first Sect. (9 items) includes general information. The second uses of ICT (18 items), which are categorized as academics and recreational. The third section includes the benefits of ICT (16 items) that are classified as academic, psychological, and social. The survey had to be answered with a five-point Likert scale, with values ranging from 1 to 5 (never-always) (Likert 1932), as shown in Table 1.

4.3 Sample and Administration of the Questionnaire

Since the average grade of the previous semester was consulted, the first semester students were excluded, also, because they have a brief period in the institution, so they knew little about the impact of ICT and their uses in classes. In this sense, a multistage sampling was used (Vivanco 2005), specifically two sampling techniques are suitable: cluster sampling and simple random sampling.

The questionnaires were administered from January to March 2017, sampling a proportional number of students per career and semester of the Instituto Tecnologico de Colima. The classrooms were visited to administer the surveys with previous authorization of the institutional authorities and the class teacher.

4.4 Capturing and Validating Information

The data obtained from the questionnaire are captured in the software SPSS 24 ®, creating a database. The rows represent the different cases and the columns the variables (43) to analyze. Before the analysis, some tests were carried out to detect extreme values, values located far from the mean, by using box plots for each item. The median replaced the extreme values. In the same way, an analysis is performed to detect missing values, which represent omissions by the respondent. These missing values were also replaced by the median or second quartile of the items.

The Cronbach's alpha index is used to validate the information obtained through the questionnaire (Cronbach 1951) and its minimum cut-off value is 0.7. It is also important to mention that the index can increase if one of the items analyzed is eliminated, so it is possible that the number of items at the beginning of the analysis and the end differs.

4.5 *Descriptive Analysis*

A descriptive analysis is performed, to provide information about the uses and benefits included in the questionnaire. At this point, the median is obtained as a measure of central tendency, due to the nature of the values, which are on an ordinal scale. Also, the interquartile range (IR) of the items is presented (the difference between the third and first quartile).

4.6 *Structural Equation Modeling*

To validate the hypotheses presented in Fig. 1 a structural equation model is used. According to Avelar et al. (2014) the structural equation models (SEM) were introduced by Jöreskog in the early 70s. In this sense, a model describes the effects of causality through latent variables and assigns the explained and unexplained variance.

The model is executed in the WarpPLS 5.0 software since it is widely recommended to evaluate causal models with not-normal data. To know the efficiency of the proposed model, the following efficiency indexes are used:

- **APC**, **Average Path Coefficient**: It is obtained by calculating the arithmetic mean of all betas or relationships between the latent variables. To know their significance, the P-value is used, which must always be less than or equal to 0.05, since the model is evaluated with a confidence level of 95%.
- **ARS**, **Average R-squared**: It is a measure of the predictive validity and refers only to those R-squared presented in the dependent variables. In this sense, it is estimated the average of all the R-squared, meanwhile, the P-values are used to measure their significance.
- **AARS**, **Average Adjusted R-squared**: It measures the predictive validity of the model, but unlike R-squared, the adjusted R-squared considers the sample size. To measure its efficiency, P-value is used that must be less than or equal to 0.05.
- **AVIF**, **Average Block VIF**: It is a measure of the average collinearity of the different variables that integrate the model. On average, values recommend are lower than 3.3.
- **GoF**, **Tenenhaus index or Tenenhaus GoF**: It indicates the generic explanatory value of a model and values higher than 0.36 are recommended, indicating that models explain at least 36% of the variance contained in the variables analyzed.

5 Results Analysis and Discussion

According to the methodology described before, a total of 469 surveys were obtained. For a better understanding of the results obtained, the information is reported in the following sections.

5.1 *Descriptive Analysis of the Sample*

The 489 surveys come from participants from different bachelor's degree programs, as illustrated in Table 2, which shows the frequency and the percentage. It can be observed that the major participation is from Industrial Engineering, Architecture, and Environmental Engineering.

Likewise, Table 3 shows the distribution according to the gender of students surveyed.

5.2 *Questionnaire Validation*

Table 4 shows the result of validating the items of uses and benefits separately by using the Cronbach's alpha index. Likewise, when the general analysis is performed, that is, uses and benefits items are considered as a group. Thus, the Cronbach's alpha coefficient obtained is illustrated in the last row.

Table 2 Sample composition

Bachelor's degree program	Frequency	Percentage
Industrial engineering	74	15.8
Architecture	71	15.1
Environmental engineering	64	13.6
Engineering in business management	53	11.3
Mechatronics engineering	45	9.6
Biochemical engineering	42	9.0
Management	42	9.0
Accounting	35	7.5
Computer systems engineering	24	5.1
Informatics engineering	19	4.1
Total	469	100

Table 3 Frequency by gender

Gender	Frequency	Percentage
Male	276	58.8
Female	193	41.2

Table 4 Cronbach's alpha index

Variable	Cronbach's alpha	Items at the beginning	Items at the beginning
Uses	0.803	18	17
Benefits	0.905	16	15
Uses and benefits	0.892	34	32

Note that in each of the analyzed dimensions, uses, and benefits, an item has been eliminated, so there is a difference between the number of items at the beginning and the end of the analysis.

5.3 *Descriptive Analysis of the Items*

Table 5 illustrates the descriptive analysis of the items associated with the uses of ICT, which are sorted in descending order according to the values of their median or second quartile. It is observed that the higher academic given to ICT by students refers to searching information, sending, or receiving tasks from their teachers or collaborators, which indicates that ICT facilitates the collaborative work as is tacitly verified in the third item. Note that these three items show median values higher than 4, but it does not happen to the other last items in that category.

About the recreational uses given to ICT by students, the main ones correspond to sending photos and videos, but also to their participation in social networks. Note that only these two aspects show median values higher than three.

Table 5 Descriptive analysis of items related to uses of ICT

Uses	Q2	Q1	Q3	RI
Academic				
Do you use ICT to find information for your subject?	4.35	3.61	4.95	1.33
Do you use ICT to send or receive assignments?	4.33	3.52	4.95	1.43
Do you use ICT to collaborate with other colleagues to the subjects' assignments?	4.25	3.50	4.85	1.35
At school, do you use ICT as a means of communication?	3.93	3.08	4.69	1.60
Do ICT are useful to design or draw?	3.64	2.59	4.57	1.97
Have ICT helped you to solve mathematical problems quickly and efficiently?	3.56	2.67	4.43	1.76
How often do teachers use ICT in the teaching-learning process?	3.54	2.86	4.26	1.39
Do you use ICT to take notes (as a notebook)?	3.37	2.35	4.33	1.98
Do you use ICTs to ask questions or receive answers from your teachers?	3.30	2.38	4.24	1.86
Do you use ICT to read newspapers and/or digital magazines?	2.60	1.68	3.61	1.92

(continued)

Table 5 (continued)

Uses	Q2	Q1	Q3	RI
Do you use ICT to participate in forums and/or debates on the topics that interest you?	2.41	1.53	3.46	1.93
How often do you use the website of the library to check the online catalog, databases, or materials available?	2.40	1.51	3.39	1.88
Recreational				
Do you use ICT for sharing photos and videos?	3.58	2.60	4.47	1.87
Do you use ICT to participate in social networks?	3.48	2.44	4.48	2.03
Do you consider that ICT has more social or recreational use (games, communication, leisure) than academic use?	2.81	1.97	3.79	1.81
Do you use ICT to play online with your classmates?	1.99	1.16	3.39	2.22
How often do you listen to the radio by using ICT?	1.89	1.15	2.95	1.80
At school, do you use ICT to download music, movies, games, or other programs?	1.79	1.06	3.14	2.07

According to the benefits obtained from the uses of ICT, Table 6 shows the measures of central tendency and dispersion, sorted in descending order according to the values of their median. Regarding academic uses, the students report that ICT provide them a good and effective education, but also facilitate the timely delivery of work to their teachers. Note that these two benefits have a median value greater

Table 6 Descriptive analysis of items related to benefits

Benefits	Q2	Q1	Q3	RI
Academic				
Do you believe that ICT can provide a good and effective education?	4.24	3.47	4.87	1.39
Do you consider that ICT help to deliver assignments in a timely manner?	4.09	3.30	4.77	1.47
Do you consider that ICT make the training programs more open and flexible?	3.97	3.24	4.66	1.42
Does the ICT use facilitate understanding of contents?	3.89	3.15	4.63	1.48
Have you improved your grades by using ICT?	3.89	3.14	4.64	1.49
Has the learning process been improved by using ICT in the classroom?	3.86	3.12	4.61	1.49
Do you consider that ICT improve communication between the different agents of the teaching-learning process?	3.86	3.09	4.64	1.55
Is teaching more personalized by using ICT?	3.76	2.98	4.56	1.57
Do ICT allow you to consult your teachers without displacement?	3.75	2.91	4.58	1.67

(continued)

Table 6 (continued)

Benefits	Q2	Q1	Q3	RI
Social				
Have ICT allowed you to work as a team?	4.22	3.37	4.88	1.50
Have ICT helped to increase fellowship and collaboration between students?	3.74	2.92	4.58	1.66
Do you think that social networks play a significant role in education?	3.65	2.73	4.51	1.78
Do you consider that ICT improve communication with the teacher?	3.60	2.74	4.44	1.69
Psychological				
If you compare the results obtained when implementing ICT and not implementing, are the first ones more satisfying?	3.64	2.89	4.45	1.56
Since motivation is fundamental in the place where knowledge is exchanged, have ICT helped to motivate you?	3.59	2.86	4.36	1.50
Do you consider that the use of ICT in the classroom makes you a passive student?	3.13	2.25	3.96	1.71

than four so are the most important. The third benefits deserve to be mentioned since they refer to the education flexibility and the possibility of open education and at a distance, an innovative and recent concept.

For the social benefits, the most reported refers to the fact that ICT have allowed to the students work as a team, and as the second benefit reports increasing companionship and team collaboration. Note that in this category only one benefit has a median value greater than four. The other three benefits present values lower than four, but higher than three, which indicates that they are at least regularly obtained.

Finally, Table 6 analyzes the psychological benefits composed of three items. All median values are less than four, but greater than three, indicating that they are regularly obtained when using ICT.

5.4 *Structural Equation Modeling*

Using WarpPLS 5.0, the model proposed in Fig. 1 was evaluated and Fig. 2 shows the results obtained.

5.4.1 Model Efficiency Indexes

Before interpreting the results obtained in the model, the following efficiency indexes are analyzed:

- Average path coefficient - APC - = 0.300, $P < 0.001$
- Average R-squared– ARS - = 0.317, $P < 0.001$

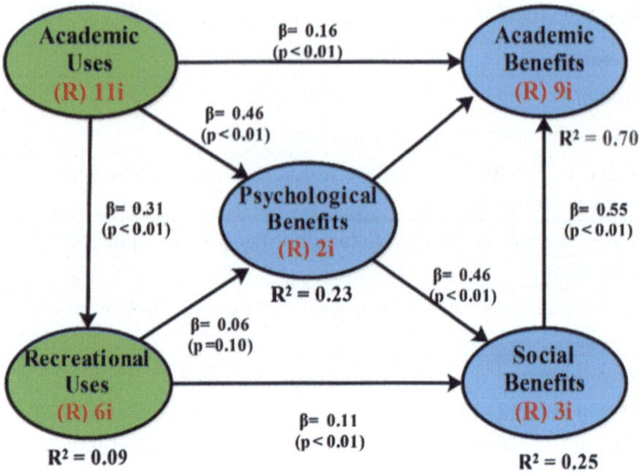

Fig. 2 Evaluated model

- Average adjusted R-squared - AARS - = 0.315, P < 0.001
- Average block VIF - AVIF - = 1.250, acceptable if <= 5, ideally <= 3.3.
- Tenenhaus GoF – GoF - = 0.409, small >= 0.1, medium >= 0.25, large >= 0.36.

5.4.2 Direct Effects

Table 7 describes the causal relationships between the variables, where a P-value is associated with each Beta value allowing rejecting or not, the hypotheses stated before. Likewise, R-squared value is illustrated for those latent variables that act as dependent variables, which are explained by at least one independent variable.

Table 7 Causal relationships between the latent variables (standardized coefficients)

Relationship	β	p	R^2	Conclusion
Academic uses → Recreational uses	0.31	<0.01	0.09	H_1: nonreject
Academic uses → Psychological benefits	0.46	<0.01	0.23	H_2: nonreject
Academic uses → Academic benefits	0.16	<0.01	0.70	H_3: nonreject
Recreational uses → Social benefits	0.11	<0.01	0.25	H_4: nonreject
Recreational uses → Psychological benefits	0.06	=0.10	0.23	H_5: Reject
Psychological benefits → Social benefits	0.46	<0.01	0.25	H_6: nonreject
Psychological benefits → Academic benefits	0.30	<0.01	0.70	H_7: nonreject
Social benefits → Academic benefits	0.55	<0.01	0.70	H_8: nonreject

5.4.3 Effect Sizes from Direct Effects

Besides, it is important to note that when more than one independent variable explains the dependent variables, the R-squared that is associated to the first variable must be decomposed into effects corresponding to each of the independent variables. Table 8 shows this process in the model proposed before. Every column represents the variance source and the rows are representing the dependent latent variables.

Table 8 Effect Sizes from Direct Effects

To	From				
	Academic uses	Recreational uses	Social benefits	Psychological benefits	R^2
Recreational uses	β = 0.308 (P < 0.001) ES = 0.095				0.095
Academic benefits	β = 0.156 (P < 0.001) ES = 0.086		β = 0.546 (P < 0.001) ES = 0.417	β = 0.299 (P < 0.001) ES = 0.191	0.694
Social benefits		β = 0.112 (P = 0.001) ES = 0.024		β = 0.459 (P < 0.001) ES = 0.222	0.246
Psychological benefits	β = 0.459 (P < 0.001) ES = 0.220	β = 0.059 (P = 0.101) ES = 0.013			0.233

5.4.4 Sum of Indirect Effects

When a variable is affected by two or three segments involving two or more independent variables, it is necessary to analyze the indirect effects. Table 9 shows

Table 9 Sum of indirect effects

To	From		
	Academic uses	Recreational uses	Psychological benefits
Academic benefits	β = 0.281 (P < 0.001)* ES = 0.156	β = 0.093 (P < 0.021)* ES = 0.016	β = 0.25 (P < 0.001)* ES = 0.160
Social benefits	β = 0.253 (P < 0.001)* ES = 0.117	β = 0.027 (P = 0.205) ES = 0.006	
Psychological benefits	β = 0.018 (P = 0.290) ES = 0.009		

*Inferences at 95% confidence level

the sum of indirect effects presented in the model, as well as the P-value associated with the test of statistical significance and the effect size, which is denoted as ES.

6 Conclusions and Implications

Although it has been statistically shown that ICT have an impact on the students' academic, social, and psychological performance, Salinas (2004) states that when introducing them in the process of educational innovation, the following aspects must be considered:

Changes in the role of the teacher: The teacher requires a training process that leads to:

- Knowledge about ICT tools.
- Interaction with the educational and social community related to the challenges of the knowledge society.
- Consciousness about of the formative needs.
- Ability to plan the development of their professional career.

Palomo, Ruiz, and Sánchez (2006) suggest that the teacher is the driver of the change of the teaching-learning system, who is a key to the success of any innovation in the teaching process. Therefore, in this new technological context, it is necessary that the teacher change his/her mentality, attitude, and role concerning teaching. In this sense, teachers should move from being content experts to being learning facilitators, which involve different tasks, such as design learning experiences for students, and provide an initial structure to encourage students to self-study.

Also, designing materials and resources adapted to the students' characteristics is another fundamental role of the teacher. The materials will not only be developed by the teacher her/himself but also in collaboration with the other partners involved in the process, thus, working in a collaborative way.

Changes in the role of the student: The student changes the traditional way of working when using ICT (computer, tablet, smartphone, etc.), it should be emphasized that they are looking for a high-quality learning and knowledge. However, it should also be considered that the ICT role in educational field it is not only an effort to copy and paste information but improving the teaching-learning process.

According to Domingo and Fuentes (2010), 90% of students emphasize that ICT imply an innovative methodological renewal that increases the motivation and participation of students, which facilitates the understanding and learning by providing new educational resources. Also, ICT increase satisfaction, motivation, and self-esteem of the teachers. Likewise, for 60% of students, ICT facilitate collaborative work, which can increase their autonomy and facilitate their evaluation and self-evaluation.

Based on the results obtained from the present investigation, it has been possible to analyze the ICT integration and their impact on the academic efficiency in undergraduate students of the Instituto Tecnologico de Colima. Also, the research has allowed investigating the academic and recreational uses that students give to the ICT and finally, has shown the academic, social, and psychological benefits that the students obtain when using them.

In this sense, the following hypotheses were verified:

- There is enough statistical evidence to state that the Academic Uses of ICT have a direct and positive impact on the Recreational Uses, since when the first latent variable increases its standard deviation by one unit, the standard deviation of the second latent variable also increases by 0.3 units.
- There is enough statistical evidence to affirm that the Academic Uses have a direct and positive impact on the Psychological Benefits. When the first variable increases its standard deviation by one unit, then the standard deviation of the second latent variable increases by 0.46 units.
- There is sufficient statistical evidence to point out that the Academic Uses of ICT have a direct and positive impact on the Academic Benefits, because when the first latent variable increases its standard deviation by one unit, the standard deviation of the second latent variable also rises by 0.16.
- There is necessary statistical evidence to declare that Recreational Uses of ICT have a direct and positive impact on the Social Benefits, because when the first latent variable increases its standard deviation by one unit, the standard deviation of the second latent variable also rises by 0.11 units.
- There is not enough statistical evidence to support that the Recreational Uses of ICT have a direct and positive impact on the Psychological Benefits, because when the first latent variable increases its standard deviation by one unit, the standard deviation of the second latent variable also rises by 0.06 units, with a P-value equal to 0.10, bigger than 0.050.
- There is sufficient statistical evidence to state that the Psychological Benefits have a direct and positive impact on the Social Benefits., because when the first latent variable increases its standard deviation by one unit, the standard deviation of the second latent variable also rises by 0.46.
- There is enough statistical evidence to state that the Psychological Benefits have a direct and positive impact on the Academic Benefits, since when the first latent variable increases its standard deviation by one unit, the standard deviation of the second latent variable also increases by 0.30 units.
- There is enough statistical evidence to affirm that the Social Benefits have a direct and positive impact on the Academic Benefits. When the first variable increases its standard deviation by one unit, then the standard deviation of the second latent variable increases by 0.55 units.

References

Aravena, et al. (2006). Investigación educativa I. Chile: AFEFCE/ECUADOR y UNIVERSIDAD ARCIS/CHILE. Retrieved January 26, 2017 from https://jrvargas.files.wordpress.com/2009/11/investigacion-educativa.pdf.

Arias, W. (1996). La innovación educativa. Retrieved January 16, 2017 from http://www.uaa.mx/direcciones/dgdp/defaa/descargas/innovacion_educativa_octubre.pdf.

Ausubel, D., & Colbs, H. (1990). Psicología educativa: Un punto de vista cognoscitivo. México: Editorial Trillas. Second edition.

Avelar, L., García, J., Cedillo, M., & Adarme, W. (2014). Effects of regional infrastructure and offered services in the supply chains performance: Case Ciudad Juarez. *DYNA, 81*(186), 208–217. https://doi.org/10.15446/dyna.n81n186.39958.

Barraza, M. (2005). Una conceptualización comprehensiva de la innovación educativa. *Innovación Educativa, 5*(28), 19–31.

Barriga, F. (2007). La innovación en la enseñanza soportada en TIC. Una mirada al futuro desde las condiciones actuales. Universidad Nacional Autónoma de México. Retrieved February 18, 2017 from www.oei.es/historico/tic/santillana/Barriga.pdf.

Belloch, C. (2012). Las Tecnologías de la Información y Comunicación (T.I.C.) en el aprendizaje. Universidad de Valencia. Retrieved March 26, 2017 from http://www.uv.es/bellochc/pdf/pwtic2.pdf.

Bisquerra, et al. (2009). *Metodología de la investigación educativa*. S.A., Madrid: La Muralla.

Bongarra, C. (2013). El uso del celular como herramienta educativa en el aula de la escuela secundaria. Instituto Juan Amos Comenio. Retrieved February 18, 2017 from http://www.fecea.edu.ar/semillero/informatica/.

Castillo, M., Larios, V., & García, O. (2010). Percepción de los docentes de la utilización de las Tecnologías de la Información y la Comunicación. Revista Iberoamérica de Educación de la OEI. Retrieved January 11, 2017 from http://rieoei.org/deloslectores/3375Castillo.pdf.

Correa, J. M. (2005). La integración de plataformas de e-learning en la docencia universitaria: Enseñanza, aprendizaje e investigación con Moodle en la formación inicial del profesorado. *Revista Latinoamericana de Tecnología Educativa, 4*(1), 37–48.

Cox, C. (2004). *Políticas educacionales en l cambio del siglo*. La reforma del sistema escolar de Chile: Universitaria, S. A, Santiago, Chile.

Cronbach, L. (1951). Coefficient alpha and the internal structure of tests. *Psychometrika, 16*, 297–334. https://doi.org/10.1007/BF02310555.

Domingo, M., & Fuentes, M. (2010). Innovación educativa: experimentar con las TIC y reflexionar sobre su uso. *Pixel-Bit. Revista de Medios y Educación, 36*, 171–180.

Fernández, I. (2010). Las TICS en el ámbito educativo. Eduinnova. Retrieved January 7, 2017, from http://www.eduinnova.es/abril2010/tic_educativo.pdf.

Fernández, R., Server, P., & Cepero, E. (2001). El aprendizaje con el uso de las nuevas Tecnologías de la Información y las Comunicaciones. OEI-Revista Iberoamericana de Educación. Retrieved February, 18, from http://rieoei.org/deloslectores/127Aedo.PDF.

Ferro, C., Martínez, A., & Otero, M. (2009). Ventajas del uso de las TICs en el proceso de enseñanza-aprendizaje desde la óptica de los docentes universitarios españoles. España: EDUTEC - Revista electrónica de tecnología educativa *29*, 1–12. https://doi.org/10.21556/edutec.2009.29.451.

Fullan, M., & Stiegelbauer, S. (1991). *The new meaning of educational change*. London: Casell.

Gallego, D., & Dulac, J. (2006). Informe final del Iberian Research Project: Madrid. Retrieved March 2006, from http://www.dulac.biz/Ibe-rian%20research/IberianResearchProject.htm.

Gallego, M., & Gámiz, V. (2011). Posibilidades de la pizarra digital para acciones de innovación educativa en la formación del profesorado en el entorno español. *Revista Electrónica Actualidades Investigativas en Educación, 11*, 1–28.

García, M., Sordo, J., & England, S. (2008). *Cómo almacenar digitalmente las explicaciones de clase en tiempo real*. Editorial Complutense, S.A., Madrid, Spain: IV Jornadas Campus Virtual UCM.

García-Valcárcel, A. (2003). Tecnología educativa -Implicaciones educativas del desarrollo tecnológico -. España: Educatio siglo XXI de la Facultad de Educación de la Universidad de Murcia. Retrieved August 7, 2014 from http://revistas.um.es/educatio/article/view/128/112.

Gómez, M. (2005). La pizarra digital. Tecnológico del Instituto de Tecnologías Educativas del Ministerio de Educación. Retrieved March 26, 2017, from http://observatorio.cnice.mec.es/.

Knight, P. (2005). *El profesorado de educación superior*. Formación para la excelencia: Narcea, S. A. de Ediciones, Madrid, Spain.

León, A. (2007). Qué es la educación. Venezuela: Educere, La revista Venezolana de Educación. Retrieved May 8, 2015, from http://www.redalyc.org/articulo.oa?id=35603903.

Likert, R. (1932). A Technique for the measurement of attitudes. *Archives of Psychology, R. S. Woodworth 22*(140), 5–55.

Marín, V., & Maldonado, G. (2010). El alumnado universitario cordobés y la plataforma virtual Moodle. *Pixel-Bit. Revista de Medios y Educación, 38,* 121–128.

Mikre, F. (2011). The roles of information communication technologies in education. *Ghana: Ethiopian Journal of Education and Sciences 6*(2), 1–16.

Muñoz-Repiso, A. (2007). Estrategias para una innovación educativa mediante el empleo de las TIC. *Revista Latinoamericana de Tecnología Educativa-RELATEC, 2*(1), 41–50.

Palomo, R., Ruiz, J., & Sánchez, J. (2006). Las TIC como agentes de innovación educativa. Sevilla: Junta de Andalucía, Consejería de Educación, Dirección General de Innovación Educativa y Formación del Profesorado. Retrieved February 9, 2017, from http://www.edubcn.cat/rcs_gene/11_TIC_como_agentes_innovacion.pdf.

Pérez, R., Rojas, J., & Paulí, G. (2008). Algunas experiencias didácticas en el entorno de la plataforma Moodle. *Revista de informática educativa y medios audiovisuales, 5*(19), 1–10.

Pérez, V., & Fernández, J. (2005). Las tecnologías de la información y la comunicación en la formación permanente del profesorado de educación física. *REIFOP, 8*(1), 1–5.

Pinto, L. (2011). No hay que prohibir los móviles en el aula, sino incorporarlos en el aula. Retrieved January 10, 2017, from www.lainformacion.com.

Salinas, J. (2004). Innovación docente y uso de las TIC en la enseñanza universitaria. *RUSC Universities and Knowledge Society Journal, 1*(1), 1–16.

Salinas, et al. (2008). Innovación educativa y uso de las TIC. Universidad Internacional de Andalucía. Retrieved February 9, 2017, from http://dspace.unia.es/bitstream/handle/10334/2524/innovacioneduc2008.pdf?seq.

Santamaría, S., Quintana, M., Milazzo, L., & Rodríguez, R. (2005) Introducción a la sociología de la educación. Retrieved February 10, 2017, from http://ubccolima.edu.mx/plataformavirtual/.

Tesouro, M., & Puijgalli, J. (2006). Las comunidades virtuales y de conocimiento en el ámbito educativo. Pixel Bit. *Revista de Medios y Educación, 28.*

Tinio, V. L. (2002). ICT in Education: UN Development Programme. Retrieved December, 2009 from http://www.eprmers.org.

Vivanco, M. (2005). *Muestreo estadístico diseño y aplicaciones*. S.A., Santiago, Chile: Universitaria.

A Series of Recommendations for Industrial Design Conceptualizing Based on Emotional Design

David Cortés Sáenz, Carlos Eduardo Díaz Domínguez, Pere Llorach-Massana, Ainoa Abella García and Juan Luis Hernández Arellano

Abstract Emotional design, since its birth in the 70s, has developed into becoming one of the main forces that drive design nowadays. The capability of understanding the user's feelings and emotions is considered to be a pivotal part of the design process. More than ever before, industrial designers need to consider the emotional side of consumers to create meaningful and successful products. The present study created a series of recommendations for Industrial Design Conceptualizing based on Emotional Design. These were developed through the analysis of several design emotion-based methodologies, conceptualization theories, and emotional design itself. The recommendations were developed to be applied in the conceptualization phase, for it is at this stage of the design process when the designer gives shape to the initial concept that will ultimately become a product. After its completion, to validate the recommendations, these were handed to a group of eight industrial design students who applied them on one of their academic projects. The results of their works were subjected to analysis to determine the impact on their projects. Students were also asked their opinion about emotional design and the recommendations they were given. Principal results evidence that only the 15% of the students knew about emotional design before being introduced by the present research. Five out of eight of the resulting projects managed to incorporate values and concepts related to emotional design. It reflects that the resulting ultimate products can be influenced by the capabilities of designers to use the recommendations. It could be concluded that the recommendations could be of great interest for industrial design to transmit emotions to products which could satisfy consumers emotional desires.

D. Cortés Sáenz (✉) · C. E. Díaz Domínguez · J. L. Hernández Arellano
Department of Design, University of Ciudad Juárez, Ciudad Juárez, México
e-mail: david.cortes@uacj.mx

P. Llorach-Massana
Barcelona School of Design and Engineering, ELISAVA, La Rambla 30-32, 08002 Barcelona, Spain

A. Abella García
Department of Statistics and Operations Research, Universitat Politècnica de Catalunya– Barcelona Tech, Barcelona, Spain

G. Cortes-Robles et al. (eds.), *Managing Innovation in Highly Restrictive Environments*, Management and Industrial Engineering,
https://doi.org/10.1007/978-3-319-93716-8_8

Keywords Conceptualization · Emotional design · Design conceptualization
Industrial design · Design process

1 Introduction

The importance of emotional design on product development has grown significantly over the past few decades. Industrial designers have gone from simple-minded functionality to a conception where their main purpose is to create products and experiences that can build emotional, deep, long, and long-lasting bonds with users (Roberts 2005). Consequently, several investigators and companies started to research and develop different ways to introduce emotional design within the design processes. Methods like Kansei Engineering, Kano Method, and Semantic Differential have been developed to understand users' emotional desires. Several categorizations of emotions along several dimensions were made. Some of them were based on arousal and valence (Hepach 2011), some others were based on the subconscious level on which they acted (Norman 2004). After the development of these methods, investigators and design professionals started to apply emotional design in their projects to create meaningful and more personal products (Norman 2004; Spillers 2004; Desmet and Fokkinga 2013). Some studies have brought up exciting and revealing news about emotions, as their deep correlation with decision making (Gray et al. 2002) or their capability to overrun cognitive barriers when it comes to product appreciation.

Nowadays, industrial designers face several difficulties when they apply emotional design into their projects. For this reason, the objective of this chapter was to provide to the industry a series of recommendations for the application of emotional design during the design process. The present research was divided into two parts. First, a series of recommendations were created to facilitate the application of emotional design during the design process. Later, these recommendations were handed to a group of industrial design students who applied them on one of their projects. As the recommendations are solely for the conceptualization phase, the resulting pieces were not graded on the emotional response they may or may not have caused; they were rather analyzed in search of concepts related to emotional design.

1.1 Introduction to Emotional Design

Emotional design, just like the industrial design and design itself, has received several different names, such as hedonic design, affective design, the design of human affective factors, and empathic design, among others. Although there is no clear or ultimate definition of what emotional design is, it seems that it is a branch of design that has made its main purpose to study the role that human emotions play as an influence on how we interact with different objects and artifacts (Aumer-Ryan 2005).

Typically, the design used to be divided into two fundamental aspects: emotive and rational. The common thinking was that they should be separated, furthermore, the rational aspect of design used to be prioritized over the emotional side. In (Gray et al. 2002), the study reinforced the theory that emotional states are associated with the different ways we process information, and in a deeper layer, with the decision-making process.

How is emotional design relevant to industrial design? Emotions have powerful effects on the decision-making process of people; hence, they play a key role when users must decide whether they like a product or not. When it comes to the appraisal of a product, emotions have the capability of overrunning the logical and cognitive walls of users, meaning that in many cases, emotions will determine if a user likes a product or not, despite cognitive factors.

The role that emotions play when deciding whether someone likes an object or not has proved to be so strong that they can be more important than functionality or usability factors. It is not whether the product works or not. The important matter is how a product makes the user feel. According to Norman (2004), on many occasions, the true value of a product is not directly related to its intrinsic characteristics. Instead, it relies on the emotional connection that bounds the product to the user. In some cases, emotional design can have an impact not only on how the user perceives the product but on how it makes the user relate to other people.

When it comes to new products, it seems that context and emotions have a direct impact on how people receive jumps on innovation. A study conducted by (Carbon et al. 2013) concluded that we are keener to appreciate innovation in a positive way when we feel safe about these innovations. For example, we are more receptive to extravagant and unconventional cars if they are presented to us in a car show as we are prepared to see new and uncommon things.

1.2 Review of Tools for Emotional Design

As it was mentioned before, several tools have been developed to achieve a design approach that envisions products that meet the user's desires. Nevertheless, these tools or methods are not specifically designed to guide the designer towards the creation of an object that triggers an emotional response. Kansei Engineering and Kano are the main emotional design tools that have been applied in the industrial sector. Some examples of the application of these methods can be found in the scientific literature, see Table 1.

Kansei Engineering provides information regarding what the user expects to see in a product, by asking a test group for their impressions of determined concepts or products. The test group gives their opinion on how they understand or like a product.

Table 1 Examples of Kansei Engineering and Kano methods application

Title	Methodology	Contribution	Reference
Satisfying emotional needs of the beer consumer through Kansei engineering study	Kansei Engineering	Describes the use of Kansei Engineering type I for the category classification of beer cans and the identification of the design elements that can satisfy emotional and sensitive needs (sense of vision)	Hirata et al. (2004)
Affective engineering in application to Bi-level human migration models	Kansei Engineering	A bi-level human migration model using the concepts of affective engineering (*Kansei* Engineering) and conjectural variations equilibrium (CVE) is developed	Kalashnikov et al. (2014)
Emotions and the urban lighting environment A Cross-cultural comparison	Kansei Engineering	The article shows the main results of an empirical research about the relation between emotions and urban lighting scenarios	Calvillo Cortés and Falcón Morales (2016)
An integrated approach to innovative product development using Kano's model and QFD	Kano model	The article analyses the notion of customer satisfaction based on the Kano model and points out the importance of product innovation in exceeding customer satisfaction	Shen et al. (2000)
Understanding customer needs through quantitative analysis of Kano's model	Kano model	A new model is developed to study the relations between customer satisfaction and the fulfillment of customer requirements	Wang and Ji (2010)
Integration of FMEA and the Kano model: An exploratory examination	Kano model	A new approach is proposed to enhance FMEA capabilities with Kano model. This includes determining the severity and "risk priority number" (RPN). The approach prevents failures at early stages of design	(Shahin 2004)
A fuzzy-AHP approach to prioritization of CS attributes in target planning for automotive product development	Kano model	The article analyzes the lack of quantitative data and undefined relationships between the attributes that make difficult to develop a quantitative model for analyzing subjective customer satisfaction (CS) attributes	Nepal et al. (2010)
Kano Model and QFD integration approach for ergonomic design improvement	Kano model	The article presents joining methods of Kano Model and Quality Function Deployment to improve the school workshop's workstation design for adolescents regarding ergonomics and users need	Hashim and Dawal (2012)

Methods like Semantic differential are based on bipolar concepts tests that, although somewhat provides information regarding how the user feels towards certain concepts, by itself it does not represent design considerations that could be used by the industrial designer who strives to trigger emotional responses with its product. Furthermore, the selection of bipolar concepts rely entirely on the design researchers, so these concepts tend to be subjective (Mondragón Donés et al. 2006).

Kano Method is another option to approach user's desires regarding product design (Beitia et al. 2009). Based primarily on assessing the degree of satisfaction that a product achieves on its users, Kano method consists of a series of multiple answer questions oriented to find out what users like better. The three main aspects that this method tries to determine are the degree of satisfaction with the functionality of a product; things that the user asks of the product; details that would make the product more pleasurable to use (Álvarez et al. 2008).

These previous methods require a considerable investment of time, resources, and knowledge, and the results are presented in the shape of statistics and data that require experience for their interpretation (Bradley and Lang 1994). Furthermore, these processes tend to be lengthy and exhaustive.

Some other tools, like the Geneva Emotion Wheel (GEW) (Sacharin et al. 2012), PrEmo (Desmet et al. 2012) and Self-Assessment Manikin Scale (SAM) (Bradley and Lang 1994), have a deeper connection to the emotional side of design. They are tools that are very effective measuring the emotional response of people towards certain objects or concepts through visual aids.

Trough anthropomorphic images, PrEmo, and SAM Scale can effectively determine whether a product triggers an emotional response, what kind of emotion, and the strength of this particular emotion. Being based on graphic tools, these methods can transcend the barrier of languages, as the emotions expressed on the images clearly show moods and sentiments.

GEW is a tool similar to the past two, though, while SAM and PrEmo use images to fulfill their purpose, GEW uses a graphic from which the surveyed can pick an emotion and how strongly they feel it, dividing these parameters on Positive/Negative emotions, and High Control/Low Control of these emotions. This allows GEW to be applied on a variety of fields, form decision making to user experiences (Sacharin et al. 2012).

Described present important knowledge about emotions and how they relate to design, yet, none of them actually tells how the novice designer should approach and conceptualize based on users emotions (Desmet and Fokkinga 2013).

1.3 Difficulties for Industrial Designers

Industrial designers who want to include the emotional aspect of design into their projects are then faced with a series of difficulties, which are given as follows:

- Although there is a fairly good amount of information regarding emotional design, most of this information is about emotional design but not about how to apply it.
- For the industrial designers that have never heard of, or know little about emotional design, it can be very difficult to apply it, as there is no established method, manual or guide that helps them on the purpose of creating objects which target is to trigger emotional responses.
- Most designers will feel lost when trying to apply emotional design, as there is no reference on how to do it.
- The methods that are the closest to emotional design (such as Kansei Engineering) require quite an amount of time, resources, and knowledge that may not be available to the grand majority of industrial designers.
- Great tools such as PrEmo are effective for measuring the emotional response, but they do not provide design guidelines for the industrial designer, their purpose is to measure results.

When comparing some of the methods used for emotional design (Table 2), we can observe that although they are successful in some ways when it refers to reach or measure emotional design, none of them was created to actually tell the designer how to design emotionally.

As a result, we can think that a possible solution to this situation is to create a series of recommendations for the industrial designer, based on emotional design, that will be capable of orienting designers on what concepts and knowledge can they rely on, in order to create objects that will capitalize on human emotions to become successful products.

Table 2 Comparison between six different tools oriented to find out consumer's preferences and emotional impact of products

	Kansei	Kano	SD	GEW	PrEmo	SAM
Based on surveys	*O*	*O*	*O*	*O*	*O*	O
Fixed answers given by surveyors	**X**	*O*	*O*	X	X	X
Limited amount of possible answers	**X**	*O*	*O*	*O*	*O*	O
Considers emotion	O	**X**	**X**	*O*	*O*	O
Considers cultural background	**X**	**X**	**X**	**X**	**X**	X
Considers functional aspects	*O*	*O*	*O*	**X**	**X**	X
Considers aesthetic aspects	*O*	**X**	*O*	*O*	*O*	O
Considers what the subject desires as design criteria	*O*	**X**	**X**	**X**	**X**	X

(continued)

Table 2 (continued)

	Kansei	Kano	SD	GEW	PrEmo	SAM
Uses visual aid to formulate answers	**X**	**X**	**X**	*O*	*O*	O
Measures emotional perception of a product	**X**	**X**	**X**	*O*	*O*	O
Aids in the design process	*O*	*O*	*O*	**X**	**X**	X
Gives EMOTIONAL based design guidelines	**X**	**X**	**X**	**X**	**X**	X
Affordable for small design studios/ independent designers	**X**	**X**	**X**	*O*	*O*	O
Requires economical investment to conduct study	High	Medium	Medium	Low	Low	Low
Requires time investment to conduct study	High	High	High	Medium	Low	Low
Requires logistics to conduct study	High	High	High	Medium	Medium	Medium

2 Methods

2.1 Creation of Recommendations to Apply Emotional Design

Even though there is not a specific guide to apply emotional design in the conceptualization process, there is plenty of information related to emotional design. Several important investigators, like Don Norman, Frank Spillers, Jodi Forlizzi, and Pieter Desmet have published various books and articles related to it (Desmet et al. 2012; Norman 2014; Spillers and Asimakopoulos 2014; Forlizzi et al. 2016).

A series of recommendations and knowledge key points were conceived from the study of diverse emotional design books, articles, reviews, and various industrial design methodologies, as well as a deep insight on how the mind works and on empirical observation of human behavior. These recommendations are focused on the conceptualization phase of the design process. It aims to give designers valuable information and tools for product conceptualization, which introduce the emotional side of users as the main criteria. The recommendations do not try to tell designers how to conceptualize. They are a gathering of concepts and knowledge about what triggers emotions on users, and how taking this knowledge into consideration might help designers reach a desirable level of emotional design. These recommendations present themselves as a serious option for designers who need a hand on understanding and designing with emotional values.

In the analytic part of the process to create the recommendations, an investigation was made about emotional design and a breakdown was started, separating its elements, both primary and secondary, to achieve a better understanding of this phenomenon. In the synthetic phase of the investigation, it is necessary to add knowledge that did not exist previously. The synthetic phase is somewhat difficult to acquire, for it is based on reflexive intuition and common sense.

The recommendations are grouped in a list and are based primarily on two factors: personal conclusions and appreciations regarding emotional design based on analysis and reflection of the gathered information, and accepted concepts within the emotional design field, proposed by various experts.

The following is a brief and simplified list of these recommendations and knowledge key points:

1. From the beginning: "Emotional design must be a fundamental part of the design process from the very beginning; it cannot be applied when the product is already finished" (Desmet et al. 2008).
2. Mood: there is evidence that a cheerful mood predisposes designer to design better solutions.
3. Eliminate preconceptions: Designers must not have preconceived ideas since the recommendations focus on the conceptualization phase. The process must start from zero.
4. Determine design possibilities: the designer must determine what the design possibility that he/she pursues, hence, the type of product that might be required is.
5. Market research based on emotional qualities: compare a similar product on the market to determine their existing emotional characteristics that designers must identify.

- Type of emotion that the product evokes: the designer must ask himself and try to determine what emotions and emotional relationships develop between the user and the product.
- Emotional level on which it acts: the designer must identify on which of the three emotional levels proposed by (Norman 2004), the product acts.

i. Visceral: exclusively based on the looks of the product. In this level, it is all about the formal and physical characteristics of the object.
ii. Behavioral: it is determined by how the product is used.
iii. Reflective: determined by knowledge and cultural baggage. It is the deeper and strongest emotional level.

6. Verbalizing is another form of design: the designer must learn to write down all of his/her ideas. It is fundamental to put down on paper the designer's ideas, for this is the first step to materialize them.

7. Mood boards: these are visual tools that support the designer when it comes to the creation or conceptualization maps and models. They help the designer to better determine the kinds of emotions that should be imprinted on the design.
8. Emotional Design Intent: this is a short and concise statement where the designer exposes the emotional level and kinds of relationships that he/she seeks to awaken in users who are using his/her product. It will serve as a future reference and guide for the designer.
9. Cultural contextualization: emotional design is deeply linked to the cultural values of the user. It is essential that the designer is aware of this and, hence, be able to identify these values (Aumer-Ryan 2005).
10. Subcultures contextualization: the so-called "protocultures" are also a differentiating factor for users, it is necessary to know them to exploit their properties for design (Lloyd et al. 2006).
11. About the aesthetics: the adage "form follows function" is being replaced by a new one: "if it looks good, it works." It is vital to understand the value, impact, and significance of the formal design. "...it requires a somewhat mystical theory of aesthetics to find any necessary connection between beauty and function" (Tractinsky 1997).
12. Aim for beauty: concepts that are considered as beautiful often have positive effects on users, affecting them on a visceral emotional level and affecting the perception of use easiness. To identify and apply these concepts to product design is vital to achieving a greater degree of acceptance among users (Tractinsky et al. 2000).
13. The personality of objects: humans tend to read each other's emotions. We have a great capacity for emotional interpretation, so much that we carry this ability to analyze the degree of inanimate objects. This feature allows the designer to give personality to objects so that they convey emotion, mood or personality.
14. Understand the effects of shape on the subconscious: certain physical characteristics often have different effects on the emotional reaction of the users. The designer must understand, for example, the calming effects of a sinuous and curved shape, or how a "heavy" or "bulky" shape can determine whether a product is attractive or not (Spillers 2004).
15. Anthropomorphizing: within the personality spectrum of objects, you may find the effects of product forms. As humans, we have a natural tendency to feel empathetic toward our species. Applying human traits and characteristics to objects has implications on the user's degree of emotional reaction. The designer should be aware of these consequences and use them in benefit of their product (Hepach 2011). An example is emotion recognition includes the capacity to read and interpret other people's emotion.
16. About objects attitudes: As with personality, it is possible to provide objects emotional states. The designer must understand how to make the object look "anxious" or "relaxed," for example. Giving emotional states to objects will cause the user to inadvertently interpret these emotional states, and feel the product differently.

17. The pride of possession: this factor has become one of the most important on nowadays product design. To make users identify themselves with products, to the point of actually feeling proud or "better" for owning them has become a factor that may determine the success or failure of these products. It is necessary to understand how to make the users acquire products, not because they need them, but because they want them.
18. Emotional memory: this is the name given to a series of concepts, memories, and experiences that the user accumulates throughout life. Emotional memory determines how the user responds to these certain concepts. Understanding these reactions and experiences, allows the designer to exploit them in order to achieve the desired emotional reactions (LeDoux 2000).
19. Establish contact with the designer's inner user: the best reference that a designer can have on the emotional impact of the objects is the designer itself. By abstracting from his/her professional skin, designers will be able to better understand the emotional impact of their own ideas (Lloyd et al. 2006).
20. Consider all aspects of design: when conceptualizing, the designer must take in account how the user should feel when holding the product in their hands, how should it smell, how much should it weight. The use experience plays a fundamental role in the emotional design.
21. Every aspect of the design must have a function: even if the sole purpose of the design is to look good. Every aspect of the design must be justified; it must have a reason to be. The consequence of adding details without purpose is having empty and meaningless products.
22. The artistic value of emotional design: design by conformity will almost always produce good and pleasurable products that most people would like. The true artistic value, however, is hardly achieved by consensus and almost always corresponds to the personal view of the designer or a small group of creative people. The designer must determine what kind of product he/she desires and intents to deliver; for it is through this contact with the user that designers can really persuade them that their vision is what the consumer wants (Crilly 2011).
23. 23.The golden ratio: also known as the golden number, Phi, Fibonacci sequence or divine proportion, the golden ratio is equal to 1.618. It has been shown that this ratio is found in many elements in nature. Its application provides balanced aesthetics and natural symmetry to objects. It is a good idea to have this proportion in mind when designing, as its correct implementation tends to generate positive affective responses on users.

2.2 *Recommendations Validation*

It becomes difficult to objectively appraise if these recommendations help designers to apply emotional concepts into their projects. An investigation could be conducted by analyzing products that resulted from conceptualizing with these recommendations and finding out if these products produce emotional responses, probably using tools such as PrEmo or Geneva Emotion Wheel. The results would not be conclusive as they would depend greatly on the individual capabilities and abilities of each designer.

Furthermore, the conceptualization phase of the design process does not include any sketching, designing or prototyping per se; it is based solely on the creation of ideas and concepts that will later become a design and, validating a process that focuses on ideas can be very subjective. Even so, there must be a way for these recommendations to be evaluated.

A study group made out of nine students from the Industrial Design Program, at the University of Ciudad Juárez, México were asked to apply these recommendations on one of their school projects and then submit these projects to analysis. It is not the purpose to determine if the design is good or not, but rather determine if the design presents characteristics that can be associated with the emotional design.

Characteristics of the study group:

- A small group of students of medium-advanced level applied the recommendations in the conceptualization phase of their projects.
- The students have never followed emotional design-based recommendations, guidelines, or whatsoever in their conceptualization processes before.
- The students did not know they were going to follow these recommendations on their projects until the day they were assigned with the design project.
- The students had approximately five days for the conceptualization phase.
- The students, after the conceptualization phase, continued with the sketching, definition, and prototyping phases of the design process.
- As a limitation, most of the projects have a requirement to be based on biomimetic, and several of them were required to be a luminaire project. Since deconstructing the resulting projects could be so subjective, another parameter must be considered to measure these recommendations.
- The resulting projects were analyzed to identify the presence or absence of concepts related to emotional design mentioned in the recommendations.
- Personal appraisal is given by the students about the recommendations. This appraisal was given through a survey oriented to find out the student's opinion regarding different aspects of the recommendations.

2.3 Designer's Appraisal on Recommendations

To find out the student opinion, a survey was applied to the study group. This was not oriented to determine the effectiveness of the recommendations, but its perceived effectiveness. The questions were made to determine how an individual perceived the recommendations, their experience using them, and their opinion.

The main aspects that were treated in the survey were the following:

1. Usefulness
2. Easiness of understanding
3. Relevance to the field
4. Effect on their design process
5. Effectiveness of recommendations when guiding trough conceptualization process
6. Effectiveness of recommendations when transmitting emotional design concepts
7. Overall opinion on the recommendations
8. Suggestions and improvements

3 Resulting Projects

As a reference of the resulting projects, graphic representations are presented; either photorealistic images (renders) or actual photographs of the prototypes. Luminaire pieces were submitted to scrutiny as a final semester by the head teacher of that assignment and by an evaluation committee to determine the quality of the design. Also, a design researcher analyzed the resulting projects in search of the emotional concepts presented in the recommendations. Non-luminaire projects were only scrutinized in search of emotional concepts.

3.1 Case One: Cuack!

The project proposed in Fig. 1 required to be a luminaire. The designer focused, during the conceptualization, on creating a design that evokes cuteness applying some recommendations, such as number 12 and 13. The recommendation 12 aims for beauty and the 13 creates personality—in this case of a duck—through simple lines and forms, this design attacks the visceral level of emotion, having its main

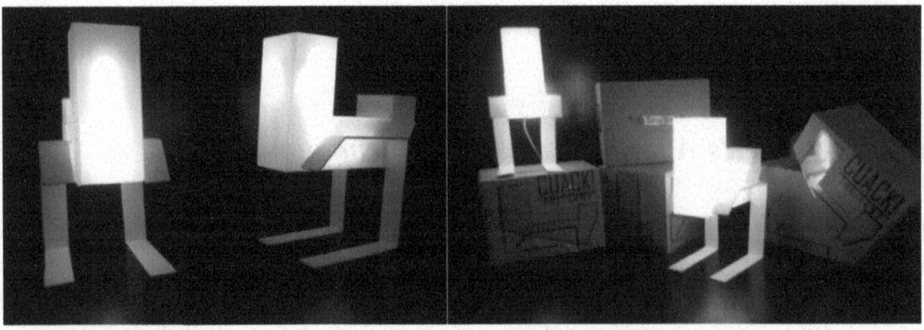

Fig. 1 "Cuack!" Luminaire, by Luis Peón

strength in the visual memories that it triggers. It is made of acrylic and steel, with an interior LED illumination.

3.2 Case Two: Nautilus Chair

As shown in Fig. 2, this chair required to be based on biomimetic. The designer was inspired by the shell of the marine creature nautilus to produce a chair made of molded urethane. In this product, the recommendation used was number 23: the application of the golden ratio.

3.3 Case Three: Periodic Table

The emotional value of Fig. 3 product relies principally on two recommendations: number 17 and 18. The first one, the pride of possessions attacks the reflective level of emotions by substituting the traditional elements with a series of music rock bands and musicians. This creates a sensation of identity that will appeal to users individually according to their musical tastes. The second recommendation evokes

Fig. 2 Concept, "Nautilus chair", by Diego Díaz

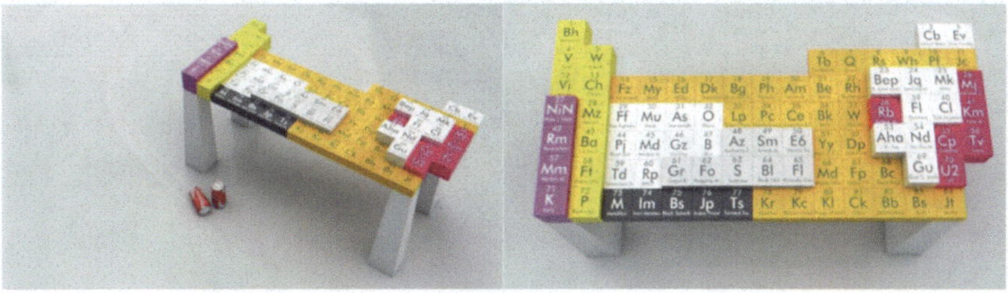

Fig. 3 Concept, "Periodic Table", by Nestor Hernández

nostalgic memories, using a periodic table that takes people back to the school and childhood. It is made of molded polypropylene.

3.4 *Case Four: USB Invaders*

In Fig. 4, the emotional design concepts of this piece are related to the recommendations 9, 17, and 18. According to the recommendation 9, cultural contextualization, this product using the famous game Space Invaders establishes a relationship with a generation that had it—the culture of gaming. It also creates a nostalgic reminiscence (recommendation 18) making those who played this game remember the 80s. At the same time, using a widely known concept aims to users to have it appealing to their sense of individuality because it is a vintage piece, as the recommendation 17 indicates.

3.5 *Case Five: Rosy*

Figure 5 illustrates a luminaire based on biomimicry of flamingos. The emotional recommendation applied in this design is number 13, giving a personality of

Fig. 4 Concept, "USB Invaders Flash memory device", by César Rodríguez

Fig. 5 "Rosy" Luminaire, by Daniela Camargo

flamingo. The shape of each piece is based on the flamingo's neck, yet, when faced one another they form the shape of a heart, alluding to the love and beauty these animals are often related to, as the recommendation 18 says. This luminaire can only be lighted when they are close to each other. It attacks two emotional levels: the behavioral level by changing the way these luminaires work (the need of being close to each other), and the reflective level by using emotional memory to recall romantic sentiments. It is made of oak and mahogany.

3.6 *Case Six: Tokay*

As can be seen in Fig. 6, this project required to be a luminaire based on biomimetic. The designer focused on a Leaf Tail Gecko and tried to imitate the tail of an animal. It is complicated to detect any of the emotional conceptualization recommendations presented in the study. It may be said that the sharpness of the lines and materials can evoke a sense of edginess, but it cannot be completely said that the designer result provokes a clear emotion. It is made of pinewood and stainless steel.

3.7 *Case Seven: EARS*

The project in Fig. 7 required to be a luminaire. The designers applied the recommendation number 13 in a rather interesting way. Understanding the cautious nature of rabbits, they gave the personality of this animal placing it on a corner in the ground. The users only observe the ears and eyes of the rabbit showing up, as if it were looking outside from safety. This gives the design a sense of aliveness that triggers emotion on users. It is made of stainless steel.

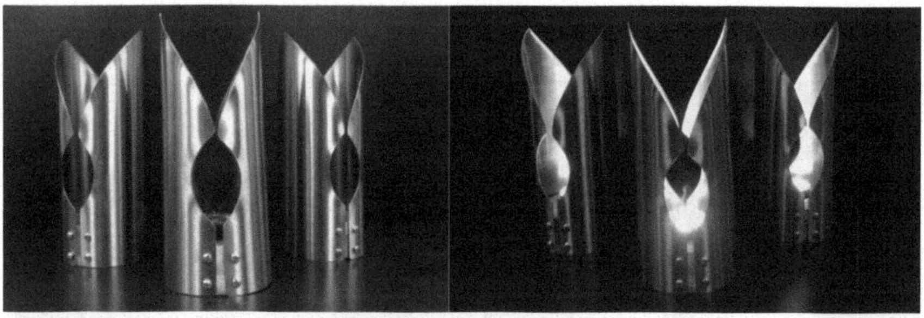

Fig. 6 "Tokay" Luminaire, by Alfonso Rojas

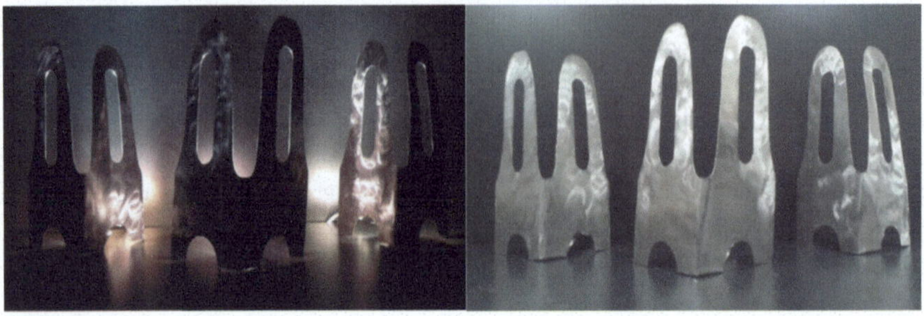

Fig. 7 "EARS" Luminaire, by Brandon Rivas

3.8 Case Eight: Colibrí

This project required to be based on biomimetic. In this case, the design of Fig. 8 chose a bumblebee as the most significant analogy. It can be said that the curviness of lines might produce some sensations, as calm, but it cannot be assured that this was the intention of the designer. So, it this product cannot be identified which recommendation used the designer in the conceptualization phase. It is made of oak and styrene plastic.

3.9 Students Survey

Based on the answers given by the study group, the following statements can be made:

1. The reading of the recommendations is entertaining and easy. Every answer in that regard qualified them as either "entertaining" or "very entertaining". According to that the recommendations can be consulted or read again easily.

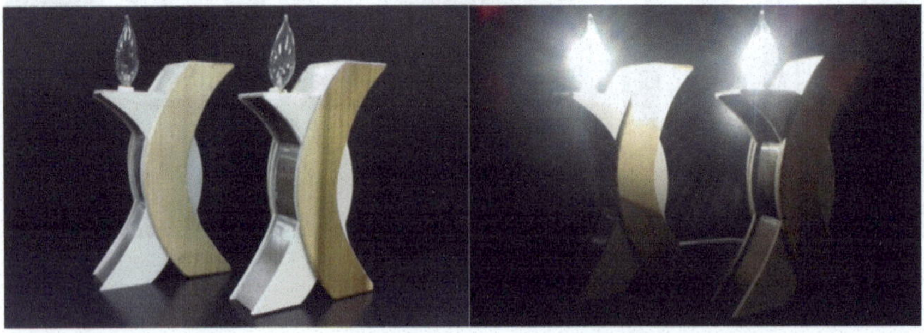

Fig. 8 "Colibrí" Luminaire, by Ailin Terrazas

This also helps to share the recommendations with other designers and engineers.

2. The designers felt the concepts were easy to understand. It means that the difficulty will not disserve the applicability of these recommendations. The doubts will not stop designers to apply them.
3. From the study group, only a 15% declared to have notions on what emotional design is; the rest admitted that it was a new concept for them. This could be a problem for the recommendations, as the designers jumped into a field they knew nothing about, yet, this also means that it was a perfect chance to test the capabilities of the recommendations when introducing designers into the world of emotional design. The recommendations present an opportunity for the designers to have an expertise in emotional design and improve their products.
4. Eighty percent of the study group feels that they now have good fundaments on how to conceptualize based on emotional design. The rest believes to have gained a medium amount of knowledge about it. Nevertheless, information related to point three underlines that these recommendations had improved the knowledge of the participants, and they felt sure about the application of them.
5. Eight out of nine designers confessed that they are not used to implement the conceptualization phase on their design processes. This might affect the study in two manners: on one side, not being familiar with the conceptualization process might make a first contact tedious or useless; on the other side, it grants the possibility to generate a fundamental working habit on the design process of these individuals. If they learn and internalize this habit, later on, they can work starting from a good base.
6. The entirety of the study group feels that the recommendations do manage to provide a good starting point for their design conceptualization. They believe that emotional concepts are well presented and grade the recommendations as either "useful" or "very useful". Furthermore, they claim that they would use it again and that they would recommend them to other designers. It is a positive result, in the sense of sharing knowledge and propose improvements in the future.
7. The study group felt that there is space for improvement, as a third of the designers spoke about the need for more and more explicit examples for each of the concepts on the recommendations. These examples can summarize all the information and promote a better comprehension, according to the different ways or manners of design.

4 Conclusions

Although the recommendations are at an early stage, five out of eight of the resulting projects managed to incorporate values and concepts related to emotional design. Several of the projects concluded in the design of a luminaire based on biomimetic. It could be considered a positive result if it is taken into account that almost all of the student designers had no real knowledge of emotional design and no real knowledge of how to conceptualize.

Students who applied some of the recommendations on their products qualified them as useful, easy to understand, easy, and fun to read. Moreover, they agree that the recommendations are a good starting point for their design processes when aiming for emotional responses.

Finally, according to resulting products designed by students, it seems that these recommendations can help industrial designers to transmit emotional concepts to products. These strategies could be of great interest for industries to develop products which could attract new consumers by addressing their emotional desires and needs.

References

Álvarez, A., Pérez, R., Aguilera, O., & Riba, C. (2008). Análisis conceptual del diseño de gradas a través del método Kano. *Rev Ciencias Técnicas Agropecu, 17,* 31–35.

Aumer-Ryan, P. (2005). *Understanding emotional design: origins, concepts and implications.*

Beitia, A., Vergara, M., González de Heredia, A., Beitia, A. (2009). Ingeniería Kansei: La Influencia de la Escala en la Aplicación del Diferencial Semántico. In: *XIII Congreso Internacional de Ingeniería de Proyectos. Badajoz*, pp. 1811–1822

Bradley, M., & Lang, P. (1994). Measuring emotion: the self-assessment manikin and the semantic differential. *Journal of Behaviour Psychiatry, 25,* 44–59.

Calvillo Cortés, A. B., & Falcón Morales, L. E. (2016). Emotions and the urban lighting environment. *SAGE Open*. https://doi.org/10.1177/2158244016629708.

Carbon, C. C., Faerber, S. J., Gerger, G., Forster, M., & Leder, H. (2013). Innovation is appreciated when we feel safe: On the Situational dependence of the appreciation of innovation. *International Journal of Design, 7,* 43–51.

Crilly, N. (2011). Do users know what designers are up to? Product experience and the inference of persuasive intentions. *International Journal of Design, 5,* 1–15.

Desmet, P., & Fokkinga, S. (2013). Ten ways to design for disgust, sadness, and other enjoyments: A design approach to enrich product experiences with negative emotions. *International Journal of Design, 7,* 19–36.

Desmet, P, van Erp J, Karlsson MA (2008) Design and Emotion Moves. Cambridge Scholars Publishing

Desmet, P., Wassink, P., & Susa Group (2012). PrEmo (Product Emotion Measurement Instrument). http://studiolab.ide.tudelft.nl/diopd/library/tools/premo-product-emotion-measurement-instrument/. Accessed 16 Feb 2017

Forlizzi, J., Saensuksopa, T., Salaets, N., Shomin, M., Mericli, T., Hoffman, G. (2016). Let's be honest: A controlled field study of ethical behavior in the presence of a robot. In: *25th IEEE International Symposium on Robot and Human Interactive Communication*, ROMAN, pp. 769–774.

Gray, J. R., Braver, T. S., & Raichle, M. E. (2002). Integration of emotion and cognition in the lateral prefrontal cortex. *Proceedings of the National Academy of Sciences of the United States of America, 99,* 4115–4120. https://doi.org/10.1073/pnas.062381899.

Hashim, A. M., & Dawal, S. Z. M. (2012). Kano model and QFD integration approach for ergonomic design improvement. *Procedia Social and Behavioural Sciences, 57,* 22–32. https://doi.org/10.1016/j.sbspro.2012.09.1153.

Hepach, R. (2011). Conceptualizing emotions along the dimensions of valence, arousal, and communicative frequency—implications for social-cognitive tests and training tools. *Frontiers in Psychology, 2,* 1–9. https://doi.org/10.3389/fpsyg.2011.00266.

Hirata, R., Nagamachi, M., Ishihara, S. (2004). Satisfying emotionals needs of the beer consumer through kansei engineering study. In: *Proceedings of the 7th International QMOD Conference*. University of LinkÖping and ITESM.

Kalashnikov, V., Kalashnykova, N. I., Acosta Sánchez, Y. G., & Kalashnikov, V. V. (2014). Affective engineering in application to bi-level human migration models. In J. Watada, H. Shiizuka, K.-P. Lee, T. Otani, & C.-P. Lim (Eds.), *Industrial Applications of Affective Engineering* (pp. 27–38). Cham: Springer International Publishing.

LeDoux, J. (2000). Emotion circuits in the Brain. *Annual Review of Neuroscience, 23,* 155–184.

Lloyd, P., Hekkert, P., van Dijk, M. (2006). *Vision in product design.*

Mondragón Donés, S., Vergara Monedero, M., & Company Calleja, P. (2006). Diferencial Semántico: Una Herramienta al Servicio del Diseño Emocional de Máquinas Herramientas. *XVIII Congreso Internacional de Ingeniería Gráfica* (pp. 124–125). Barcelona: España.

Nepal, B., Yadav, O. P., & Murat, A. (2010). A fuzzy-AHP approach to prioritization of CS attributes in target planning for automotive product development. *Expert Systems with Applications, 37,* 6775–6786. https://doi.org/10.1016/j.eswa.2010.03.048.

Norman, D.A. (2004). *Emotional design: why we love (or hate) everyday things.*

Norman, D.A. (2014). Emotional design: People and things. In: jnd.org. https://www.jnd.org/dn.mss/emotional_design_pe.html. Accessed 1 May 2014

Roberts, K. (2005). *Love marks, the future beyond brands*. Power House Books

Sacharin, V., Schlegel, K., Scherer, K. (2012). *Geneva emotion wheel rating study*. Geneva

Shahin, A. (2004). Integration of FMEA and the Kano model: An exploratory examination. *International Journal of Quality and Reliability Management* 21, 731–746. https://doi.org/10.1108/02656710410549082

Shen, X. X., Tan, K. C., & Xie, M. (2000). An integrated approach to innovative product development using Kano's model and QFD. *European Journal of Innovation Management, 3,* 91–99. https://doi.org/10.1108/14601060010298435.

Spillers, F. (2004). Emotion as a cognitive artifact and the design implications for products that are perceived as pleasurable. *Experience Dynamics* https://www.experiencedynamics.com/about. Accessed 13 June 2014

Spillers, F., & Asimakopoulos, S. (2014). Does social user experience improve motivation for runners? *A Diary Study Comparing Mobile Health Applications*. Lecture Notes in Computer Science (including subseries Lecture Notes in Artificial Intelligence and Lecture Notes in Bioinformatics) 8520 LNCS pp. 358–369. https://doi.org/10.1007/978-3-319-07638-6_35

Tractinsky, N. (1997). Aesthetics and apparent usability: Empirically assessing cultural and methodological issues. In: *Proceedings of the ACM SIGCHI Conference on Human Factors in Computing Systems*. ACM, Atlanta, pp. 115–122

Tractinsky, N., Katz, A. S., & Ikar, D. (2000). What is beautiful is usable. *Interacting with Computers, 13,* 127–145. https://doi.org/10.1016/S0953-5438(00)00031-X.

Wang, T., Ji, P. (2010). Understanding customer needs through quantitative analysis of Kano's model. *International Journal of Quality and Reliability Management* 27, 173–184. https://doi.org/10.1108/02656711011014294

Indexing and Mapping Examples of Heuristics Compiled from TRIZ

Emanuela L. Silveira, Marco A. de Carvalho and Júlio C. de Carvalho

Abstract TRIZ is recognized as an effective and systematic approach for inventive thinking and includes several tools, including Inventive Principles, Standard Solutions and Separation Principles. These tools are what we term Inventive Heuristics. Heuristics are cognitive strategies that direct problem solvers to areas of the solution space where promising solutions are more likely to be found. Examples of Inventive Heuristics are the Inventive Principle 'Dynamization' and the Standard Solution '1.2.2'. TRIZ contains a relatively high number of Inventive Heuristics (469 were found in a previous study), which requires significant time to be understood, selected and applied. Another problem concerns the existing repetitions of heuristics, which entails significant practical demands, especially for newcomers to TRIZ. To address the last issue, a compilation of Inventive Heuristics was created, thus reducing the total number of Inventive Heuristics to 263, leaving only the original heuristics and eliminating overlaps. This chapter presents an index and examples build on the created compilation, which is structured in a catalogue, thus facilitating its learning and use. This chapter describes the research strategy, the mapping examples, index structuring and the practical use of the Heuristics Catalogue in a case study that is related to solving classical TRIZ problems. The authors expect that the use of the catalogue will simplify the process of applying TRIZ and will result in faster invention and innovation cycles.

Keywords Problem-solving · Ideation · Inventive heuristics · TRIZ

E. L. Silveira · M. A. de Carvalho (✉)
Postgraduate Program in Mechanical and Materials Engineering, Federal University of Technology—Paraná (UTFPR), Av. Sete de Setembro 3165, 80230-901 Curitiba, Paraná, Brazil
e-mail: marcoaurelio@utfpr.edu.br

J. C. de Carvalho
Bioprocess Engineering and Biotechnology Department, Federal University of Paraná (UFPR), Caixa Postal 19.011, 81531-990 Curitiba, Paraná, Brazil

G. Cortes-Robles et al. (eds.), *Managing Innovation in Highly Restrictive Environments*, Management and Industrial Engineering,
https://doi.org/10.1007/978-3-319-93716-8_9

1 Introduction

TRIZ methodology is recognized for its efficient, systematic product ideation and inventive problem-solving (Ilevbare et al. 2013). The central purpose of TRIZ is to foster inventive thought; thus, this methodology has a logical and systematic structure that provides repeatability, predictability and reliability while creating solutions to a problem (Rousselot et al. 2012). Therefore, this method is distinct from intuitive methods, which remain widely used in the ideation process in industry and scientific research, despite the fact that their use often leads to unpredictable or unrepeatable results (Barry et al. 2010).

Ilevbare et al. (2013) emphasize that the TRIZ methodology has been expanding and applied in several universities in more than 35 countries, in addition to a growing number of global companies and organizations, such as General Electric, Procter and Gamble, Intel, Samsung, among others, who find in TRIZ a useful method to stimulate the development of innovative products in a systematic way (Souchkov 2008). The effectiveness of its application in global organizations demonstrates its potential for replication and effectiveness. Betting on its application within the context of Latin America can generate ideas that foster innovation and assist in the economic growth of markets still under development.

According to (De Oliveira 2001), technological training includes the individual ability, as well as the capacity of integration and synergy of an organization, involving technical, managerial and institutional specialties. It depends, among other things, on education and training. It is through training that a nation or region can assimilate the new knowledge and apply it in production processes, as well as creating a favourable environment to technological innovation.

The most commonly used way to envisage the result of investments in technology and technological innovation is through Total Factor Productivity (TFP), since the adoption of new methods and techniques of production, as well as investment in human and physical capital, can generate externalities capable of adding positive effects on economies of scale (CEPAL 2010). In Latin America, the rate of TFP, according to CEPAL, is historically low compared to developed countries.

De Oliveira (2001) emphasizes that to encourage the continuous innovation of the policies of Latin American countries, the following fronts should be emphasized: technology and applied science; reorganization and modernization of the productive sector; viability of permanent instruments, the incorporation of new technologies and international cooperation. In this context, the use of tools that promote and encourage innovation, such as TRIZ, which provide global examples of technological innovation as a reference can become an important step for economic development.

TRIZ application entails the use of many tools and concepts that include inventive principles and combinations, patterns and templates, which can be termed inventive heuristics. The main objective of a heuristic is to aid in problem-solving by making the process more efficient through the sharing of previous experience, and thus to guide the generation of solutions to new problems (De Carvalho et al. 2003).

1.1 Problem

Tessari and De Carvalho (2015) noted that even the inventive heuristics of TRIZ that are used as tools for troubleshooting require significant time to be understood, selected and applied due to their high number (approximately 470). This large number represents an obstacle for newcomers to TRIZ.

To overcome this problem, the authors performed a compilation of heuristics, finally reducing their number significantly (of the 469 heuristics analyzed, 263 remained). However, the authors recognize the need for future studies that satisfy not only the goals of compilation but also the exemplification and indexing of these heuristics.

1.2 Goal

Based on the above discussion, the following study aims to develop a catalogue containing the 263 inventive heuristics of TRIZ that were previously compiled by Tessari and De Carvalho (2015) by indexing and exemplifying these heuristics. The purpose is to contribute to the application of the heuristics by developing a more didactic model while providing textual and visual descriptions for the examples to aid in their understanding and effectiveness when applied by designers who are new to TRIZ.

1.3 Rationale

TRIZ tools allow the generation of better ideas while solving problems effectively and creatively with continuous improvement (Rantanen and Domb 2010). Thus, the use of TRIZ is an excellent way in which companies seeking innovation can consolidate their positions in the market and differentiate themselves from their competitors (Marques 2014). One of the core competencies for training and generating creative ideas is related to the knowledge and skills that are acquired deliberately from outside specific designer practice areas (Epstein et al. 2013; Kwon et al. 2015).

Therefore, it is appropriate to exploit the universal features of TRIZ and its heuristics by mapping examples from other areas such as chemistry, physics, biology and medicine to ensure the most efficient problem resolution process by sharing experience and inspiring designers to explore new directions.

2 Research Method

The research is focused on the qualitative analysis of interpretative characteristics (Caleffe and Moreira 2006). It follows an exploratory approach that enables research on a given issue or situation, which then provides insights for further research (Reis 2008). To validate this procedure, a comparative method is used to analyze and interpret existing similar material in the literature (Marconi and Lakatos 2007).

The comparison process used to validate the catalogue is analogous to the method of numerical model validation. The numerical method produces some results, which are then the subject of a benchmarking process. Then, the results are compared to the results of a pre-established reference model, such as a physical, experimental or other numerical models (Jauregui and Silva 2011). Thus, the strategy to validate the catalogue relies on a reference model that corresponds to previous problems using inventive solutions that are recognized for their effectiveness and innovation within the Classic TRIZ approach.

The chapter has four sections: (1) A bibliographical analysis of the TRIZ heuristics groups compiled by Tessari and De Carvalho (2015). (2) Some mapping examples and analysis of existing indexes. (3) The results of structuring the unified index and examples within the Heuristic Catalogue. (4) A case study to validate the Heuristic Catalogue and demonstrate the guidelines for their use.

2.1 Heuristic Compilation by Tessari and de Carvalho

Tessari and De Carvalho (2015) performed a literature review following a historical line to clarify which of the selected heuristics groups were originally designed in Classic TRIZ. Then to evaluate all others that were developed as 'improvements' of the original groups.

The research conducted by the authors was focused on six main groups of heuristics: (1) The 40 Inventive Principles (PIs) that were developed by Altschuller from 1956 through 1971 (Altshuller, 1998). (2) The 76 Standard Solutions (Altshuller et al. 1999). (3) The 121 Heuristics (De Carvalho et al. 2003). (4) Six heuristics published as 'New Sub-Principles' of the original 40 PIs (Savransky 2000). (5) The combined principles, including 37 combinations of Inventive Principles (Mann 2002). Finally, (6) The 63 templates of the General Theory of Innovation (GTI) (Yezersky 2006).

Before starting the confrontation and compilation of the 469 heuristics present in these six groups, all heuristics were rewritten according to the following syntactic functional pattern to facilitate comparison: <verb> + <object> + <complement>.

In the above standard use, the infinitive answers the question 'what to do?', the object answers 'do it with what?' and the addition answers 'how, when and where to?' Thus, the central intent of each heuristic was immediately assessed, which the authors understand to facilitate the comparison. Tessari and De Carvalho (2015) defined two exclusion criteria for their comparison: i. Heuristics with the same meaning in their descriptions should be eliminated. ii. Heuristics with common examples should be removed. Accordingly, more heuristics are eliminated in each group, and the remaining 263 heuristics are considered unique.

2.2 The Importance of Examples for TRIZ Heuristics

With an analysis of the six groups compiled, it was possible to identify the presence or absence of examples. The lack of an example, or the little care used in its description, makes the process of assimilating the meaning and relevance of heuristics much slower and more difficult for the designer.

Problem-solving through analogies can be defined as a strategy of transferring knowledge from past problem solutions to new problems. Thus, it is possible to relate past experience of resolution and transfer knowledge to the construction of specific problem solutions (Mayer 1992). The examples presented below are intended to demonstrate the importance of examples for the applicability of heuristics and the relevance of their analogies for solving new problems.

The software Innovation Workbench Software System™ (Zlotin and Zusman 2005), presents textual/illustrated examples of solutions for some heuristics. The heuristic 'Inversion—Replace one action with a counteraction' is illustrated in Fig. 1.

The example relates to controlling the athlete's running rhythm during training: Correct control of the race rhythm occurs when the athlete has an understanding of when to run at a moderate or fast pace. The most efficient method of training is then, instead of the athlete running on a track, to have the athlete run on a treadmill while the trainer varies the speed, thus being able to control the athlete's pace precisely.

Fig. 1 Example of the 'Inversion' heuristic (Zlotin and Zusman 2005)

2.3 Researching Examples

The following procedure was applied to achieve the goal of mapping examples for each of the heuristics:

- Identification of examples existing among the references discussed by the precursor authors of each of the six heuristics groups;
- Analysis and selection of complementary examples within the literature;
- Searching for new examples or improved examples when (1) The existing examples do not clarify the relationship between the example and the heuristics and/or when the description is very superficial; (2) When existing examples are too complex and/or difficult to understand due to insufficient textual and/or visual explanations; (3) When no examples are available for the heuristics in the literature.

For heuristics relating to Combined Principles, no examples were found in the literature. Thus, this group covering '46 Combined Principles' required a search targeting new examples.

The method used to find new examples included the use of symbolic analogy (Back et al. 2008). This method involves the use of keywords that express the condensed form, function and attributes of each heuristic. The search procedure for new examples can be described in the following steps:

- Choosing keywords representing the function and attribute that are directly associated with the terms used in the heuristics description.
- Using synonyms, as well as more specific and more general terms and attributes, to perform a new query and reach satisfactory results. For instance, for the heuristic regarding the Principle Combined '41—Reducing the weight of the individual parts', a keyword that was directly related to a function used while searching for the examples was 'reduced weight' (Reduce weight); further synonyms and/or words with similar meanings were explored such as 'Decrease weight', 'Reduce dimensions' and 'lighten'. The attribute function 'Structural Optimization' was also used to search for examples—Fig. 2;

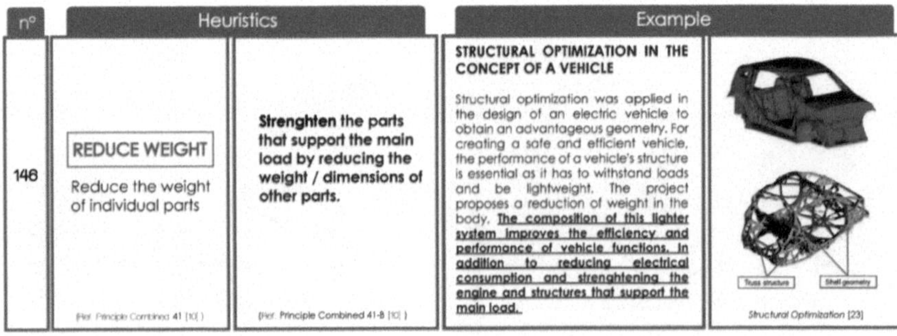

Fig. 2 Catalogue's heuristic structures with example

- Mapping and selecting potential examples relating to the heuristics;
- Writing a description of the example that emphasizes the relevant features of the system and their connection to the heuristics;
- Finding visual examples (images, graphics, photographs, etc.) that correspond to the selected example and are easily assimilated.

The research process was focused on situations, processes and real products in which enhanced creativity and greatly improved competitiveness due to the innovation were evident. The research sources including online databases, patents, online magazines and books, among others.

2.4 *Analysis of Indexes*

The compiled heuristic groups are fundamental to the composition of the unified heuristic catalogue. Thus, we compared the proposed indexes in five of the groups described above, excluding only the group of 'Six Principles' by (Savransky 2000), which has no index indication, as shown in Table 1.

This analysis allowed the identification of existing common indexes in different heuristics groups as follows:

- The classes 'Transformation in Space' and 'Transformation in time' are presented in three different groups: i. 121 Heuristics (De Carvalho et al. 2003); ii. Templates (Yezersky 2006); and iii. The inventive principles index for physical contradictions (Mann 2002);

Table 1 Comparison between existing indexes

40 inventive principles	Standard solutions	121 Heuristics	Combined principles	Templates
1. Transformations in space	1. Building and destroying su-field models	1. Shape transformations	1. Physical performance	1. Addressing a problem at a higher system level
2. Transformations in time	2. Enhancing su-field models	2. Structure transformations	2. Efficiency	2. Problems related to time
3. Separation on condition	3. Transition to the super-system and micro-levels	3. Transformations in space	3. Ility	3. Problems related to space
4. Transition to an alternative system	4. Standard solutions for detection and measurement	4. Transformations in time	4. Manufacture/ cost	4. Prevent a negative event from happening

(continued)

Table 1 (continued)

40 inventive principles	Standard solutions	121 Heuristics	Combined principles	Templates
4.1 Transition to a subsystem	5. Standards for applying standard solutions	5. Transformations of movements and forces	5. Measurement	5. Incorrect range/output function
4.2 Transition to a super-system		6. Transformations of a material		6. Poorly controlled action
4.3 Transition to an alternative system		7. Expedients of differentiation		7. Standard restriction on a potential solution
4.4 Transition to an inverse system		8. Quantitative changes		

- The class 'Transition base system for a super-system or a micro-level (subsystem)' can also be found with similar nomenclature in three groups: i. Inventive Principles (Mann 2002); ii. Standard Solutions (Altshuller et al. 1999); and iii. Templates (Yezersky 2006);
- The class 'Measurement and Detection' is suggested in two groups: i. Standard Solutions (Altshuller et al. 1999) and ii. Combined Principles (Mann et al., 2010).

(De Carvalho et al. 2003), also, compare and classify 40 Inventive Principles within the proposed classes for the 121 heuristics. The results obtained show that these heuristics are not directly related but are complementary, emphasizing the importance of the existence and interaction of each heuristic group.

2.5 Previous Similar Studies

Calle-Escobar et al. (2014) Proposed a methodology that consolidates heuristic rules and application examples from different areas (e.g. engineering, industrial design and biology). This approach encompasses three phases. The first one, structuring, is related to the understanding of the problem itself (decomposition of the problem into functional blocks). The second, the formalization of the problem, where the subsystem to be redesigned is selected, and the cause and effect method is used to analyze the selected subsystem. The third, resolution, is where the implementation of the 78 heuristic rules occurs. It is suggested to carry out a brainstorming later. The methodology proposed allegedly enables designers to understand how the different principles of solution can be integrated into their creative processes.

Daly et al. (2012) proposed the 'Heuristic Project'. The study carried out by the authors started with the analysis of the ideation process of engineers and designers. The focus is on developing the ability of students and professionals to design innovative solutions to the increasingly complex problems in today's world (DUDERSTADT 2008; Sheppard et al. 2009). According to the authors, when considering a diverse set of potential concepts, the designer is more likely to come up with new and innovative solutions (Brophy 2001; Liu et al. 2003). However, designers often focus on specific options generated early in the design process. In this context, the set of heuristics identified by the authors in previous empirical studies provided instructions for the development of 'Cards' to assist designers in search of new ideas. The strategy used in the 'Cards' includes the description of the heuristic, represented by an abstract image which describes its application, and two examples of products that show how the heuristic is evident in existing consumer products.

According to Daly et al. (2012), the cards developed in the 'Heuristic Project' are not design rules that should be followed, but strategies to facilitate the process of ideation itself, increasing the variety of combinations and developing ideas in unusual directions. As a result, it was observed that the designers who used the cards explored different alternatives, generating much more creative concepts. The authors consider that the uses of heuristics, together with their examples, have a high potential for generating new and creative solutions.

2.6 Considerations on the Theoretical Framework

Considering the heuristics compiled by Tessari and De Carvalho (2015), some examples were found for five of the six groups studied. However, in many cases, existing examples had a superficial description, not enough to illustrate the heuristic and even some of these examples were too complex, and then, difficult to conceptualize simply. So, the need to search for more didactic examples became clear.

The two models analyzed for similar studies, in Sect. 2.6.1, allowed an understanding of how such an approach could be effective. However, it is important to emphasize that none of the approaches demonstrates the use of an index to organize heuristics, which can make their selection and application difficult in a practical way. Besides, the heuristics studied by these authors do not encompass all compiled inventive heuristics of TRIZ.

The previous sections allowed the identification of points for improvement in the inventive heuristics of TRIZ because they require significant time to be understood, selected and applied due to their high number. It is evident the importance of standardizing an index that unifies the heuristics compiled by Tessari and De Carvalho (2015), while at the same time making its application accessible even to less experienced designers. The new examples found proved to be of great relevance by making the understanding of heuristics clearer.

It is understood that the use of examples allows a better comprehension of heuristics and the generation of analogies with processes and products already implemented. Although the examples available are not of the specific area of the designer, they can foster and stimulate creativity and then, to facilitate the ideation process in another project.

2.7 Results

This section presents the final structure of the unified index and images that graphically demonstrate some classes and examples within the proposed Heuristic Catalogue.

2.8 Unified Index

In the analysis described above, it was considered that the structure of an index with more specific categories, focused on individual functions of each heuristic. Although relevant, could hinder the connection and classification process of different groups due to the high number of heuristics. Furthermore, it was observed that more specific classes, containing special features of each heuristic, would form an index exhibiting a wide variety of parameters, but containing a smaller number of these heuristics. This factor could hinder a designer's decision due to the variety of classes to choose from, and the risk remains of not finding within the class the best heuristics to solve a problem. Hence, the decision was to adopt a uniform index model with broader classes.

However, with a more embracing index, the number of heuristics to be analyzed for each class is greater. This makes the selection process more time-consuming for the designer but increases the chances of finding a more appropriate solution. Thus, ten classes were defined for the 263 compiled heuristics. Figure 3 presents the proposed unified index and the associated classes:

Table 2 contains a description of each Heuristic Class.

2.9 Directing the Heuristics for Each Class

The targeting process by which each heuristic was compiled into classes involved an individual analysis of their features and functions, thus assigning them to classes of greater affinity. Also, the original classifications proposed by Altshuller et al. (1999), Mann (2002), De Carvalho et al. (2003), Yezersky (2006), Mann et al. (2010), was considered. That is, the classes that were shared between indexes described by different authors were included in a combined index, and their

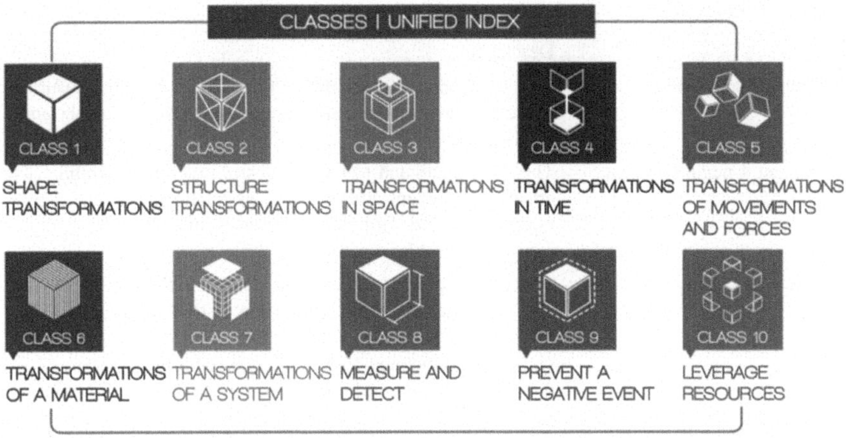

Fig. 3 Heuristic classes and unified index

Table 2 Description of heuristic classes

Classes	Description
1. Shape transformations	Considers the physical properties of the external geometry and the internal characteristics of the system. This class is aimed at improving ergonomics, optimizing the use of the raw material, and increasing safety, among others
2. Structure transformations	This class refers to the provision and organization of the essential elements that comprise the system and includes system optimization, standardization and improving reliability, among others
3. Transformations in space	This 'space' comprises the area occupied by a system in one dimension, distance, area or volume. The class is aimed at eliminating the need for an object, the use of unoccupied space and the use of another dimension, among others
4. Transformations in time	Heuristics in this class suggest ways of solving problems that relate to the need for additional time, time control and the postponement of action, among others
5. Transformations of movements and forces	This class is related to the action of a system in terms of carrying out tasks and includes the reduction of effort and energy losses and altering the frequency of movement by eliminating needless actions, among others
6. Transformations of a material	This class is related to the resolution of problems related to the incompatibility, waste, inefficiency and low reliability of the materials that comprise each system
7. Transformations of a system	This class proposes the systematic analysis of problematic situations by considering the relationships within and between systems at different levels and aims to modify parts to improve function and reliability and differentiating between systems, among others

(continued)

Table 2 (continued)

Classes	Description
8. Measure and detect	This class relates to the difficulty of detecting and measuring processes such as the inspection or analysis of complex and costly operations
9. Prevent a negative event	This class is related to solving issues involving harmful effects, insufficient and/or excessive relationships, unnecessary and dangerous influences, among other interaction problems within the project
10. Leverage resources	This class focuses on restrictions related to the introduction of new substances or fields within the system and proposes the use of existing resources and environmental fields, such as phase transitions and adding elements and fields that decompose, among others

heuristics were grouped. Certain heuristic relationships represent more than one category and were therefore repeated; the same example was used in the different classes.

In Fig. 4, the distribution percentage of the heuristics in each class is shown.

Each class grouped from 15 to 37 Main Heuristics. Groups which presented repeated heuristics within the classes include the Inventive Standards, Combined Principles, Templates, and especially, the Inventive Principles. This is mainly due to the more general and comprehensive character of this heuristic group.

In Fig. 5 The graphic structure of the Heuristics Catalogue is illustrated.

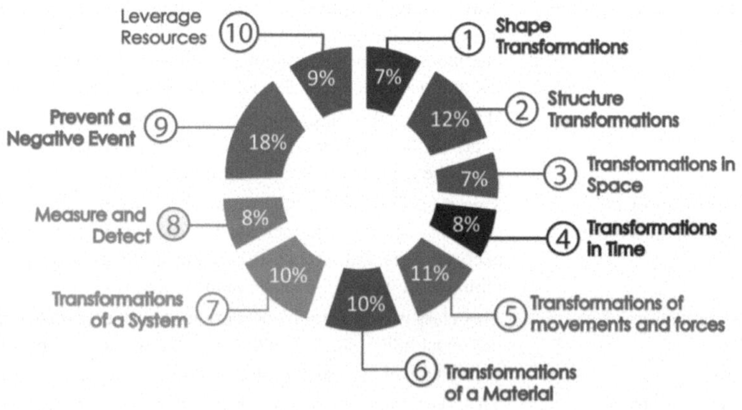

Fig. 4 Heuristic distribution per class

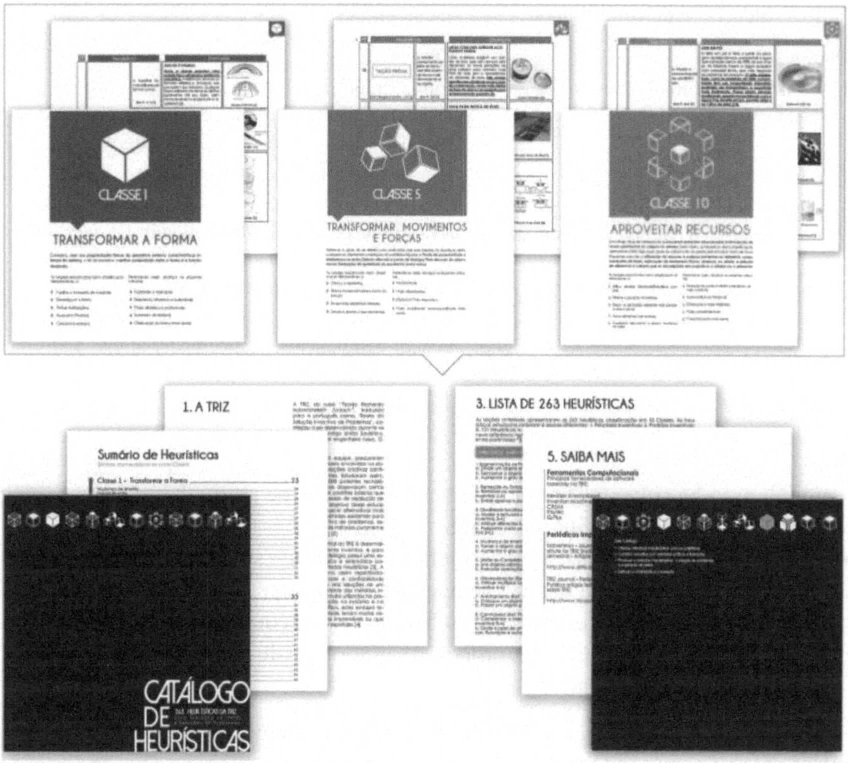

Fig. 5 Graphical structure of the heuristic catalogue

2.9.1 Heuristics Catalogue Guidelines

The following steps demonstrate how designers and problem solvers can optimize the use of the compiled heuristics catalogue, and select the class most appropriately within the unified index. The application of steps 1 to 3 can be seen in the case study of this chapter, in Sect. 4.

- Step 1: Define the Problem: The guideline proposes a form that can support the description of the Problem (Table 3). The form is based on TRIZ tools such as: 'Problem Hierarchical Explorer' (Basadur 1996), 'Window Operator' or 'Ideal End Result' (Mann 2002);
- Step 2: Solution search: Define 'What is the cause of the problem?' And 'Where is the solution opportunity?' The response should be directed to select one of the unified index classes.
- Step 3: Selection of Heuristics and Particularization: after class selection, designers will have access to heuristics with examples of specific solutions. These should be analyzed abstractly to find functional affinities with the specific

Table 3 Step 1—problem definition

Questions	Response
1. What is the original problem (be specific)?	Curved pipe pieces (elbows) in shot-blasting machines (steel) are damaged due to the speed at which the grit passes through the elbows
2. What is the general problem?	Elements that rub and generate wear
3. What is preventing us from solving the problem? What else is a problem?	The use of more resistant pipes would increase the cost. Input protection elements will be damaged by grit when travelling at speed. However, the use of grit increases blasting efficiency
4. What is needed to solve this problem?	Increase the blasting efficiency, prevent damage to system components, reduce material loss and reduce costs
5. What is the ideal final result?	The pipe protects itself

problem to be solved. The selected heuristics must be particularized, generating ideas to solve the specific problem;

- Step 4: Generated Ideas: If the solutions found are still not satisfactory, it is recommended to define the problem again and/or select a new class or heuristic. However, if ideas are considered sufficient, it is recommended to apply evaluation methods to identify the proposed solution with the greatest potential for subsequent implementation.

The systematic process depicted by the guidelines indicates how to proceed to obtain an efficient and creative concept for problem-solving. The process of abstraction and particularization assists in the generation of analogies to solve the specific problem, allowing a more efficient compared with the solutions of other well-solved similar problems.

3 Case Study

The following case study proposes a comparison to validate the notion that the ideas generated by the catalogue are consistent and resemble reference solutions that were previously proposed in the Classic TRIZ literature.

3.1 Case Study 'Pipe for Blasting Grit'

Grit blasting is a classic method for cleaning or polishing metal. However, curved sections known as 'elbows' in pipes made of polymeric material that are present in certain steel shot-blasting machines are damaged due to the speed and pressure with which the grit passes through the pipe. To protect the pipes, some coating should be

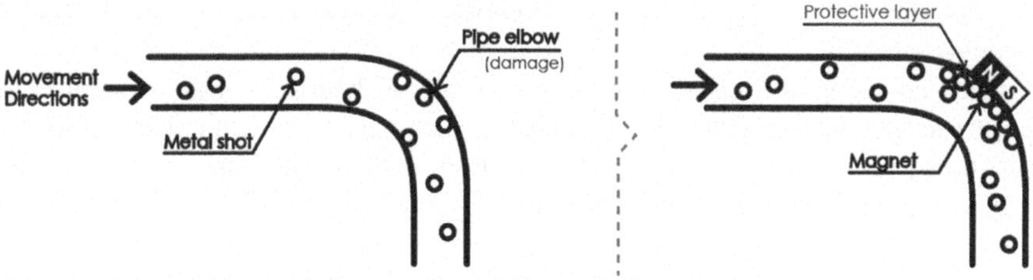

Fig. 6 Grit being transported in pipe elbow (Fey and Rivin 2005)

used; absent such a need, no coating should be used because such coatings would be eventually damaged and increase the cost. This physical contradiction needs to be addressed (Savransky 2000).

Classic TRIZ suggests the following inventive solution: The grit can itself serve as a coating. A magnet placed on the outside of each curve such that it covers the wear region can extract some of the grit particles from the stream, thus forming a layer of protection inside the elbow (Fig. 6).

To reach this solution, the use of the Standard Solutions method covering the following heuristics (Standard Solutions 1.2.2) is proposed: 'Eliminate harmful interactions by introducing a 'new' substance that is the modification of an existing substance'. Thus, the 'new' substance added to protect the elbow is not a foreign object but is part of the system itself (the grit) (Fey and Rivin 2005).

In this case study, the solution suggested by TRIZ is used as a reference due to the features achieved by these systems, such as simple implementation, low cost, longer system life, ease of maintenance, and greater controllability, among others.

3.2 Heuristics Catalogue Application

Considering the solution reached using Classic TRIZ for solving the wear problem of grit-blasting pipes as a reference, the application of the Heuristics Catalogue was tested to ascertain whether its use could lead the user to converge to the same solution.

To ensure the proper use of the Heuristic Catalogue, usage guidelines were developed. These guidelines are divided into three steps and are exemplified using this case study.

Step 1 refers to problem definition. To achieve this, a questionnaire was proposed. The questions are listed in Table 3, together with some answers that are appropriate for the case study.

By describing the specific problem, it is possible to understand that the granular component, although causing the problem, should be maintained in the system to ensure the efficiency of the blasting process. Thus, it is understood that for the problem to be resolved, is necessary to prevent the harmful interaction that occurs.

Thus, in step 2 (choosing the class), we chose to analyze heuristic class 9: Avoid negative events.

After choosing the class, designers can access heuristics with examples of specific solutions. These should be analyzed abstractly to find functional affinities with the specific problem. In step 3, it is proposed to individualize the selected heuristics, thus generating ideas to solve a specific problem set.

Three heuristics were selected for problem solution generation. These were considered potential solutions because they allow the generation of coherent ideas based on the proposals of Classic TRIZ and encourage new ideas.

The first heuristic selected, '28—Mechanical Media Replacement' (refer to Inventive Principle 28) suggests the use of electromagnetic fields, among others, to interact with objects. The elbow and grit were considered as mechanical objects in the problem. In this case, the index suggests the creation of an electromagnetic field. For this, a possible solution would be to introduce a magnetic elbow to attract the beads to the pipe wall, thus forming a granular blanket that would absorb the impact energy of other metal particles; this solution is consistent with that presented by Classic TRIZ in Fig. 7.

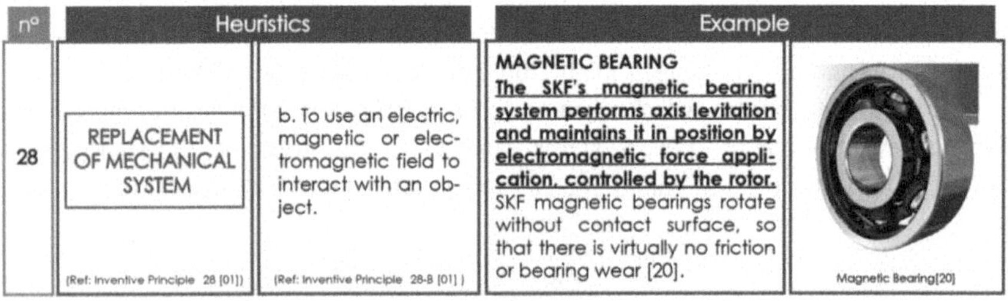

Fig. 7 Heuristic 28—replacement of mechanical system

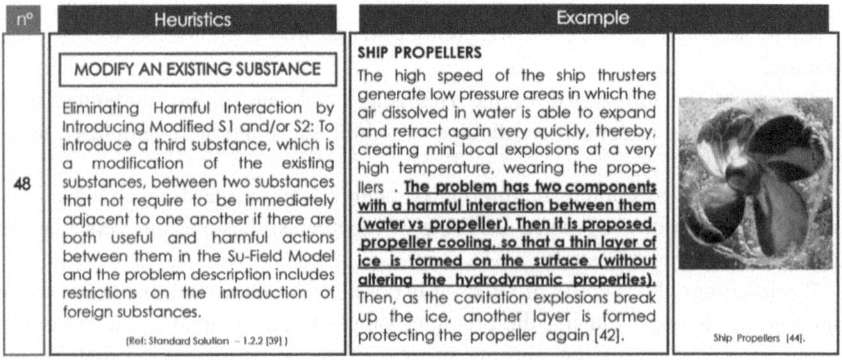

Fig. 8 Heuristic 48—modification of an existing substance

n°	Heuristics	Example	
56	USE MAGNETIC ADDITIVES Complex Ferro-Field Models: To use additives (such as a coating) to give a non-magnetic object magnetic properties. May be temporary or permanent. (Ref:Standard Solutions – 2.4.5 [39])	**MAGNETIC PAINT** A ferrous product applied to the surface of non-magnetic object to give magnetic properties to it. Magnetic attraction is the result of tiny iron particles that are mixed with paint. The higher the concentration of ferrous particles, the greater the magnetic flux and attraction. It can be applied to various surfaces of non-magnetic materials such as wood, metal, masonry, drywall, plaster, among others [53].	 Magnamagic [53]

Fig. 9 Heuristic 46—using magnetic additives

The second heuristic, '48—Modify an existing substance' (refer to Standard Solution 1.2.2), Fig. 8, proposes the elimination of harmful interactions by introducing a 'new' substance that represents a modification of existing substances. This heuristic matches the same solution like that proposed by Classic TRIZ, showing a possible convergence of solutions. It is worth noting that the solution for 'ship thrusters', an idea that relates to the generation of an ice layer, has been renewed; that is, the avoidance of contact and damage. This idea could be implemented in the piping system as a whole or just the elbows of a sandblasting machine.

The third heuristic selected, '56—Using magnetic additives as a coating on non-magnetic objects' (refer to Standard Solution 2.4.5), suggests the use of additives in the form of a coating to confer magnetic properties (temporary or permanent) on a non-magnetic object. This heuristic may also promote ideas that are similar to those proposed by Classic TRIZ; i.e. although a polymer-based pipe shows no magnetic properties, the introduction of a magnet to its outer surface would provide the magnetic properties needed to attract grit, thus forming a protective layer that may be temporary (Fig. 9).

Moreover, the example of 'Magnetic Coatings' proposed in Heuristic 56 could easily promote the generation of similar ideas, such as the painting of pipe elbows with magnetic ink, which, given its concentration of ferrous microparticles, would attract the grit to the elbows, thereby protecting it from subsequent abrasive contact.

3.3 *Discussion*

In the case study, the Heuristics Catalogue generated ideas that were similar to the reference proposals generated by Classic TRIZ. In addition to similar ideas, alternatives were found, such as the suggestion of 'elbow freezing' (forming a protective ice layer that is renewed during the process). Thus, it is reinforced not only the potential of the catalogue in the ideation but of the TRIZ itself, since the heuristics present in the catalogue came from it.

Based on the Heuristics Catalogue, the heuristics of the different groups could be viewed in a unified manner, thus enabling the beginner designer using TRIZ to become aware of the potential of six different heuristics groups.

During the validation process, the textual and visual description of the present catalogue example assisted in understanding the heuristic and stimulated the generation of analogies. However, the abstraction process must be carefully performed such that potential heuristics are not incorrectly discarded by designers when they do not perceive a direct relationship between the example and a specific problem.

4 Conclusion

In the context of the constant search for innovative solutions, this study has demonstrated the inventive and systematic potential of TRIZ, thus emphasizing the cognitive potential of the associated inventive heuristics and the generation of ideas and the space exploration of different solutions.

The process of mapping examples played a key role in the understanding of certain complex heuristics, thus creating a more didactic presentation and stimulating the analogous reasoning of designers starting to work with TRIZ. Therefore, this study is believed to have assisted the lack of systematization and exemplification in supporting the collected heuristics.

Thinking in the context of Latin America, the application of such a catalogue can encourage technological development, and this can create opportunities to promote economic development. It is understood that the study is limited to one aspect of innovation, i.e. that of the invention. However, it represents an initial step, which in conjunction with public policies and research incentive programs can assist in the development of the region.

Acknowledgements The authors thank UTFPR, FUNTEF-PR, Fundação Araucária, CAPES, and CNPq for providing the infrastructure and financial support for this research.

References

Altshuller, G. (1998). *40 principles: TRIZ keys to technical innovation*. Worcester, MA: Technical Innovation Center.

Altshuller, G., Zlotin, B., Zusman, A., & Philatov, V. (1999). *Tools of classical TRIZ*. Southfield, MI: Ideation International.

Back, N., Ogliari, A., Dias, A., & Silva, J. D. (2008). *Projeto integrado de produtos: planejamento, concepção e modelagem*. Barueri (in Portuguese): Manole.

Barry, K., Domb, E., Slocum, M.S. (2010). *Triz Journal*. https://triz-journal.com/404.asp?action=print&404. Accessed 13 Feb 2016.

Basadur, M. (1996). *The power of innovation: how to make innovation a way of life and put creative solutions to work*. London: Pitman Publishing.

Brophy, D. R. (2001). Comparing the attributes, activities, and performance of divergent, convergent, and combination thinkers. *Creativity Research Journal, 13*(3–4), 439–455.

Caleffe, L. G., & Moreira, H. (2006). *Metodologia da pesquisa para o professor pesquisador*. Rio de Janeiro (In Portuguese): DP&A.

Calle-Escobar, M., Mejía-Gutiérrez, R., Nadeau, J.P., Pailhes, J. (2014). Methodology for the implementation of heuristics in the design process. in *Virtual Concept Colombian International Workshop*, Mar 2014, (pp. 1–7) (in Spanish) Medellin, Colombia.

CEPAL (Comisión Económica para América Latina y el Caribe) (2010), La hora de la igualdad: brechas por cerrar, caminos por abrir (LC/G.2432(SES.33/3), Santiago de Chile (In Spanish).

Daly, S. R., Yilmaz, S., Christian, J. L., Seifert, C. M., & Gonzalez, R. (2012). Design heuristics in engineering concept generation. *Journal of Engineering Education, 101*(4), 601–629.

De Carvalho, M. A., Savransky, S. D., & Wei, T. C. (2003). *121 Heuristics for solving problems*. Morrisville, NC: Lulu.

De Oliveira, G.B. (2001). Algumas considerações sobre inovação tecnológica, crescimento econômico e sistemas nacionais de inovação. Revista da FAE, 4(3) (In Portuguese).

DUDERSTADT, J. (2008). Engineering for a changing world: A roadmap to the future of engineering practice, research, and education. Holistic Engineering Education. Ann Arbor, MI: University of Michigan.

Epstein, R., Kaminaka, K., Phan, V., & Uda, R. (2013). How is creativity best managed? Some empirical and theoretical guidelines. *Creativity and Innovation Management, 22*(4), 359–374.

Fey, V., & Rivin, E. (2005). *Innovation on demand: new product development using TRIZ*. Cambridge, UK: Cambridge University Press.

Ilevbare, I. M., Probert, D., & Phaal, R. (2013). A review of TRIZ, and its benefits and challenges in practice. *Technovation, 33*(2), 30–37.

Jauregui, R., & Silva, F. (2011). Numerical validation methods. *Numerical analysis—theory and application* (pp. 155–174). InTech: Rijeka.

Kwon, J., Lee, M., & Kim, H. R. (2015). Does a creative designer necessarily translate into the creative design of a product? Exploring factors facilitating the creativity of a new product. *Creativity and Innovation Management, 24*(4), 675–692.

Liu, Y. C., Chakrabarti, A., & Bligh, T. (2003). Towards an 'ideal' approach for concept generation. *Design Studies, 24*(4), 341–355.

Mann, D. (2002). *Hands-on systematic innovation*. Clevedon: Creax Press.

Mann, D. L., Dewulf, S., Zlotin, B., & Zusman, A. (2010). *Matrix 2010: updating the TRIZ contradiction matrix*. Ieper: Creax Press.

Marconi, M.D.A., Lakatos, E.M. (2007). Metodologia científica, Vol. 4. Atlas, São Paulo (In Portuguese).

Marques, J.F. (2014). Aplicação da metodologia TRIZ e da manutenção autónoma em atividades de manutenção industrial. Dissertation, Faculdade de Ciências e Tecnologia da Universidade Nova de Lisboa (In Portuguese).

Mayer, R.E. (1992). *Thinking, problem solving, cognition*. WH Freeman/Times Books/Henry Holt & Co.

Rantanen, K., & Domb, E. (2010). *Simplified TRIZ: new problem solving applications for engineers and manufacturing professionals*. Boca Raton: CRC Press.

Reis, L. G. (2008). *Produção de monografia da teoria à prática—o método educar pela pesquisa (MEP)*. Brasília (In Portuguese): SENAC.

Rousselot, F., Zanni-Merk, C., & Cavallucci, D. (2012). Towards a formal definition of contradiction in inventive design. *Computers in Industry, 63*(3), 231–242.

Savransky, S. D. (2000). *Engineering of creativity: introduction to TRIZ methodology of inventive problem solving*. Boca Raton: CRC Press.

Sheppard, S. D., Macatangay, K., Colby, A., & Sullivan, W. M. (2009). *Educating engineers: Designing for the future of the field*. San Francisco: Jossey-Bass.

Souchkov, V. (2008). *Breakthrough thinking with TRIZ for business and management: An overview*. ICG Training & Consulting, pp. 1–8.

Tessari, R. K., & De Carvalho, M. A. (2015). Compilation of heuristics for inventive problem solving. *Procedia Engineering, 131,* 50–70.

Yezersky, G. (2006). *Creating successful innovations: General theory of innovation and its applications*. Vinci: Institute of Professional Innovators/Università Degli Studi di Firenze.

Zlotin, B., Zusman, A. (2005). Theoretical and practical aspects of development of TRIZ-based software systems. *Paper Presented at Conference TRIZ Future*. Graz, November 2005.

The Use of Affective Computing in the Conceptual Design Stage of New Products

Agustín Job Hernández-Montes, Guillermo Cortés-Robles, Giner Alor-Hernández and Jorge Luis García-Alcaraz

Abstract The innovation process, seen as a set of problem-solving activities needs novel approaches to assist decision-making in the evaluation of different potential solutions. Typical perspective as the Analytic Hierarchy Process, similarity-based techniques, and essential performance indicators find their limits when it is necessary to take into account a crucial factor: emotions. The emotional response to unique product attributes is in fact, a determinant element to succeed in a market. The affective computing paradigm is a recent technology that allows knowing the person's mood by using different strategies: the use of the camera to recognize an image, the tone voice of a person, and the use of different sensors and wearable devices. Some common features are the use of sensors that measure heart rate, the excitement of an individual, and the detection of the bioelectric activity of the brain at the instant of someone sees something specific. The affective computing has become more relevant in the past years. Emotions are fundamental to human experience, influencing cognition, perception, and everyday tasks such as learning, communication, and even rational decision-making. Industry asks for a different effective mechanism to select the best alternative during design and evaluation of products or services, particularly in the conceptual design stage. Decisions at this

A. J. Hernández-Montes (✉) · G. Cortés-Robles · G. Alor-Hernández
Tecnologico Nacional de México-Instituto Tecnológico de Orizaba,
Av. Oriente 9 No. 852, 94320 Orizaba, VER, Mexico
e-mail: ahernandezm@ito-depi.edumx

G. Cortés-Robles
e-mail: gcortes@ito-depi.edu.mx

G. Alor-Hernández
e-mail: galor@ito-depi.edu.mx

J. L. García-Alcaraz
Department of Industrial Engineering and Manufacturing, Universidad Autónoma de Ciudad Juárez, Chihuahua, Mexico
e-mail: jorge.garcia@uacj.mx

G. Cortes-Robles et al. (eds.), *Managing Innovation in Highly Restrictive Environments*, Management and Industrial Engineering,
https://doi.org/10.1007/978-3-319-93716-8_10

stage will determine the primary attributes of a product. The affective computing paradigm can show what characteristics of the product are more attractive to the design stakeholder and allow a review before launching a product.

Keywords Conceptual design · Affective computing · Industrial design Decision-making process

1 Introduction

Computers have begun to acquire the ability to express and recognize emotions. Recent neurological studies underline that emotions play an essential role in human cognition and perception. Thus, research on affective computing not only indicates that it has the potential to assist people but also to enable computers to acquire decision-making skills (Picard 1995). There are several techniques by which affective computing can get the feeling of a person; some useful resources are using a camera to capture one or several images of the user, and from their gestures or reactions to identify, or infer their feeling. Some educational sites use this technique frequently (Montes 2016). The voice is another way of knowing the emotions. This method is occupied by the customer service centers to mention some (Chorianopoulou 2016). The use of video recordings to obtain the feeling of the users in different moments is also a technique used, although not so recommendable in web developments due to the consumption of bandwidth when storing the video (Bond 2017). By using various sensors that can measure the heart rate, arousal level of a person, among other parameters. An example of sensors is the galvanic skin response (GSR), which measures the skin conductance, which correlates with the function of the sweat gland, which, in turn, is controlled by the nervous system. When an individual is frightened or experiences anxiety, there is a rapid increase in skin conductance due to the increased activity of the sweat glands (Vera-Munoz 2008). These are the most commonly used sensors. However, there are some others like electromyogram (EMG), respiration rate (RR), which provide other functionalities and technical resources. Affective computing initially emerged as an educational approach. Nevertheless, today is a research opportunity in other technical areas, particularly interesting in the context of this chapter as a tool for assessing the conceptual product design. With the use of affective computing, it is possible to measure the responses that an individual present when observing a product, and know which characteristics of that product are the most relevant to users. This function is essential since currently, measuring the experience of customers with the goal to predict or merely to identify their preferences with the goal to envelop them in the new products or services design is a complex activity or carried out later when there is a prototype. Focus groups and self-reports where respondents ask about what they like or what features they want in a product are some of the most common strategies to deal with this problem. Despite the usefulness of typical approaches, these methods lack reliability, since they do not capture actual momentary feelings during the evaluation (Hyung-il Ahn 2014). The affective computation paradigm offers new resources to evaluate the different conceptual designs of a product or service. With this information will be possible to know what attributes are in the customer's taste or which create is the one that has the better reception. The implementation of such mechanism estimates an increase of confidence during the selection of a final design even before launching a test with a focus group or launching a product to the marked.

Simultaneously, in the past years a discipline called Computer-Aided Innovation—CAI have appeared, which combines several disciplines with the purpose of increasing the effectiveness in the development of new products (Réne Lopez Flores 2015; Zaripova 2014). CAI approach represents an improvement that responds to the demands of the industry to propose tools that facilitate the development of new products. Initially, the concept of CAI did focus on helping product designers in the early stages of the design process. Currently, CAI systems are perceived as an element capable of transforming business opportunities and customer demands to crystallize these intentions into new products or processes. CAI's methods and tools inspired by different techniques and methodologies such as the theory of inventive problem-solving (TRIZ), the deployment of the QFD quality function, axiomatic design, synectics, mind mapping, brainstorming, lateral thinking, and Kansei engineering, among others.

The goal of these new tools is to help CAI creative performance, with the expectation paradigm shifts using this new category of software tools (Leon 2009). However, the CAI approach currently has several opportunity areas. In the context of this chapter two stand out:

1. CAI systems must implement problem-solving processes based on multi-user collaboration.
2. It is necessary to assimilate different tools to facilitate the evaluation process, to explore the best alternative in the conceptual design of products.

These opportunities are the central theme of this chapter and the essential elements for proposing a research hypothesis: Incorporating affective computation into a collaborative problem-solving process can improve the performance of a CAI-type solution. To explore and demonstrate this proposition, the rest of this chapter includes six sections: A description of the Affective Computing, a brief analysis of the Computer-Aided Innovation approach, an architecture, the evaluation tools, a methodology, a case study, and finally, the conclusion and future work. The second section describes the basic concepts of affective computing; the third section contains a summary of some affective computing works with product evaluation, the fourth element has the main elements of the CAI. The fifth element describes the prototype of an architecture of emotions detection and the sixth part deal with the evaluation tools. The seventh section describes a basic methodology, and finally, eighth section shows a case study. The last section contains the conclusion and the future work.

2 Affective Computing

Affective computing is the computation that relates, arises from, or deliberately influences emotion or other affective phenomena. Emotion is fundamental to human experience, affecting cognition, perception, and everyday tasks such as learning, communication, and even making rational decisions. However, technologists have considerably ignored emotion and created an experience often frustrating for people, partly because the effect has been misunderstood and difficult to measure. Research in this discipline prompted the development of new technologies and theories that deal with the crucial understanding of feelings and their role in the human experience. Affective computing combines engineering and computing with psychology, cognitive sciences, neurosciences, sociology, education, psychophysiology, value-centered design, ethics, and more (Picard n.d.).

Currently, affective computing covers several fields of study, and as mentioned above, affective computing emerged from the need to endow computers with "emotions". In turn, a computer will be able to identify the emotions of users. One of the disciplines where more study takes place is in the educational context. Over the years, this field has evolved and adapted to discoveries and technologies.

The use of affective computing has increased, as well as the various techniques used to obtain the emotion of a person. Day by day all these mechanisms evolve, and new algorithms are developed to improve the accuracy when it is necessary to get information about the emotion of a person. Table 1 lists the various techniques used to gain the feeling of a person.

Several techniques are available to obtain the feeling of a person in a context of product evaluation (Hyung-il Ahn 2014; Hoonhout 2008; Boucsein 2008). The purpose of these applications is to know what alternative, among a set of potential solutions, causes the best impression to the customer. Also, the application aims to

Table 1 Techniques for obtaining the feeling of a person

Technical	Description	Applications
Pictures	To identify the user's emotion by taking a set of photographs (face)	HTC and spotify mood player
Videos	As with photography, one obtains the feeling of the user through his face and posture. The difference is that it is possible to know the emotion of the person in the different periods of time that the video lasts	EmoVu player
Voice tone	One of the alternatives is using a microphone to record the sound and from the audio and voice tone to get the emotion of the user	EMO speech
Sensors	There are several sensors to obtain human data such as heart rate, skin conductance, sweating, the stress of an object, which determine the degree of anxiety of a person, the level of excitement of an individual to see something specific that attract your attention	Galvactivator

uncover what specific features are received better by the user. Each technique has its advantages and disadvantages. For example, the skin conductance sensor determines the degree of excitation of a person and has high accuracy, but results will be obtained only for those using the sensor and with the appropriate equipment. On the other hand, techniques of facial photographs and videos have lower accuracy than the sensors. However, these techniques can be occupied in a massive way among the users using the collective intelligence and then, obtain a higher amount of results and evaluations. Intensification in the use of mobile devices facilitates the adoption of this technology. Hence, a website can provide affective computing tools to registered users.

The application of affective computing occurs in different approaches such as educational, medical, and in recent years, to the business domain. The latter being the least exploited. Since affective computing is becoming more relevant in this area, it is necessary to combine diverse ways for better results, a greater diversity of application domains including among them the Computer-Aided Innovation (CAI). Given the importance of CAI in the context of this chapter, the following section describes their key features.

3 State of the Art

Currently, the use of affective computing has grown in different areas; one of them is the evaluation of some products. This section presents some important works.

Patwardhan (2013) mention that consumer often reacts expressively to products such as food samples, perfume, jewelry, sunglasses, and clothing accessories. Thus, the author proposes a multimodal affect recognition system to classify whether a consumer likes or dislikes a product tested at a counter or kiosk. The information about the consumer is obtained through the analysis of the consumer's facial expression, the body posture, hand gestures, and the voice after testing the product. The recognition system was developed in the c# language. Also, it uses the Windows Kinect to connect the camera and the microphone and thus to collect the images of the user and their voice. The system offers the user the possibility to enable or disable the different devices according to the stimulus that the system will sensate. The system works with five basic emotions: angry, disgust, happiness, scare, and sadness. According to the author, the most frequent emotions were angry, disgust, and happiness. Finally, this work concludes that a multimodal system gives better results than a unimodal system for the products' feedback through a system's evaluation and with the Kinect they got better results and more accuracy in individuals' modalities in real time. The work of Damien Dupré (2015) establishes that the recent interest in the affective computing approach will have a positive impact for the consumers and the products; especially in emphatic applications and application that simulates to feel and that understand the client. Hence, to materialize such functionalities, it is necessary to measure the user's emotional experience. To accomplish this objective, the author developed an emotion measure platform. This platform proved that the user's emotional experience could be measured discreetly, and simultaneously to obtain information about the emotions in the user. In the work of Westerink (2009), the author underlines that the usability of a product or application is an essential indicator of the user's satisfaction. This indicator directly reflects the user's acceptance of the product. As a consequence, the usability acceptance is a relevant issue for almost any companies. The author executes some basic operation to process the user experience and to analyze and map the relationship between the users' actions and their emotions. Thus, the author combines some theories of emotion generation with the affective computing to finally explain that the affective computing could be used as a method for determining the usability test of a product and the better form to execute this test. According to Tian (2008), the works where the user's experiences are linked to products, systems or services is attracting the interest of the industry, particularly in some companies that have the goal to anticipate the commercial success of a product. The author proposes a project with two objectives. The first one is to provide an approach to identify emotions through a multi-device based on a physical view. The second is to offer some solutions for the affective computing and then, to evaluate the user's emotions and experiences. Also in this work, the authors did an experiment where ask some users if they want to use a classic mouse (this was the control group) or a hand-gesture interaction device for some activities. The author had the hypothesis that the new product will trigger stronger emotions than the classics products. The author developed two emotional evaluation tools. The first one Oudjat "Verbal Self-Annotation of Emotions" and the second EmoLyse "Nonverbal Self-Annotation of the Motivational component of Emotions." Finally, the author concludes that emotions are a potential success indicator for the products in the market. Samer Schaat (2015) assumes that emotions and social norms play a key role in consumers' decisions. The authors developed a sociocognitive agent to examine the psychological and sociological factors influencing consumer decision-making. The author created a decision model that integrates motivation, emotion, and normative mechanisms using a unified activation and valuation framework. To accomplish this purpose, the authors use Java, H2 database, Hibernate for the mapping object-relational, the software MASON for simulations, and Java FX for the users' interface. The authors executed different simulations with the model and proved that results were consistent. Christophe Vaudable (2009) performs a study focusing on the consumers' emotions during an interview about some products. This research is based on the analysis of some annotations of some dialogs. The author made some tests with 40 subjects and collected a database. During the interviews, the subjects revealed several "real-life" complex emotions. Also, in the interviews, the user manipulated many products, one of them was a food product, and in this particular case, the authors observed the expressions of many emotions. Also, the work underlines the relationship between the eating habits and some emotions. Thus, in the end, the study allowed the creation of a protocol allowing gather emotions information that could be used for the creation of a model to detect emotions automatically.

4 Computer-Aided Innovation

Innovation is fundamental for the development of society, for the renewal and growth of companies, and is a critical goal for the survival of organizations. Originality can emerge from the connection of other systems or object, such as technologies, structures, markets, culture, strategies, products, services, or another set of functions that have a common purpose. From this perspective, the innovation process is typically iterative, interactive, with a specific context, multitasking, uncertain, path-dependent, and is the result of a new combination of extremes and means (Stefan Hüsing 2009).

In the initial studies on computer-aided innovation, the central objective was to help process engineers during the initial stage of the design process, also called diffuse front-end. Subsequently, the scope expanded, then CAI systems are now effective support to the innovation process, from the initial stage or idea generation passing by the detailed design and development stage to the reverse logistic or recycling of a product. Recent advances in information and communications technology and theoretical approaches to innovation, provide a new environment for the next evolutionary step of CAI and allows users to propose inventive solutions and improve the efficiency while solving complex problems (Réne Lopez Flores 2015).

The CAI approach has evolved and has been complemented by other technologies that currently allow covering various aspects of the innovation process. It is necessary to highlight the role of the TRIZ theory, the QFD, and recently the collective intelligence, and the semantic web. It is from the assimilation of new technologies that CAI evolved to what is now called Open CAI 2.0, which aims to generate an open innovation process between organizations to implement a continuous or systematic innovation process.

4.1 CAI Classification

The CAI software includes simple applications for specific activities of the innovation process as model conflicts or basic semantic search and more complex solutions such as the software that supports the most significant stages of the innovation process. Consequently, CAI proposes a description for the different explications and delimits the scope of these tools (Réne Lopez Flores 2015), the classification is given by Stefan Hüsing (2009) and has three classes:

(a) Strategy management: Innovation managers help address strategic issues such as portfolio or scenario management.
(b) Managing ideas: It helps to direct the diffuse front-end of the innovation process, from the generation of concepts to the evaluation of ideas.

(c) Patent management: This type of tools is used in projects of inventions and in the search and analysis of patents as an approach to stimulate creativity.

In Fig. 1, we can see the categories of CAI tools divided into sections.

It is important to mention that it is possible to find different tools that contain some of the subcategories shown in Fig. 1, and it is expected that these categories will change over time since the CAI is in a constant evolution process. One of the most relevant trends is the use of collective intelligence, and due to its importance, the following section addresses this issue.

4.2 Collective Intelligence in CAI

The concept of collective intelligence in the computer field is better known as Web 2.0. It has a substantial effect on consumer behavior and has contributed to customer empowerment. The consequences are far-reaching, not only in the area of technological development but also in the domains of business and marketing strategies (Efthymios Constantinides 2008). The term known as Web 2.0 gave by Tim O'Reilly in a conference in 2005 mentioned that Web 1.0 or websites with static information, offered only a unidirectional relation. That is, a user could just consume the information provided on a website. The arrival of the Web 2.0 with the introduction of websites with a dynamic description, where users could already upload their information produced a two-way relationship between users and Internet sites (O'Relly 2005).

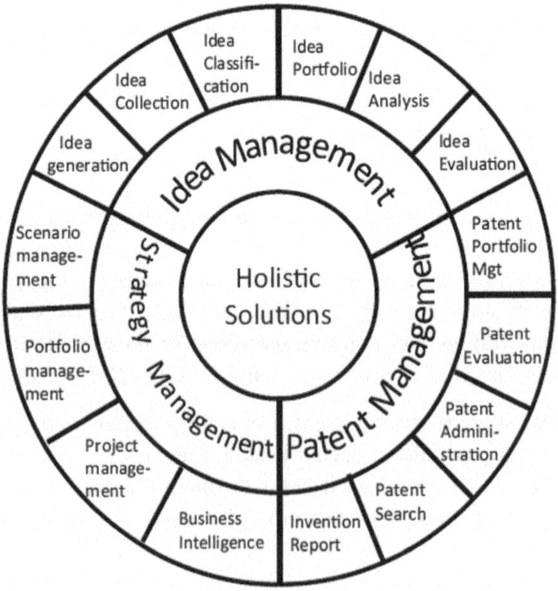

Fig. 1 Detailed CAI software categories (Réne Lopez Flores 2015)

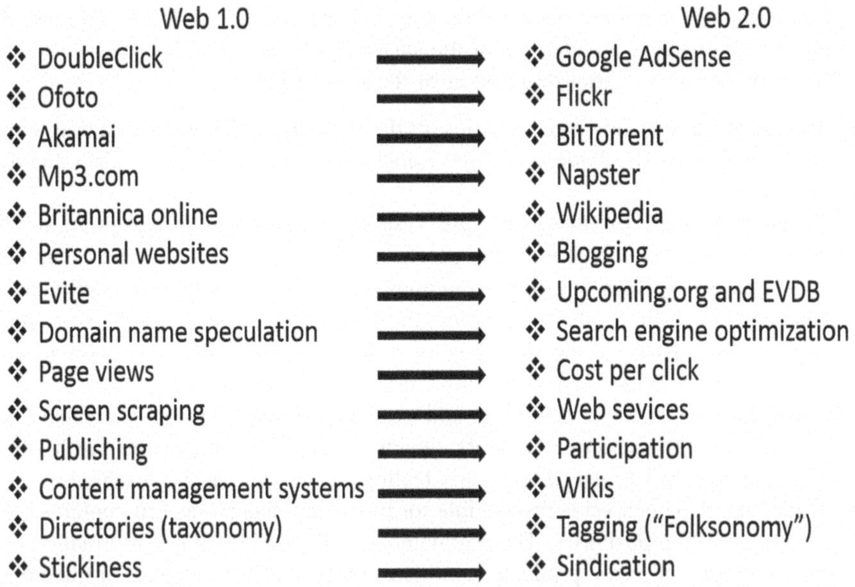

Fig. 2 Some transitions from web 1.0 to the web 2.0 (O'Reilly 2005)

Figure 2 shows some examples of the transition from Web 1.0 to 2.0.

At present, the CAI has made use of collective intelligence or Web 2.0 for the advantages it offers, creating an approach called Open CAI 2.0. The aim is to apply processes where knowledge has a free flow, so it is possible to implement an open innovation process. The study of emerging intelligence in groups of individuals doing things together is not new, but in recent years particular attention has been received with the emergence of applications in Web 2.0. This new way to interact helps to unlock the potential of collective intelligence due to its architecture centralized in the participation of the user, while simultaneously allowing the connectivity. As a platform for collaboration, the Web 2.0 is useful for implementing different collaborative patterns, for example Réne Lopez Flores (2015): Temporal, synchronous, asynchronous, and multi-synchronous; Space: Locally and distributed, and rules: rules of work, rules, and restrictions.

In this chapter, it has been developed a prototype that with the use of the affective computing get the person´s emotion, that is, watching images or videos of a specific product. Next, the section describes the prototype architecture.

5 Architecture

A layer-based architecture was chosen for the tool's development since it allows having the different functionalities of the tool separated. Similarly, several modules collaborate to produce more than one feature or function in the software. The use of layers facilitates the maintenance and the use of resources (Schmidt 1996). Figure 3 shows the layer-based architecture of the system under the CAI approach.

Next, points offer a brief description of the system layers:

- **Interface Layer**: This level contains all the logic that rules the interaction with the user. It uses the JQuery Mobile framework to make the page adaptable to mobile devices.
- **Services Layer**: In this layer, the system contains the "Configuration of Services" process, which is in charge of selecting the different services that the user will use. The tier has direct communication with the collective intelligence module and the affective computing module, which will use the Skybiometry API for the detection of emotions. The collective intelligence module contains the logic that allows users to collaborate with other entities/users/teams to generate new ideas and evaluate products. On the other hand, a section of the affective module is in charge of interacting with the Skybiometry API to get the person's feeling and store it in the database.
- **Data Layer**: This level is responsible for providing data storage. It contains the necessary data repositories. The user database will contain all the information of the two types of users, products, comment, ideas, and the database of affective material that incorporates all the interrelationships between an object or concept and the evaluations offered by all users.

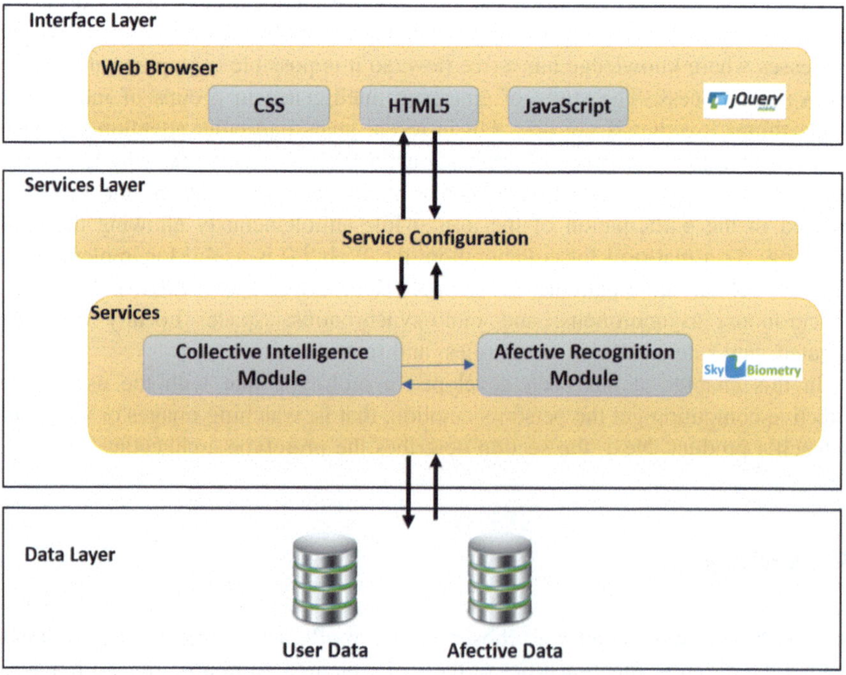

Fig. 3 Layered system architecture

6 Evaluation Tools

One of the industry's most frequently used assessment tools is the self-report. In the context of this chapter, it is significant to identify the strengths and limitations of this mechanism, as well as continually strive to improve its quality (Sallis 2000). The use of self-reports in the evaluation of design, and features of a product has one crucial advantage: It can be applied to a large number of people, who can answer the questions and hence, to obtain a huge volume of valuable. These mechanisms of communication are simple to implement and easy to understand by the user. However, the main disadvantage of self-reports is the lack of some mechanism to collect the intrinsic information that generates during the interaction or utilization of a product. Consequently, the feeling generates during the interaction with the product stop being so real at the time that the user answers some questions. Also, answer some questions before trying a product or considering a concept is tedious for the user, who can moreover lie in their answers, in addition to other limitations such as time demanding.

A very dynamic research area is the design of tools to evaluate products and services with the purpose to generate more real assessments. The goal is to try to predict the success of rapidly changing products in the market and improve the methodologies for product development. The field of affective computing is an alternative solution to evaluate more accurately the conceptual designs, prototypes, or characteristics of a product through the observation of the actual emotion presented by the user when he or she interacts with some of the attributes of a product. It also has the advantage of offering a wide range of applications in the market, derived from the use of mobile devices, which already have integrated a webcam. As a result, it is possible to reach a large number of customers to obtain a considerable amount of data. Available data allows organizations to evaluate alternative solutions with which the consumer is most identified. Finally, it is meaningful to generate a massive evaluation tools to have a bigger number of options for companies and try to increase the probability of success for launching new products to the market.

7 Methodology

The increasing market pressure, the rapid technological development, and shorter product life cycles are forcing corporations to promote new products systematically. The services industry is facing an even more critical situation (Chung 2017). As a result, companies need to develop the ability to manage innovation on an ongoing basis. The requirement to market a product that meets the needs of consumers or that surprise a customer forces organizations to develop a series of processes to increase the likelihood of success of a product. Figure 4 shows the methodology with the elementary operations for the development of a product and the area of

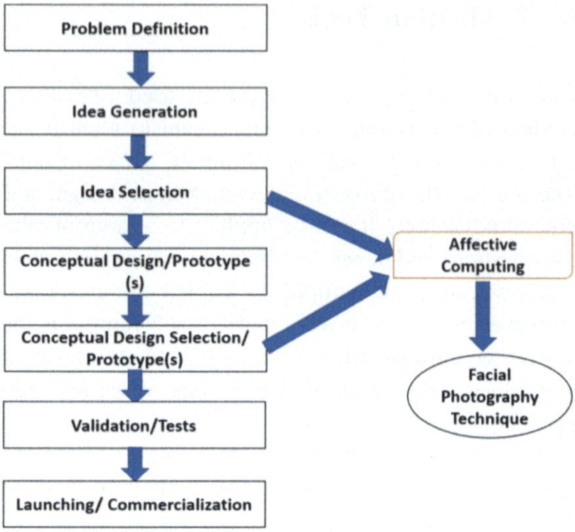

Fig. 4 Methodology for the development of a new product, areas of opportunity for affective computing

opportunity that has the affective computing to evaluate the different alternatives in a particular stage (Edgett).

As can be seen in Fig. 4, the development of a new product for an organization represents an area of opportunity for affective computing and more specifically, during the evaluation of the best alternatives to fit the user needs. Here are the steps to follow:

(a) **Problem statement**: This activity focuses on defining the scope and purpose of the design process.
(b) **Idea generation**: This phase consists of obtaining internal and external ideas that guide the process of product design. These ideas are stored and evaluated at a later stage.
(c) **Idea selection**: This point defines some qualitative and quantitative criteria to evaluate and filter collected ideas. Subsequently, the most promising or best-received concept selected by an expert panel, a focus group or by the target population will move to the next step of the process. At this stage, it is possible to use affective computing to filter a lot of ideas presented to the target population or a work team. Then, those with the better affective assessment will pass to the next stage of the process. In the case where none of the concepts have enough value to accept it, then users will have to choose more ideas and repeat the process until at least one concept is selected.
(d) **Conceptual design/Prototype (s)**: In this stage, the initial concept moves to the conceptual design of some potential solutions. In the context of a service, we suggest performing the conceptual design of the service production process and then to model the interaction with the user.
(e) **Conceptual design selection/Prototype(s)**: Once the conceptual design concludes, the most promising alternatives get materialized in prototypes. In the

case of services, a test scale implements the service production process. The design of a product is another area of opportunity for affective computing. The purpose is to capture the emotions generated by the prototypes in the target population and then to select those that have the best reception in the clients or users.

(f) **Validation/Test**: The Company executes some test on the product to validate its functionality and to provide feedback to improve the design methodology. This stage includes the validation of the production process.

(g) **Launching/Commercialization**: The last step, once that product design is validated, is to execute the marketing strategies that support a successful introduction to the market. Again, this stage is another opportunity area for the use of the affective computing about the client emotions when the users interact with the product or hire a service.

Figure 4 shows how the role of affective computing to facilitate the decision-making process in the stages of ideas selection, conceptual designs evaluations, and in the user's feedback.

7.1 Process of Selecting Ideas and Selecting Designs/ Prototype

Figure 5 depicts the process to get the users evaluations of ideas or conceptual designs of new products. At the moment, detecting emotions from the use of a product or service is temporarily outside the scope of this chapter, we focus the effort in the initial design stages.

Skybiometry API was selected for their advantages compared to other APIs on the web. Among the most relevant are: its usability, a free license which includes some calls per month, extensive documentation, and their degree of accuracy in recognizing the emotions of a person in a photograph. Nonetheless, to assess the accuracy of the Skybiometry API, we use a database of faces called Radboud. The database contains images of faces. A set of images of 67 models having 8 emotional expressions. The results obtained by comparing seven images from seven patterns in the Skybiometry API has a great performance. It is important to notice that a larger quantity produces a timeout error. The final result of the evaluation shows that by making use of the Radboud database images, the API is 100% accurate in predicting the feeling of the selected photograph in the database, which proved that the Skybiometry API is an excellent option for development. However, it is necessary to comment that the photographs of the database took in the best way and people try to simulate the different feelings, but despite this, the API offers very good results, and that is why it was chosen to have other APIs such as Microsoft, Google, and EmoVu, among others.

The information generated can be treated graphically. It is important to mention that this feature allows a more accurate evaluation since the Skybiometry

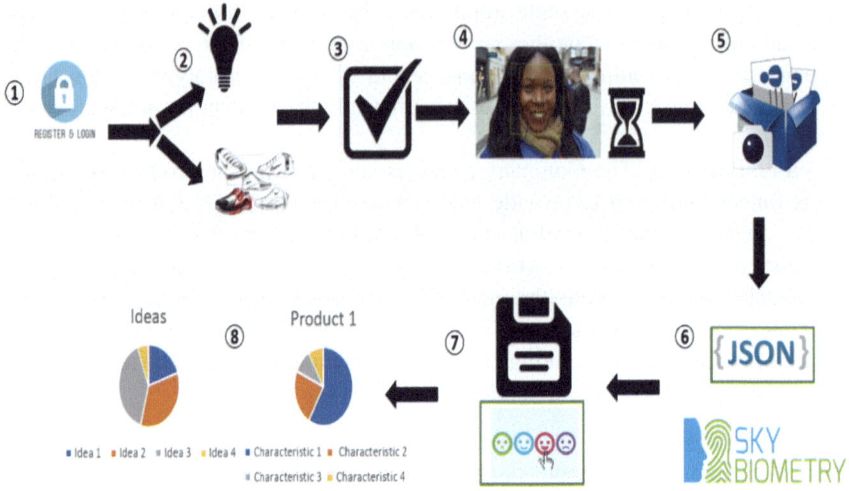

Fig. 5 Process of selecting ideas and selecting designs/ prototype

application captures photographs and analyzes the user's emotions, while evaluating ideas or designs of a product represented as images or videos. Described below is the process to perform the identification of emotions:

(1) Login: This function enables the resources of the system, so the user can access the properties of the application and use them.
(2) Alternatives selection: The user can select between the list of ideas and the list of designs/prototypes of a company. Also, the company or organization interested can suggest to a user some assessment considering their priorities. The purpose is to make use of collective intelligence for users to obtain an acceptable amount of ideas and then apply some filters and work with the most promising.
(3) Product selection/ Idea selection: Once the user selects an alternative from the list of ideas or design/prototypes of a company, he can visualize it in the system in the form of a set of images, audio, or video files.
(4) Take a photo: The system will take some pictures of the face of the person for a period; the time determined by the users depends on their needs.
(5) Store photo: Photographs get stored on a server. The Skybiometry API processes these images files to obtain the emotion and each photograph will be related to each characteristic of a product observed by the user. These relationships will be stored in the database to obtain the graphs of each product from this information.
(6) Getting the feeling: All stored photographs were analyzed for a specific event called the Skybiometry API. The result of each photograph produced by the API is also stored in a database. It is imperative to notice

that the use of the API goes through the use of a web service. Hence, the system makes a request and obtains a result, in this case, a result in a specific format called JSON. Then, this information needs to be decoded. The affective module is responsible for carrying out the necessary processes to obtain the emotion.

(7) Store Feeling: The API stores the feelings obtained in the database for further retrieval, as well as the gender of the person and the approximate age. The API provides this information. If the system cannot take a successful photograph of the user's face, the assigned emotions will be considered as disinterest because it indicates that the person was not observing the material loaded by the company of the product or was turning over. Therefore, the API cannot detect the face of the person, for that reason no emotions will be obtained.

(8) Generate Graphics: Once the user's emotions are stored in the database, the company can access the specific data of each design or idea to obtain graphs about the users' behavior to the ideas or conceptual designs they selected to visualize. The company knows how many users have seen a particular characteristic, how many have presented a positive emotion and how many have a negative state of mind.

8 Case Study

In this section, a case study is presented as proof of concept of the system for the detection of people's emotions with the purpose of evaluating different ideas or conceptual designs of new products. The case study describes the operation of the system and explains each one of the elementary functions offered for classifying a set of ideas and evaluate different product designs. Thus, companies will be able to collect information about the way people feel when they see their models. Obtaining this information and its use in the decision-making process is crucial since it represents a feedback loop that can assist a team or an expert in the process of selecting ideas and conceptual designs.

In the case of the study, we used six assorted designs of a product. The product has widespread use. It has multiple electrical contacts (multiplug) that contain four to six plugs. It may or may not have USB ports. The case study contains different conceptual designs for the product, which are observed by different users on the web. Through this mechanism, the company collects information and statistics related to the emotions that users show when analyzing a conceptual design or idea. In Fig. 6, it is possible to observe login of the system where the user enters their email and password and, in case of not being registered, choose the option "Register here" to save a profile and be able to access the system's functionalities.

If it is the first time in the site, then the option "Register here" is the first step. The user will introduce their data and will select the type of user in the system.

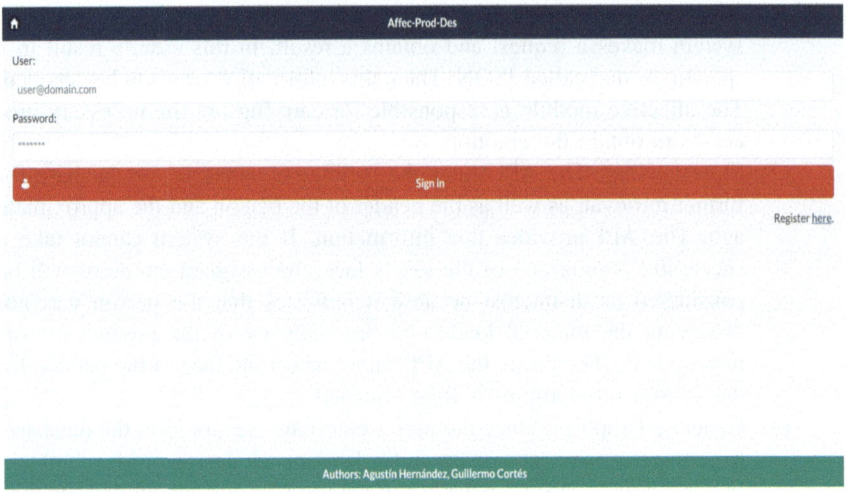

Fig. 6 Login

Depending on the option there will be different features available. Figure 7 depicts the registration form to enter the user's data. It is also possible to provide access to a company or stakeholder. A user will have to the bundle of ideas or designs of products of the different businesses registered in the system. If the record corresponds to a business, then the business will have all the information of the products, ideas, or designs in an image or video format so that users can watch them. They can also generate graphics about each product, observing the emotions presented by users when they see their products designs or ideas.

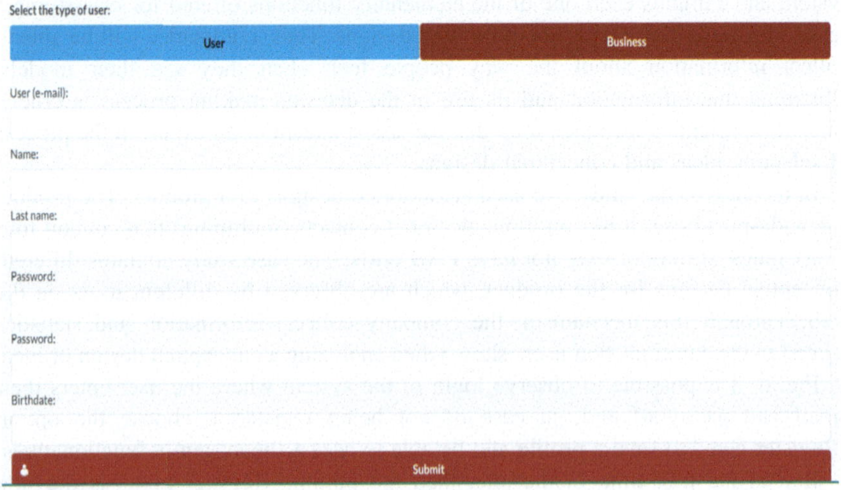

Fig. 7 Form user registration

Once the access stage is granted, the system stores the user's data in the database, and with the username and password, a user will be able to log into the system. The system also has validation of the data of users or business stakeholders. Once verified the access, the system shows a welcome message to users with their name. Right before that, the next step is to complete a form that asks for some information to launch a search for products that are in the database of the system (Item registered by companies or proponents). Figure 8 shows some results that match the search criterion defined by users.

As seen in Fig. 8 when a user enters some search criteria, all products that meet this criterion are listed with a brief description of the product, the category to which it corresponds, and the name with which the product was registered. The user selects the one that best matches their needs. In this case, the result "Just one touch unplugs easily" was selected from the category "Electronics". If a company wants a private evaluation, it is possible to send an invitation to some specific users. Otherwise, the idea, product or initial design is open to any user.

Once the product is selected, a screen is displayed where the user chooses at least one option that can be the conceptual design of a product or an idea. Figure 9 shows the basic information of this step.

When the users select the category of conceptual designs, in the present case they can observe six conceptual models that one or different companies have uploaded. Before watching each of the conceptual designs, the system displays a message to ask the user for their permission to use the webcam. Figure 10 shows the authorization to use the webcam in the Mozilla Firefox browser. It is necessary to point out that if the user does not have a webcam, it will not be possible to get the emotion and therefore the system will not register any information during the evaluation of the different concepts of the product.

Once the system can use the webcam, all the designs that the company uploaded are visible to users. In this case, there are six designs, which are evaluated by the

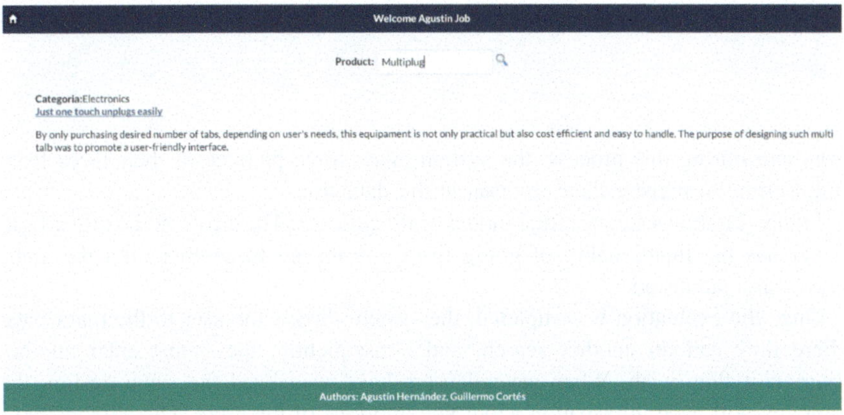

Fig. 8 Results from the criterion entered by the user

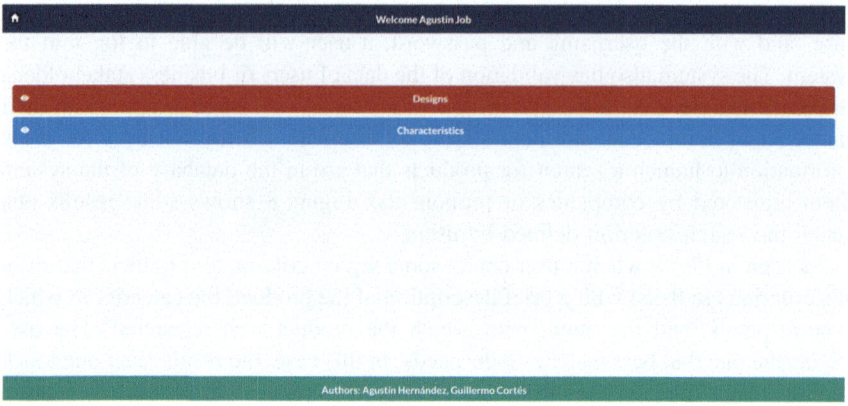

Fig. 9 Selecting content to display

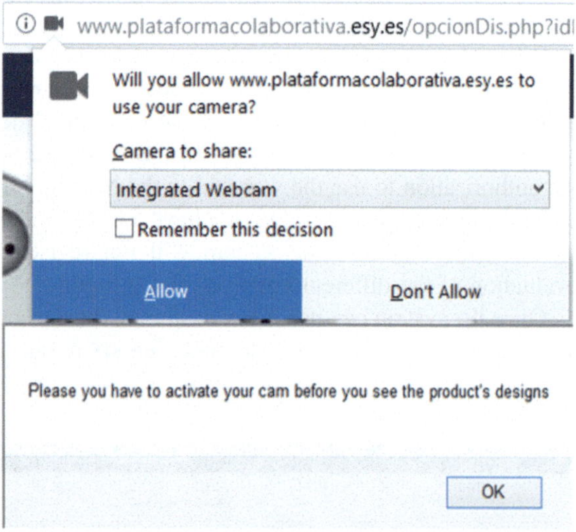

Fig. 10 A request messages by the system to have access to the webcam

users and during this process, the system takes some pictures of their faces from time to time to register their emotion in the database.

Figure 11 shows a conceptual model of the product. The arrow in the upper right corner has the functionality of going from one design to another until the alternatives are exhausted.

Once the evaluation is completed, the system directs the user to the main page where they can do another search, and consequently, they must enter another criterion if they wish. When several users have visualized the same design, the company will have information about the emotions of the users. The administrator of the company accounts must log into the system. Once the session started, the

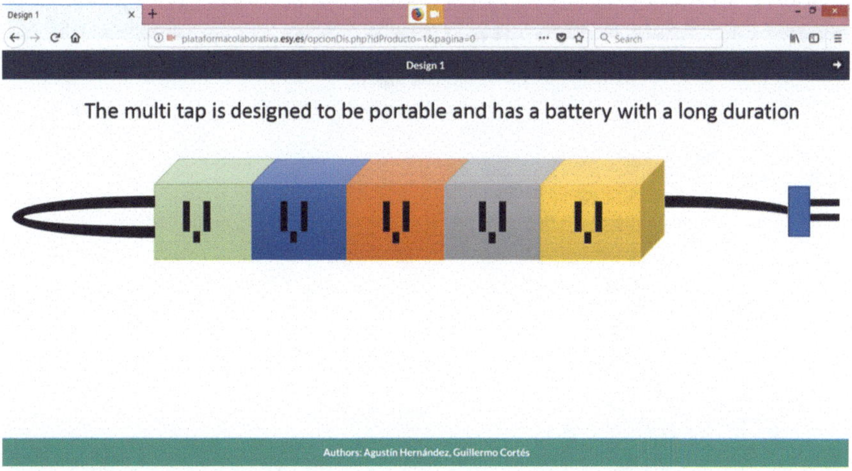

Fig. 11 Presentation of the design of the product

administrator can observe all the products they have registered, as well as to add other products, design alternatives, ideas, or specific features of that product in different formats for the user's visualization. After the user participation for a period, the business will reward the customer by sending him a free product to your home to stimulate the evaluation and motivate other users.

By selecting the option Graphics Results, the administrator can see the information collected from several users. For this case study, the administrator visualizes the graph that accounts the emotions in each design presented to the user. Figure 12 shows the graph of each emotion that the Skybiometry API can recognize, or the absence of emotions if not photographs are available. The six designs are presented each with a specific color where it can be seen that the model that produces a surprise to most people was the design three with 31 users in this category.

As it was possible to observe, the use of affective computing as an alternative solution and support for the decision-making is feasible. It provides relevant information and the expressions that users offer when they see a conceptual design or idea. This information allows companies to have an idea of whether such a product is interesting and have some attributes that are relevant to a potential user.

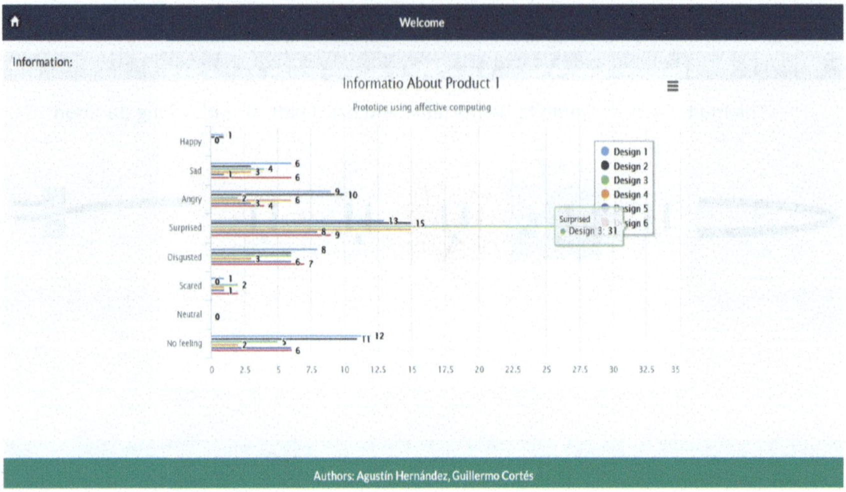

Fig. 12 Graph of the emotions presented by different users

9 Conclusion and Future Work

The object of this chapter is to explore the potential that has the affective computing to assist the decision-making process during the evaluation of ideas or the conceptual design of products. The ability to acquire the emotions of the user during the evaluation is an area of opportunity in the computer-aided innovation systems or to the open innovation systems.

Affective computing provides capabilities not previously considered in the CAI approach, which represent a research opportunity in this field. A CAI system uses less performant techniques to evaluate different solutions and alternatives for a problem. Consequently, the affective computing approach can help in the decision-making, and provide to a CAI system with new techniques to carry out the innovation processes. Also, by developing a system that can detect the emotion of a person when observing conceptual designs, features, and ideas of a new product will produce valuable information before the detailed design stage. It is possible to suppose that with such information, the probability of success of a product will increase at the time of launching it to the market.

As future work of this project, we have the following points:

- To add other technologies that will meet the diverse needs of a company with the purpose to improve the different stages of the innovation process.
- To validate other technique of affective computing and improve performance to obtain more accurate results.
- To explore the use of affective computing in other tools and different approaches provided by the CAI for a better structure of the system.

- To allow companies to obtain valuable information from the data collected and search to extrapolate this information to other domains.
- To use the Skybiometry API to obtain another kind of information, since the API provides an approximate age and the gender of the user, with this information will be possible to analyze a market and identify non-evident restrictions.

10 Acknowledgement

The National Council of Science and Technology (CONACYT), the Public Education Secretary (SEP) through PRODEP, and the Tecnologico Nacional de Mexico sponsored this work. Additionally, the ROPRIN working group (Network of Optimization in Industrial Processes) supported this work.

References

Bond, R. R. (2017). SenseCare: Using affective computing to manage and care for the emotional wellbeing of older people. *eHealth 360*, 352–356.

Boucsein, W. et al. (2008). *Objective emotional assessment of industrial products.* Springer Netherlands.

Chorianopoulou, A. K. (2016). Speech emotion recognition using affective saliency. *Interspeech,* 500–504.

Christophe Vaudable, L. D. (2009). Study of consumer's emotions product interviews. In *3rd International Conference on Affective Computing and Intelligent Interaction and Workshops. ACII.* Amsterdam, Netherlands.

Chung, L. et al. (2017). The unique chinese innovation pathways: Lessons from chinese small and medium sized manufacturing firms. *International Journal of Production Economics* 190, 80–87.

Damien Dupré, A. T. (2015). Emotions triggered by innovative products: A multi-componential approach of emotions for user experience tools. In *International Conference on Affective Computing and Intelligent Interaction (ACII).* Xi'an, China.

Edgett, S. J. (n.d.). *Stage-Gate.* Retrieved 01 20, 2017, from dea-to-Launch (Stage-Gate®) Model: An Overview.

Efthymios Constantinides, S. J. (2008). Web 2.0: Conceptual foundatiosn and marketing issues. *Journal of direct CONSTANTINIDES, Efthymios; Fountain, Stefan J. Web 2.0 Journal of Direct, Data and Digital Marketing Practice, 9*(3), 231–244.

Hoonhout, J. (2008). *Inquiring about people'S affective product judgements. Probing Experience.* Springer.

Hyung-il Ahn, R. W. (2014, April-June). Measuring Affective-Cognitive Experience. *IEEE Transactions on Affective Computing, 5*(2), 173–186

Leon, N. (2009). The future of computer-aided innovation. *Computers in Industry, 60*, 539–550

Montes, A. J. (2016). Sistema de detección de emociones para la recomendación de recursos educativos. *Tendencias de la Ingeniería de Software.*

O'Reilly, T. (2005). *What is web 2.0: Design patterns and business models for the next generation of software.* Retrieved July 20, 2017, from http://www.oreilly.com/.

O'Relly, T. (2005, September 30). *What is web 2.0*. Retrieved July 20, 2017, from www.oreilly.com.

Patwardhan, A. S. (2013). Multimodal Affect Analysis for Product Feedback Assessment. In *IIE Annual Conference. Proceedings. Institute of Industrial Engineers-Publisher.*.

Picard, R. (n.d.). *Affective Computing*. Retrieved October 10, 2017, from http://affect.media.mit.edu/index.php.

Picard, R. W. (1995). Affective Computing. *Affective Computing-MIT Media Laboratory Perceptual Computing Section Technical Report No. 321, 2139.*

Réne Lopez Flores, J. P. (2015). Open computer aided innovation to promote innovation in process engineering. *Chemical Engineering Research and Design, 103*, 90–107.

Rene Lopez Flores, J. P.-M. (2015). Usin the Collective Intelligence for Inventive problem solving: A contribution for Open Compuer Aides Innovation. *Expert Systems with Applications, 42*, 9340–9352.

Sallis, J. F. (2000). Assessment of physical activity by self-report: status, limitations, and future directions. *Research quarterly for exercise and sport, 71*(suppl. 2), 1–14.

Samer Schaat, S. W. (2015). Modelling Emotion and Social Norms for Consumer Simulations Exemplified in Social Media. In *2015 International Conference on Affective Computing and Intelligent Interaction (ACII),* Xi'an, China.

Schmidt, D. S. (1996). Pattern-Oriented Software Architecture. Wiley.

Stefan Hüsing, S. K. (2009). Computer aided innovation—state of the art from a new product development perspective. *Computers in Industry, 60*, 551–562.

Tian, X. W. (2008). A Study on the Method of Satisfaction Measurement Based on Emotion Space. In *9th International Conference on Computer-Aided Industrial Design and Conceptual Design, CAID/CD, IEEE.* Kunming, China.

Vera-Munoz, C. P.-S. (2008). A wearable EMG monitoring system for emotions assessment. *Probing Experience*, 139–148.

Westerink, J. O. (2009, September). Emotion measurement platform for daily life situations. In *3rd International Conference on Affective Computing and Intelligent Interaction and Workshops. ACII,* (pp. 1–6). Amsterdam, Netherlands.

Zaripova, V. (2014). Computer Aided Innovation model for university-enterprise cooperation. *ICTIC—Proceedings in Conference of Informatics and Management Sciences*, Vol. 3, pp 377–382.

Taking Advantage of the Innovation in Die Service Design

María Blanca Becerra-Rodríguez, Diana Jiménez-Juan, Arturo Hernández-Hernández, Ángel Adad Franco-Baltazar, José Marcos Zea-Pérez and Isabel Ernestina López-Navarro

Abstract The present work has been planned to satisfy the needs of the metal-mechanic industry in Mexico, San Juan del Río, which generates 16% of the Gross Domestic Product and counts with 8,932 economic units in Mexico. The metal-mechanic sector is a pillar for other industrial activities. This work focuses on the thinking skills for designing dies as tooling that are used to punch metal parts with high or low production volumes. Design skills, due to a lack of investment and development, have been lagging in Mexico, compared to foreign markets, which export tooling to Mexico mainly from Asia, for 2,600 Million Dollars (MDD) per year. Consequently, the objective of this chapter is to demonstrate that the service is feasible from different points of view. Based on the scientific method, this research work is structured to reach next objectives: (1) To detect the need to locate the market that allows the creation of the company and (2) To deploy the product lifecycle method to determine the stages that a product must follow from its conception to the manufacturing process. Results show that the market will accept the new service with the expected profitability of 195% the first year. The product also

M. B. Becerra-Rodríguez (✉) · D. Jiménez-Juan · Á.A. Franco-Baltazar
I. E. López-Navarro
Instituto Tecnológico de San Juan Del Río. Av. Tecnológico #2,
C.P. 76800 San Juan Del Río Querétaro, Mexico
e-mail: industrialblanca@gmail.com

D. Jiménez-Juan
e-mail: yanis_jim10@hotmail.com

Á. A. Franco-Baltazar
e-mail: aasub_academica@itsanjuan.edu.mx

I. E. López-Navarro
e-mail: isalopmx@yahoo.com.mx

A. Hernández-Hernández · J. M. Zea-Pérez
Universidad Politécnica de Querétaro, Carr. Estatal 420 S/N, El Rosario,
C.P. 76240 Querétaro, Mexico
e-mail: arturo.hernandez@upq.mx

J. M. Zea-Pérez
e-mail: marcos.zea@upq.mx

G. Cortes-Robles et al. (eds.), *Managing Innovation in Highly Restrictive Environments*, Management and Industrial Engineering,
https://doi.org/10.1007/978-3-319-93716-8_11

provides opportunities for growth and improvement with a competitive service price that has the potential to solve some market needs that would help the development of the industry in Mexico.

1 Introduction

The technology advances rapidly, which opens the doors to different areas for improvement. However, to compete in the service market and, more important yet, to remain competitive, it is necessary to mobilize the experience of highly trained and experienced people to be part of the technological tools used in the improvement of transformation, extraction, and services activities.

In this case, NX®, Siemens software, offers several characteristics that with a time of study and dedication, it is possible to master it and apply it in projects such as the services design of dies. With such preparation, it is possible to achieve a reduction of up to 30% of the time in die design, a basic level of experience, under the supervision of experts of the subject in dies and services design.

Die industry supports two industries that nowadays have an important growth in Mexico: aeronautics and automotive. With companies like Honda, Audi, Volkswagen, Mazda, Volvo, Nissan, GM, and Airbus, operating in the central region of Mexico this service has a dynamic market. However, to respond to such growth, which is forecasted to be very large, the new company must endure strong investments to have a promising future.

In the company where this project was developed, there were several restrictions: the lack of capacity, not enough facilities and machines, and the necessity to understand technology, but the most critical issue is to have a set of experts and designing tools. Due to these needs, most of the service for designing dies is made abroad because in Mexico there are not specialized people in this area.

Today, in Mexico there is no exact information about the number of manufacturers that are producing molds and dies, and it is estimated that there are between 1000 and 1500 companies (Shop 2015).

A study by the Scientific and Technological Advisory Forum (FCCyT, by its Spanish acronym) titled "Design and Development of High Value-Added Products and Processes" reports that Mexico imports more than $18 billion in manufacturing machinery and production processes, and more than $2.6 billions of dollars in molds, dies, and tools.

An emergent national industry of molds, dies, and tools explains the high level of imports and the absence of experienced suppliers. Here is the need for designing new services and creating commercial relations with companies that may have the capacity for manufacturing from domestic parts to the realization of assemblies of die cut pieces.

Also, the National Council for Science and Technology (CONACYT, by its Spanish acronym) indicates that Mexican companies only provide between 5% and 10% of the molds, dies, and tools required in the country. Thus, most of these components come from the United States, Canada, Germany, Portugal, Spain, and Asia (Shop 2015).

Following a similar direction, a report of the firm VSI Consulting says that Mexico has problems producing molds and dies at the national level, because of the lack of special high-strength steels, the lack of specialized labor, and the lack of specialized die-cutting careers.

The report also underlines that approximately 85% of Mexican mold and die companies are workshops without any certifications, and also that national companies are specialized more in the repair of molds, which do not give them any added value (Shop 2015). "The main problem is repairs: Mexican companies send their molds to Europe to be repaired, and if it is not the first option, foreign engineers travel to Mexico to make the repairs." Also, "The FCCyT points out that national engineering and design competencies, together with experience in machining and transformation processes, can take advantage of and further develop to design and manufacture these types of products nationwide."

"This will generate greater value and allow maintaining control over capital goods, increasing competitiveness and reducing the technological dependence of Mexican manufacturing companies in the strategic sectors to develop local capacities for the design, engineering, and manufacture of capital goods" (Shop 2015).

The Technological Development Program in Advanced Manufacturing aims to transform Mexico into the leading manufacturing center of Latin America. To reach this objective, the program seeks to increase the levels of the import volume; in the first stage, it focuses on the development of capacities for the manufacture of molds and tools in Mexico. ProMexico, which is part of the program, emphasizes that the plan for the country is to produce molds for the injection of advanced plastics, high-pressure aluminum, as well as design, engineering, and manufacture of dies, attending industries in the national territory such as automotive, aerospace, metalworking, and electrical, among others.

The Department of Economy also works in this direction, through the Support Program for Technological Improvement of the High Technology Industry, which aims to help high-tech industries increase their capacity to adopt new technologies, and to promote the participation of companies in activities of higher added value, through specialization of human capital, certification of productive and human capacities, generation of specialized information, and also technological development and transfer.

The program attends suppliers, from companies of high technology industries that offer specialized services in the maintenance of molds and dies, repair and maintenance of aeronautical parts, aircraft, or boat manufacturing, among others.

According to data from the Sustainable Development Department, the approximate investment for the construction of the first engineering and technological development center for the molds and dies field in the State of Queretaro was 36 million pesos (with federal, state, and private resources). This investment was made in collaboration with the National Chamber of the Transformation Industry (CANACINTRA, by its Spanish acronym) of Queretaro City.

"The molds importation in the state, in areas related to plastics injection molds, is approximately 200 million dollars a year. This represents an important development area" according to the Secretary of Sustainable Development of Queretaro.

The National President of CANACINTRA proposed the creation of a cluster of molds and dies in the state of Hidalgo. Two reasons are behind this decision: (1) The region has a metalworking vocation and (2) It has several logistic advantages because it is located near to those states that have strong metal mechanical industries (automotive, aerospace, and appliances, among others), which demand molds and dies.

Hence, "Molds and dies are a business segment that is wasted in the country and with high potential. It is a link in the productive chain of the country that is deteriorated badly. Besides, another initiative is the creation of the Mexican Association of Molds, Matrices and Dies located in Santiago de Queretaro and whose objectives are: to create technology transfer programs with other countries, develop clusters, as well as expand the domestic mold production market."

There are some studies of aluminum dies associated with the melting processes, where porosity is one of the most frequent defects. Porosity formation limits the use of casting in critical high-strength applications. In this study, the use of the Taguchi approach was conceptualized to obtain optimum adjustments of the parameters of casting under pressure, to increase the quality and efficiency of the casting under pressure. The goal was to increase the quality and efficiency of the alloy of aluminum (EN AC-46500) reducing the formation of porosity (Apparao and Birru 2017).

A study, about the analysis of the destructive mechanisms for die inserts used in the first operation of wheel forging process, showed that the dominant destructive mechanisms in die inserts are thermal fatigue occurring in the initial exploitation stage and abrasive wear (Hawryluk et al. 2017).

A study of different materials, such as aluminum alloys and steels, has become one of the means to lighten the weight of a vehicle's body. Due to the formation of hard brittle intermetallic compounds, fusion welds or steel welds by resistance points are used very little. Self-piercing riveting is an example of this type of welding; the lap shear strength was found linearly correlated with the undercut for a fixed top sheet thickness and an equation was fitted to predict the lap shear strength regarding undercut and top sheet thickness (Shop 2015).

Riveted diagrams were used to illustrate the riveted ranges and joint quality under each set of rivet and die combinations. It was found that the softer rivets and larger dies could improve the riveted range, and meanwhile decrease the joint strength. The longer rivets and smaller dies could narrow down the riveted range but increase the joint strength (Ma et al. 2018).

Thermoplastic forming (TPF) is an efficient process for the fabrication of metallic glass (MG) components. However, the die and the glass adhesion has been one of the crucial issues in the TPF production (Monfared et al. 2017); comprehensive experimental and theoretical studies on the adhesion between two typical MGs (La-based and Zr-Based) and various die materials including electroless Ni–P, Si, polytetrafluoroethylene (PTFE), sapphire, and SiC were carried out. It was found that of the above die materials investigated, PTFE and sapphire were the best in preventing adhesion followed by SiC, electroless Ni–P, and Si.

Further, theoretical prediction also indicated that the work of adhesion of PTFE and sapphire is the lowest among the employed dies, which agrees very well with the experimental results. However, the low melting point of PTFE makes it not a suitable die material for the TPF of some MGs. By considering some of forming requirements and conditions, this study concluded that sapphire is the best die material for the thermoplastic forming of MG components (Monfared et al. 2017).

The term "adaptable die design" is used for the methodology in which the tooling shape is determined or modified to produce some optimal property in either product or process. The adaptable die design method, used in conjunction with an upper bound model, allows the rapid evaluation of a considerable number of die shapes and the discovery of the one that produces the desired outcome. A variety of criteria can be used in the adaptable die design method, for example, dies which produce minimal distortion in the product. A double optimization process is used to determine the values for the flexible variables in the velocity field and second to determine the die shape that best meets the given criteria (Gordon et al. 2007).

Because of the above, it is clear that technology is the knowledge that will boost the innovation in the die industry. Thus, it will allow the improvement of the die designing process, quality service, delivery time savings, and reducing operating costs.

2 Problem Description

In Mexico, there are manufacturing sectors with more development than the Latin America average. These sectors have an enormous economic impact. As a result, the Mexican economy is one of the 15 largest economies of manufacturing exports worldwide (Mexican Council of International Affairs, Mexico in the world, 2010). The economic report of the export manufacturing industry up to May 2016 had a remarkable performance with an annual growth of 2.31%.

A report by the National Institute of Geography and Statistics (INEGI, by its Spanish acronym) mentioned that the manufacturing industry contributed in the year 2014 with 17.5% of the Gross Domestic Product (GDP) of the country, being one of the economic sectors with greater potential growth.

On the other hand, Mexico reported economic growth of the manufacturing industry from 2013 to 2015 of 1.8% per year, a percentage that is far from the rhythm that has a substantial impact on the economic growth of a country and which is estimated at an annual average of 7 and 8%.

According to the report about World Development Outlook for the Organization for Economic Co-operation and Development (OECD, 2014), Mexico has made great achievements regarding export diversification and raised the sophistication of its manufacturing industry. However, the achievements mentioned are based on imported that are re-exported with low levels of added value and little use of national suppliers.

This phenomenon is reflected in the dynamics of the automotive, auto parts, electrical appliances, and metallurgical industries, which are related to the lack of capacities in specialized national suppliers. There is an export model with relatively low labor costs, which raises the national capacity to compete in international markets. However, to dynamically sustain growth, it is necessary to create and strengthen the capacities to design, develop, and introduce to the market new products or technological processes. In a few words, to foster innovation.

Mexico exports metalworking products mainly to the United States and Canada. ProMexico is a Mexican program that offers several resources to the industry, and it has identified different opportunities for impelling the scientific and technological development in 15 metal mechanical processes in the automotive, auto parts, electrical, consumer electronics, and appliances sectors, as well as the aeronautics sector.

Among the eight processes with the greatest market opportunity (in millions of dollars, mdd) are (1) stamping ($12,724 mdd); (2) casting ($11,225 mdd); (3) forge ($10,291 mdd); (4) machining ($8,969 mdd); (5) injection and extrusion of plastic ($6,796 mdd); (6) molds, dies, and tooling ($4,131 mdd); (7) surface treatments ($2,305 mdd); and (8) thermal treatments ($1,450 mdd) (Proméxico 2017).

Therefore, it is necessary to encourage the national supply of molds, dies, and tools through the creation of infrastructure, knowledge development, and exchanges of experiences with the productive sector for their design, testing, and construction.

Consequently, the creation and strengthening of the technological infrastructure in molds, dies, and tools, with an incentive to applied research, and the development of specialized human resources, will establish the technological bases for the competitive and productive growth of the national companies.

The design, development, and manufacture of molds, dies, and tooling require a high investment in equipment for advanced manufacturing as well as the updating of production equipment.

This context demands a multidisciplinary collaboration among different companies, higher education institutions, and research centers that have previous experience and related infrastructure and who, through the integration of a group of firms or associations, can address the current problems and manage to generate significant technical expertise to consolidate the regional integration of supply chains.

3 Objectives

Create a company with highly trained and specialized Mexican people in the design of dies for covering the five percent of the national market. This company will reduce the import and the technological dependency of Mexico and simultaneously will generate employment and well-paid labor directly and indirectly.

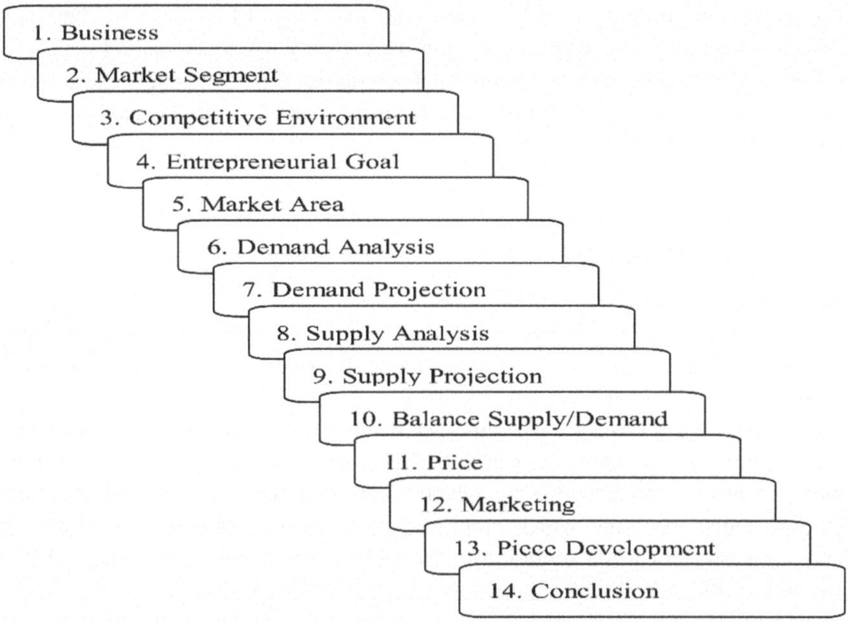

Fig. 1 Entrepreneurship methodology of a company, specialized in dies innovation

4 Methodology

The methodology followed by this project is shown in Fig. 1, and this methodology is followed in the course of formulation and evaluation of projects in the National Technological Institute of Mexico, where the students will be able to formulate, evaluate, and manage investment projects which allow them to set out companies of goods and services based on the criteria of competitiveness and sustainability.

5 Development

5.1 Business

The Torueles Mexicanos company has as main activity to provide specialized in die design in San Juan del Río, Queretaro, using specialized software, including Solid Works ® and NX ®. The company will provide the design and analysis of competencies to enterprises that require a tooling. As a local supplier, the enterprise can offer service with several advantages and thus to reduce their operating costs and achieve quality in their final product. The design activities will be made by the members of the company, who already have experience in the use of software, as well as previous training in dies design with two types of experts: teachers and external consultants in companies, having the capacity to satisfy market needs.

The design analysis will be an important activity for the company, because it will offer some solutions to problems coming from the concept of the pieces, with tools such as finite element.

5.2 Market Segment

The potential markets are compounded of all companies dedicated wholly or partly to the manufacture of dies. It is estimated that at the national level the metalworking industry has 8,932 economic units by 2015.

This stage of the project will focus specifically on the central and lowland zone of the country that includes the states of Guanajuato, Jalisco, Queretaro, San Luis Potosí, Zacatecas, and Mexico City, which according to INEGI as a whole had total of 3,052 units economic which employed more than 18,000 people with total salaries greater than 1,705 million Mexican pesos. It is worth mentioning that it is expected to offer the service throughout the country in the future.

The service will also be provided to companies that require metal parts so that their products achieve the objectives of operation/cost. Within these companies is the automotive and aeronautics, among others.

5.3 Competitive Environment

The metalworking industry in Mexico generates a great contribution to the country's economy. However, it has not had enough development or investment to cover the national demand, which continues to grow each year because of the impact that it has on their industries.

Therefore, the Mexican industry demands additional resources to cover the needs of hundreds of companies. In the Mexican market, domestic suppliers only offer between 5 and 10% of the molds, dies, and tools required. The rest is covered with products imported from the United States, Canada, Germany, Portugal, Spain, and Asia, according to CONACYT (Shop 2015). For instance, the Company Huilyee stamping die, the largest exponent in China, is reaching up to 2,600 million dollars in 2015, according to the FCCyT study. Thus, the market share is important. It is necessary to introduce to the market a service of quality and price comparable with the Asians, which will allow a competitive advantage in Mexico.

5.4 Entrepreneurial Goal

The goal is to offer the dies service design to companies located in the central region of Mexico that works with any matrix.

5.5 Market Area

Due to the location of the company, the central and lowland zone of the Mexican Republic has been selected as the main market at this stage of the project, not closing the options to other areas of the country. The states that belong to this zone area are Aguascalientes, Mexico City, Guanajuato, Jalisco, Puebla, Queretaro, San Luis Potosi, and Zacatecas.

5.6 Demand Analysis

It is crucial to recognize that the demand of the metalworking industry has increased moderately in recent years with an annual average of 7%, as can be seen in Fig. 2.

The metalworking industry stands out the activity, machining of metal parts for machinery, and equipment in general, where the punching is located. This activity contributes 88% of the economic activity of this sector as shown in Fig. 3, with a total of 8,932 companies.

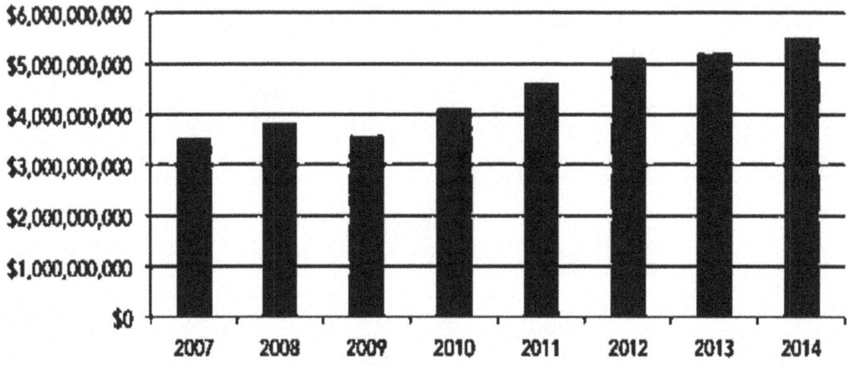

Fig. 2 Total manufacturing production in Mexico (INEGI 2017)

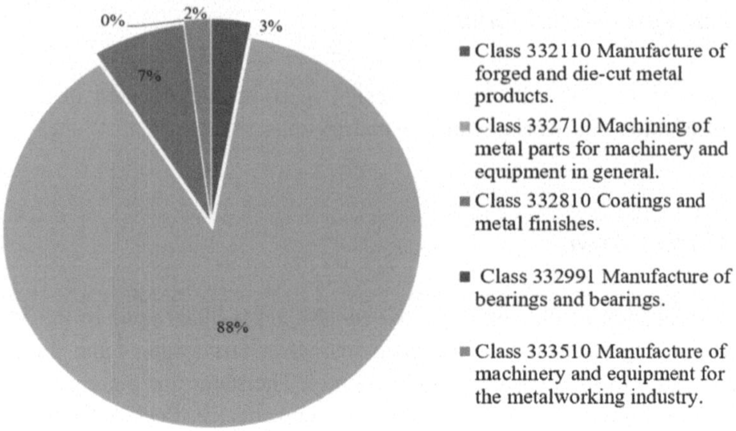

Fig. 3 Distribution of economic activity of the metalworking industry (INEGI 2017)

5.7 Demand Projection

Figures 4 and 5 show the current demand situation and also a comparison of data at the national level with the preferred market zone (INEGI 2017).

There are more than 3,000 economic units in the selected market. Among them are selected those that generate more than 1,705 million pesos, and it is expected to give attention at least for 15% of them in the first year, which are approximately 450 companies. Using this information and demand, it is possible to propose a forecast. Figure 6 depicts the tendency.

- 3000 = number of economic entities in the year 2015.
- 7% = average growth of the metalworking industry.
- 10 = years required to estimate.

Fig. 4 Comparison between total national and total regional economic INEGI 2017

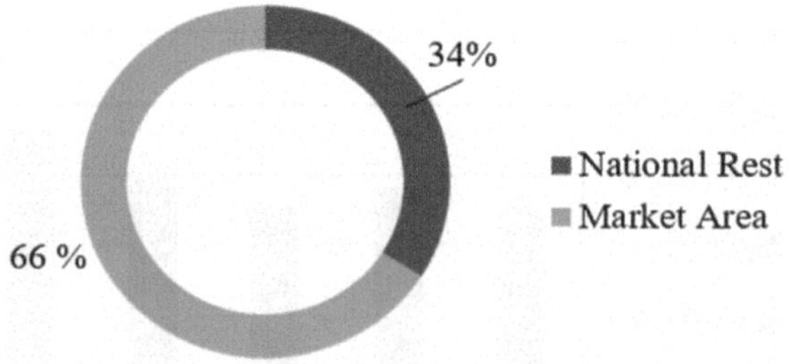

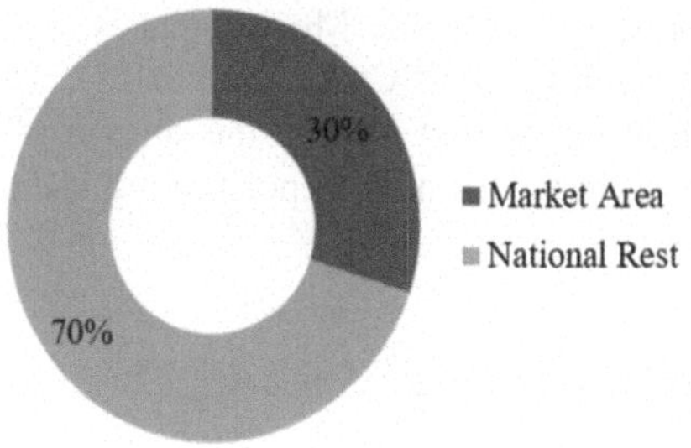

Fig. 5 Number of persons hired directly in the potential market (INEGI 2017)

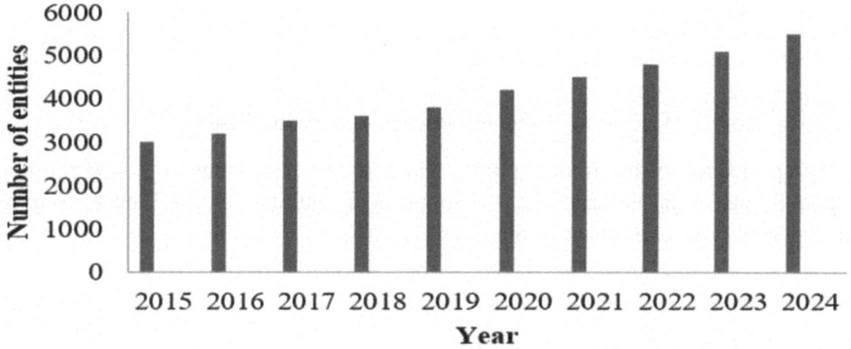

Fig. 6 Demand estimation in 10 years (INEGI 2017)

This demand projection shows about 80 economic entities for this year, which is the demand of the market sector, ensuring growth that guarantees the operations in the market.

5.8 Supply Analysis

The main market offer is abroad according to CONACYT. Local suppliers in Mexico only satisfy between 5 and 10% of the molds, dies, and tools that are required.

In Mexico, there are also companies that offer the die design. For instance, ACM metal, dies, and industrial matrices. Akiyama is another example. However, according to the VSI Consulting, at least 85% of them do not have any certification or specialization. The lack of this specialization is a problem because they cannot generate an added value to the product. Nevertheless, their experience and time in the market is a point to consider as a competitive advantage.

There is information on imports that Mexico makes abroad, which exceeds 2, 600 million dollars, according to FCCyT (Shop 2015). This volume of service together with the work of the current market, the support of large companies, and the local factor can be an immense help in positioning itself in the market.

5.9 Supply Projection

The incertitude in available information, particularly in demand for the new service, makes difficult to estimate the service demand. Nevertheless, taking into account imports in the Mexican market makes feasible to estimate demand. Figure 7 contains this information.

- 2, 600 = importation in mdd of dies.
- 7% = average annual growth of the metalworking industry.
- 10 = number of years to compare between 2015 and 2024.

The projection of the future supply, which in this case is about 3, 000 million dollars for next year, indicates that the supply will continue to grow. Hence, there is an opportunity to participate in this market.

5.10 Balance Supply/Demand

Table 1 shows the estimation of the supply and the demand for the current year. In this table, it can be observed that the growth is upward when compared with the historical data registered in the INEGI.

Based on this analysis, the information provides security to the company since there is a market in which it can compete with a high demand risk, leaving the level of growth in the marketing strategy.

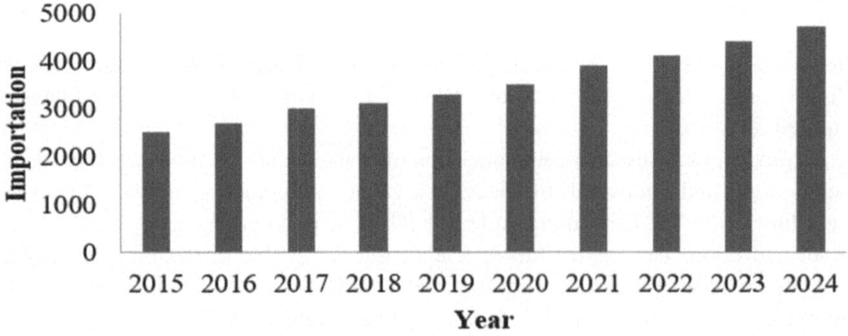

Fig. 7 Supply for the service (INEGI 2017)

Table 1 Balance supply/demand of the company

Balance supply/demand at the end of 2017	
Supply	3,500 Economic units
Demand	$3,000,000

5.11 Price

To estimate the hourly price service, it is necessary to perform a cost and expense analysis. The cost–benefit relations are prepared and the expense per month provides a profit of 40%. Likewise, to calculate the annual profit, it was considered a 1-year projection with equal sales and an equal number of projects. The relation shows an amount of 85% of the capacity. The estimation considers ten hours per project, a total of 40 projects, and a team of three experts. Tables 2, 3, 4, and 5 as well as in Fig. 8 show the analysis performed.

With this information, it is possible to calculate the profit per month and estimate the annual profit considering inflations. According to a group of experts consulted by the Bank of Mexico, which is also similar to the estimation of the Forbes magazine, the forecast is about 5.24% for this year.

The available information moves the profit to a value of 195%, which translate the project into a promising service that will have an opportunity in the Mexican market.

This analysis was determined with a service price per hour of 557.25 Mexican pesos, which gives a financial estimation of 132,776.92 Mexican pesos per project. Under such conditions, the profit will be near to 2,143, 024.82 Mexican pesos per year. The investment costs and the profit produce an index of 95%, comparing the profit with the investment.

Table 2 Company's expenses and costs

State of expenses and costs		
Element	Per month	Per hour
Software SolidWorks ®	$ 21,538.46	$ 134.62
Software NX ®	$ 15,384.62	$ 96.15
Rent	$ 5,000.00	$ 31.25
Electricity service	$ 500.00	$ 3.13
Water service	$ 500.00	$ 200.00
Salary payments	$ 32,000.00	$ 6.25
Logo rent	$ 1,000.00	$ 6.25
Paper and office supplies	$ 1,000.00	$ 6.25
Phone service	$ 1,200.00	$ 7.50
Publicity	$ 10,000.00	$ 62.50
Internet	$ 1,500.00	$ 9.38

Table 3 Benchmarking among similar company operation costs

Operation costs	$68,084.62	$ 428.65
Expected utility	30%	
Net price per hour:	$ 557.25	

Table 4 Company's utility estimation per month

Expected utility per month		
Total estimated sales		$222,900.00
Operation costs	$ 234.38	
Administrative expenses	$ 328.89	
Total operation expenses		$ 90,123.08
Net utility		$ 132,776.92

Table 5 Company's inflation estimation

Inflation estimation	
Period (Month)	Utility
1	$132,776.92
2	$139,734.43
3	$147,056.52
4	$154,762.28
5	$162,871.82
6	$171,406.31
7	$180,388.00
8	$189,840.33
9	$199,787.96
10	$210,256.85
11	$221,274.31
12	$232,869.08
Annual utility	$ 2,143,024.82
Annual average	$178,585.40

5.12 Marketing

The budget for the project is estimated at 10, 000.00 Mexican pesos per month for specialized advertising, although it should be noted that the community participation would be the most important.

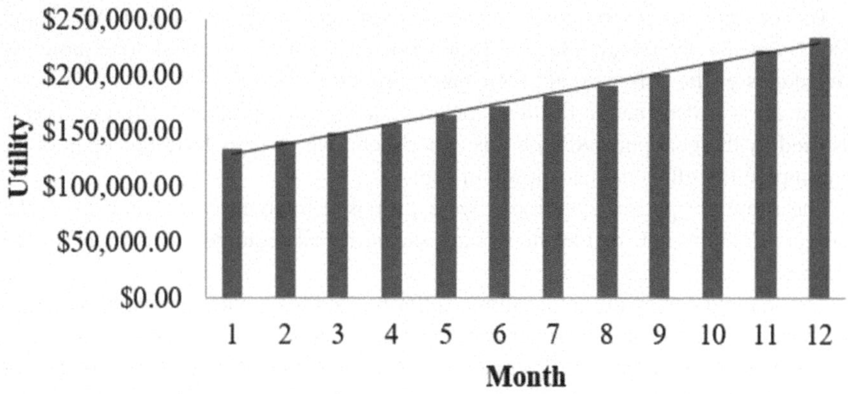

Fig. 8 Company's annual projection utility (INEGI 2017)

5.13 Piece Development

5.13.1 Specialized Software

Technological advance is becoming more intuitive, making easier their implementation in industrial and service sectors. Such is the case of Siemens, where the implementation of the innovative software NX ® has made a big contribution to the design of dies and molds.

In Mexico, there is a saying: "Mexico is good making more of the same thing, but in the disruptive" (Castillo 2001); this phrase calls the attention because there are few specialists, many tools, and no guidelines on how to use them.

5.13.2 Dies Design Efficiency

Tooling design has passed from being a complex process based on complex formulas that require specialized people to a much easier and more efficient process, aided by the development and implementation of specialized software, which allows the design of a toll in few hours.

Regarding this industrial need, we designed a die using both, AutoCAD ® which is general design software and NX ® Siemens, based on blocks specialized software, which through a list of steps guides the user during the design process allowing a non-experienced person to handle it. However, the non-experience does not mean a lack of knowledge about the process and the parameters to be introduced. The user must be capable of understanding the process, as well the inputs and the outputs to be able to take complete advantage of the software.

To compare the performance of the two software packages (AutoCAD ® and NX® Siemens), the research team selected a piece belonging to a lock to explore the advantages of the software and their basic structure.

The die was designed to fit into a lock, due to its basic structure. This decision is justified by the goal pursued with this experiment, which is to design piece in order to compare the efficiency as shown in Fig. 9.

The same design and conditions were proposed for both cases, mainly on the software domain and for the die design, including all required calculations in the process.

The loading forces die dimensions, suffering, guides, punch retainers, punches, and "required steps", among other parameters, which were calculated to understand the fact that once the piece is designed, the software proceeds with the steps, except for some required data. Material thickness tables and opposition to cutting are considered, among other variables.

On the other hand, it is worth mentioning that when the specialized software is not available, there are people who perform the design calculations by hand to have a more significant certainty of the die parameters before the manufacturing process. With the specialized software, this activity is done faster. The user just needs to feed the data requested by the software, as shown in Fig. 10.

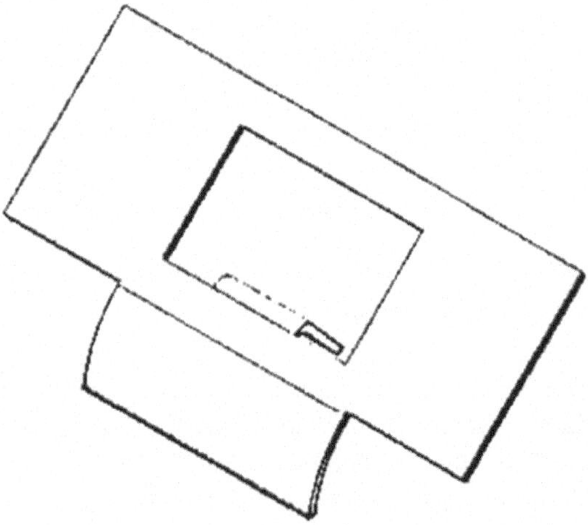

Fig. 9 Metal lock/part for which the die will be designed

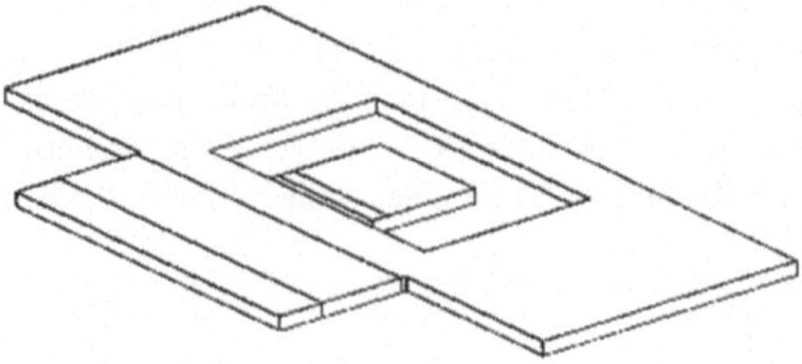

Fig. 10 NX ® Part/Software deployment

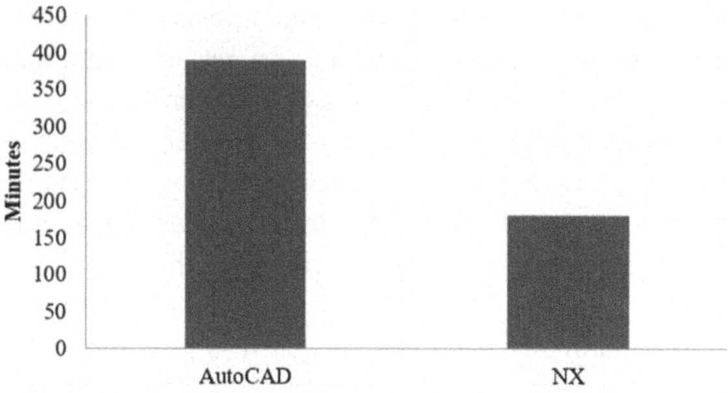

Fig. 11 Spent time (minutes) for die design in each of the chosen softwares

A meaningful reduction of time in the die design was observed among the software. Figure 11 explains this difference.

In the case of AutoCAD®, 390 min were required to perform the design, while the specialized software (NX® Siemens) required 185 min. Based on these results and considering this specific design, an improvement of 52% is observed; however, the improvements will vary according to the complexity of the design.

5.14 Remarks

Results of the case study for the service in the die design show several advantages such as follows:

- High-speed production and greater number of equal parts at lower cost;
- Savings in the use of raw material, less scrap, and a bigger number of parts per roll;
- More safety in the operation;
- Flexibility in the operation due to a fewer operators and presses;
- Space saving;
- Automatic operation;
- Saving time in the die adjustments;
- More safety and comfort in the operation; and
- Reduction of fixed operating costs.

6 Conclusion

In Mexico, the companies with the need to work with high-speed dies face the problem of training personnel. This situation demands the development of several skills related to the design, adjustment, maintenance, and refurbish of dies.

It is overwhelming the fact that, even when there are new technologies available in the market or new developments every day, in Mexico there is still a lack of interest from several actors (the industry, research centers, and universities to mention but a few) to learn how to get an advantage of this situation. As a result, there is an economic and industrial dependency to other countries.

Consequently, it is highly recommended that the universities and research centers work together to propose training programs for satisfying the need for specialized people in the die design sector. This action will enhance the capacities of the country, and Mexico will have the technical capabilities in this industrial field.

It has already been shown that it is possible to carry out a work of such magnitude, without too much experience in the use of the software. However, there are more to develop and investigate to take advantage of available software and thus, to move the design competencies to the next level of the product, the management of the lifecycle and the manufacturing process, for finally, develop functional and active tooling.

As revealed in the study, the services offered by the company are a business opportunity for proposing new services in the design of dies, but also, for proposing new training services. Finally, the technical resources involved in this chapter reveal that it is possible to reduce the time while designing dies and conditions that can foster the innovation in some national businesses due to a fast response not available now in the Mexican market.

References

Apparao, K.C., Birru, A.K. (2017). Optimization of die casting process based on taguchi approach. *Materials Today: Proceedings* 4 (2):1852–1859. https://doi.org/10.1016/j.matpr.2017.02.029.

Castillo, J.C. (2001). Desarrollo de una técnica de diseño de troqueles de alta velocidad para producir laminación rotor-estator grapado. Universidad Autónoma de Nuevo León, San Nicolás de los Garza, N. L. México.

Gordon, W.A., Van Tyne, C.J., Moon, Y.H. (2007). Overview of adaptable die design for extrusions. *Journal of Materials Processing Technology* 187–188 (Supplement C):662–667. https://doi.org/10.1016/j.jmatprotec.2006.11.158.

Hawryluk, M., Zwierzchowski, M., Marciniak, M., Sadowski, P. (2017). Phenomena and degradation mechanisms in the surface layer of die inserts used in the hot forging processes. *Engineering Failure Analysis* 79:313–329. https://doi.org/10.1016/j.engfailanal.2017.04.036.

INEGI (2017) Instituto Nacional de Estadística y Geografía. http://www.inegi.org.mx/. Accessed 14 Mar 2017.

Ma Y, Lou M, Li Y, Lin Z (2018) Effect of rivet and die on self-piercing rivetability of AA6061-T6 and mild steel CR4 of different gauges. *Journal of Materials Processing Technology* 251:282–294. https://doi.org/10.1016/j.jmatprotec.2017.08.020.

Monfared, A., Liu, W., Zhang, L. (2017). On the adhesion between metallic glass and dies during thermoplastic forming. *Journal of Alloys and Compounds* 711:235–242. https://doi.org/10.1016/j.jallcom.2017.04.005.

Proméxico (2017). PRO México, Trade and investment. www.promexico.gob.mx/en/mx/home.

Shop, M.M. (2015). Industria de moldes y troqueles en México. http://www.mmsmexico.com/art%C3%ADculos/industria-de-moldes-y-troqueles-en-mxico. Accessed 15 Jul 2017.

TRIZ Evolution Trends as an Approach for Predicting the Future Development of the Technological Systems in the Food Industry

Jesus-Manuel Barragan-Ferrer, Stéphane Negny, Jonas Damasius, Diana Barragan-Ferrer and Dalia Cizeikiene

Abstract Innovation is regarded as a vital element for the development of new technologies to create a competitive advantage in the food industry. The innovation process in the food companies is principally driven by two strategies: the technological change (technology push) and the market orientation (market pull). Although both strategies provide the elements to encourage innovation, they can be only applied to solve specific problems or needs that are easily identified limiting their applicability. A new alternative strategy to innovate is to try to anticipate the future developments of technological food systems from an evolutionary perspective based on the TRIZ Evolution Trends (TETs). Thus, to improve the innovation process in the food industry, the present research proposes to combine elements of the TRIZ Evolution Trends, the technology push and the market pull strategies to create an approach to suggest recommendation of the most favorable transformation to improve or develop technological food systems. An example is presented to demonstrate the usefulness of this approach.

Keywords TRIZ · Evolution trends · Technology push · Market pull
Food industry

J.-M. Barragan-Ferrer (✉) · J. Damasius · D. Cizeikiene
Faculty of Chemical Technology, Department of Food Science and Technology, Kaunas University of Technology, Kaunas, Lithuania
e-mail: barraganferrer@gmail.com

J.-M. Barragan-Ferrer · D. Barragan-Ferrer
Faculty of Medicine, University of Applied Sciences, Kaunas, Lithuania

S. Negny
Laboratoire de Génie Chimique UMR CNRS/INPT/UPS 5503, Université de Toulouse, Toulouse, France

G. Cortes-Robles et al. (eds.), *Managing Innovation in Highly Restrictive Environments*, Management and Industrial Engineering,
https://doi.org/10.1007/978-3-319-93716-8_12

1 Introduction: Innovation in the Food Industry

Innovation in the industrial sector is recognized as a strategic factor for the economic growth and the company's success in which knowledge is used to solve technological problems or generate value by using new or improved methods, materials, and devices (Galanakis 2006). In particular, the food industry, that is highly competitive and one of the largest manufacturing sectors, is in continual change, and therefore requires constantly innovating to propose inventive products.[1] In this sector, the innovation entails the need to improve or develop new technologies to create successful products with high quality, to satisfy the changing demands of the consumer, and at the same time to reduce the time to market. A successful innovative product will, in turn, create a business opportunity to retain existing customers, and to attract new customers to obtain an advantage over competitors, as well as a way to add brand value for the company and ensure better sales. However, the innovativeness of firms in the food industry, most of which are SME's, is restricted to the proposition of solutions with a low level of inventiveness as well as to the use of mature technologies that usually are developed outside the food sector (Bigliardi and Galati 2013a, b). Also, radical innovations in this industry are very scarce (Traitler et al. 2014). This limited innovative capacity is reflected by the reduced amount of patents in the food technology compared with other sectors (Bigliardi and Galati 2013a). Another cause for this low innovation level, according to (Capitanio et al. 2010), is that the consumer has a conservative behavior, and refuses to accept new food products, thus creating a "consumer inertia" in which the industry only introduces a small variety of food products. All the situations mentioned above give the impression that the food industry is a slow-growing sector with a low technological level, and low R&D intensity (Hirsch-Kreinsen et al. 2006; Mäkimattila et al. 2013). It is therefore obvious the need to start changing this perception by managing the innovation process to enhance the innovative capacity in the food industry to develop new technologies with the potential to create a competitive advantage.

The innovation process in the food industry is a complex process that is led by two principal "forces": the technological development (the technology-push), in which the companies can develop or acquire technologies, and the market pull, in which consumers and external factors exert pressure on the food companies (Galanakis 2006). While both strategies share a common purpose of developing the technology to satisfy the customer needs with new or enhanced products, the idea of simplifying the innovation process to only two strategies seem to be very restrictive to anticipate the future business opportunities in the food industry (Mäkimattila et al. 2013). Therefore, food companies have a growing need for developing or integrating new approaches to enhance the innovation process and then, simplify the generation and commercialization of new technologies. An alternative

[1]The term product includes a product, process or service at different hierarchical structure levels (component, system, super-system or environment).

possibility is to complement these two strategies with the TRIZ Evolution TRENDS (TETs) as an additional source of innovation to provide new technological opportunities in the product development. These TETs have been applied to technological analysis as a support for a decision making tool to predict further improvements of technological products (Park et al. 2013). Thus, the purpose of this chapter is to introduce an approach to apply the TETs in food industry.

2 Innovation Strategies in Food Technology

Next paragraphs describe the most common strategies for managing innovation in the food industry (Del Giudice et al. 2012).

The **technology push strategy** is based on the premise that the advances in science and technology govern the evolution of technological innovations (Nemet 2009; Di Stefano et al. 2012). These technological opportunities emerge thanks to an idea or a scientific discovery (scientific research) that brings the technological change necessary to create an innovative product (an invention that is commercialized). In other words, a scientific discovery that becomes an invention and creates a technological opportunity that is transformed into a new product that pushes the market and impulses the demand (Godin and Lane 2013). Here the demand means the acquisition and consumption of products. The principal activity in technology push is the discovery of new knowledge and/or the transfer of know-how to solve different needs or problems that could be later crystallized in innovative products, processes, and services. The innovation process following this strategy starts with the scientific research stage, then the creative stage, next, the technology generation or invention stage (development), and ends with the diffusion stage (market).

From a different perspective, the **market pull strategy**, considered the major influence in the innovation process, states that the market conditions determine the evolution of innovation. These conditions are based on the needs (also called demands) of the people and the market, which "push" for a demand of inventions that are then transformed into commercial products (Nemet 2009). A need can be interpreted as a wish, a problem, an unsatisfactory solution, requirements, scarcity of resources, regulations (legal, political, sociocultural or environmental) or a human desire (e.g., convenience, safety, health, and taste) in which the science must seek a solution. This strategy is based on the principle that "*the majority of successful innovations arise in response to a specific need*" (Godin and Lane 2013). The market pull strategy also includes the demand pull that occurs with the need to improve a new technology after its introduction into the market (Galanakis 2006). This demand pull transforms the gaining experience during the use of this new technology into incremental or radical innovations. This experience is the knowledge created in a progressive manner during the different stages of development, production, and usage. The innovation process of this strategy is similar to the technology push, but it adds a stage of exploring market needs at the beginning of the process.

By comparing the level of innovativeness of these two strategies, when the market pull is associated with incremental innovations and mature technologies, technology push are related to technological breakthroughs (radical innovations) (Brem 2008; Peters et al. 2011). The choice between one strategy and another may reduce innovativeness of firms. For example, using only the market pull, we could propose a replacement for the existing product based on consumers needs, while the competitor could introduce a new technology and then threaten our competitive advantage (Brem 2008). Thus, both strategies can be considered complementary rather than contradictory. Therefore, they can coexist sequentially or combined as proposed by various authors (Kleinknecht and Verspagen 1990; Arthur 2007; Nemet 2009; Bigliardi and Galati 2013b) For instance, in combination, the introduction of a new technology with high potential for market success requires stimulating the market, and the market-needs sometimes requires competences and knowledge that goes beyond the scientific frontier of the existing fields of technology (Mowery and Rosenberg 1991; Di Stefano et al. 2012). In sequential synergy, both strategies can create a sustainable competitive advantage in a company. This means that the technology push is a long-term strategy focused on radical innovation creating radical technological solutions, major improvements, or reinventing the market. While the market pull strategy is focused on incremental innovation through the creation of continuous improvements or substitutions to meet the actual market demands to retain the customer, and ensure the financial health in the short-term (Ferrer et al. 2012). Thus, the challenge for the firms in the food industry is to find the right balance between the technology push, to get the right technology, and the market pull, to understand the market.

The problem is not to choose the right or the wrong way to innovate using one of these strategies, but the multiple factors that influence the product development which are difficult to identify using these strategies. The "new needs" is one factor that is hidden or unrevealed. Therefore, this factor would require the correct market demand from the infinite possibilities of human needs, the identification of the product that the consumers hadn't known they needed, or the identification of the current or future value of the product (Nemet 2009). The market opportunity for a new discovery is another factor that would require the identification of a need for a product that did not exist, or the identification of a specific need (or problem) that is going to be resolved with the new scientific discovery or technology.

Also, these strategies have shown limitations to get a real competitive advantage. To create a technological change with a completely new system, technology push requires the discovery of a new phenomenon that is very rare in science (Altshuller 1999). Furthermore, there is a resistance to invest in a new, unproven technology due to the high level of risks and uncertainty. On the other side, market pull fails to develop innovations that create a real dominance in the marketplace (Galanakis 2006). The market pull is merely interested in satisfying the market conditions without considering whether the technological capabilities exist or not. To face the limitations of the above-mentioned factors, it is possible to include an evolutionary perspective with the TRIZ Evolution Trends (TETs).

3 TRIZ Evolution Trends of Technological System in Food Industry

An alternative strategy to address the problems from an evolutionary perspective for predicting the future development of the technological systems and anticipate market demands in the food industry is the use of the **TRIZ Evolution Trends** (TETs), also referred in the literature as patterns of evolution (Terninko et al. 1998) or laws of evolution (Altshuller 1984). Through the development of the TRIZ theory, Altsuller, its creator, analyzed the historical evolution of numerous technological systems and noted that they evolve not randomly, but follow certain principles, which have repeatedly been used to solve problems during their development. These principles were generalized into objective regularities or "*Patterns of Technological Evolution*" (Zlotin and Zusman 2013) that can be used to propose recommendations of the most favorable transformations to create the next generation of a technological system (Altshuller 1984). As Fey et al. explain (Fey and Rivin 2005): the TETs "*describe significant, stable, and repeatable interactions*" between the system, their elements, and the elements in the environment during its progressive evolution process. Hence, solutions proposed for resolving the contradictions or improving any technological system must be consistent with these TETs (Cavallucci and Khomenko 2007).

The TETs also suggest ways of how various resources from different sources such as the parts of the system, adjacent systems or the environment can be discovered, generated, consumed or even reduced (because are inefficient or redundant) to continue developing the system itself or other systems. Therefore, the TETs allow the control of the evolution of a given technological system by predicting how it would evolve (Brem 2008).

Initially only eight trends were proposed by Altshuller (1984), but over the time, the TETs have been expanded, improved, and strengthened to the point that various trends and lines of evolution are now available to explore different scenarios to improve technological systems. A **line of evolution** or path, describes the sequence of incremental evolutionary changes or transitions that a system undergoes within a trend. Therefore, we can identify the actual state of a system and know in advance the next steps in its evolution. For example, the line of evolution of the trend "substance segmentation" is as follows: solid, powder, liquid, gas, and field. We can observe this trend in the commercial sugar: sugar blocks, powder sugar, syrup, and aerosol.

The ability of the TETs to predict the evolution of technological systems, that is often difficult with the market pull and technology push strategies could bring new opportunities in the development of food technologies during the conceptual design phase. We have therefore performed an analysis of previous works about TETs, including Altshuller's work, to arrange TETs in such a way as to facilitate the application of the existing patterns to analyze and predict the potential evolution states of particular food technology. There are many possible ways to organize the TETs. According to Park et al. (2013), the TETs cannot be equally applied in all the current lifecycle of a technological system. Thus, the trends should be arranged and used regarding the current lifecycle stage of the system under analysis. For Mann (2002), the trends are grouped in three interface situations: time, space or interface. In this research, we structured the trends depending on the technological change of the system: general evolution trends, evolution trends at the system level, evolution trends at the higher-level system, evolution trends at micro-system, and other additional relevant trends, as described below.

3.1 General Evolution of Technological Systems

The trends included in this category describe the necessary conditions at the structural and functional level of a technological system.

3.1.1 Trend of Increasing the Degree of Idealness of the Technological System

The evolutionary development of a technological system tends toward the ideality. This ideality is acquired as a result of increasing and improving its useful features (functions and properties), and reduces its disadvantages (harmful actions/effects, cost, noise, non-needed properties, complexity, and problems, etc.). This trend is used to recognize what functions, effects, elements or properties in a technological system need to be improved, and which ones must be removed with the possibility of applying other trends. Furthermore, the ideality serves to evaluate the result of any solution used to improve a technological system.

The line of evolution in this pattern includes several ways to improve useful features, and to reduce disadvantages (Petrov 2002; Zlotin and Zusman 2013):

- Improve the existing functions/properties/parameters in the system
- Increase the use of resources present in the system and its environment
- Absorb a useful function of adjacent systems or the environment
- Add new functions/properties in the system
- Reduce/prevent/eliminate the harmful factors
- Inventing new applications for harmful factors

These general ways to increase ideality are directly related to the other evolution trends. For example, to improve the existing function in a system can be achieved by the resources trend.

3.1.2 The Trend of Non-uniform Evolution of Technological Systems

The parts in a technological system develop unevenly. Considering the trend of ideality, when we identify the useful feature, or the disadvantage and its associated element in the system or environment, we focus our efforts to improve this part causing an irregular development of the other parts in the system. This irregular development will inevitably cause conflicts (problems, bottlenecks, and contradictions), thus impeding its future development (Savransky 2000; Rantanen and Domb 2010). This trend also reveals that the parts of a system evolve independently and at different rates. Then, we need to identify the parts "underdeveloped" and try to apply the others TETs.

3.2 Evolution at System Level

The trends in this category describe the different ways to improve a technological system at the system level, including its components. These trends aim to generate incremental changes in the system through the use of resources, dynamization, coordination, and the enhancement of interactions and controllability.

3.2.1 Use of Resources Trend

It is well known that in the technological system, multiple resources are present, but not used, or we do not know how to use them. This "hidden resource" can be a source of new functions or properties (Negny et al. 2012); thus it should be identified and try to use it creatively. Resources include material, substances and their state, fields, spaces, and time available in the system and its environment. The resources in a technological system are divided into two general groups: substances and fields. Consequently, two sub-trends, namely evolution in substance and evolution of field are proposed as a way to reach better outcomes in a technological system. In each sub-trend, a series of lines of evolution are also proposed, as described in the Annex 1. This annex has 12 tables. Each table describes a particular line of evolution (Table 12.1).

3.2.2 Dynamization Trend

A technological system evolves towards more adaptable, moveable, adjustable, flexible (more degrees of freedom) structure capable of adjusting to the changing environment. Its structure can also be configured in different ways, having more variety of states and functions or the ability to adapt to multiple requirements. Thus, in this trend, the system started as a solid substance, followed by increasing its segmentation (becoming smaller and smaller) and, finally, the structure becomes a

Table 12.1 The possible links between nodes in a network of problems. *Source* (Cavallucci et al. 2010; Becattini et al. 2015)

Nodes	Relation	Description	Representation
System → System/ component	Can be subdivided into	The technological system can be subdivided into sub-systems or components to facilitate its description	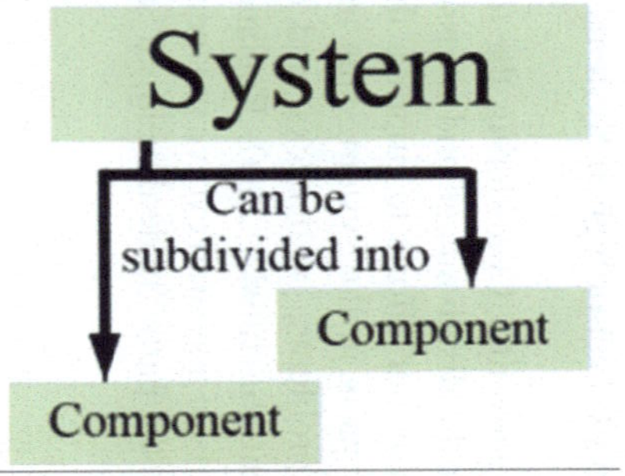
Goal → Subgoal/ problem	Can be subdivided into	Goals to be achieved can be subdivided into sub goals or problems to simplify its description	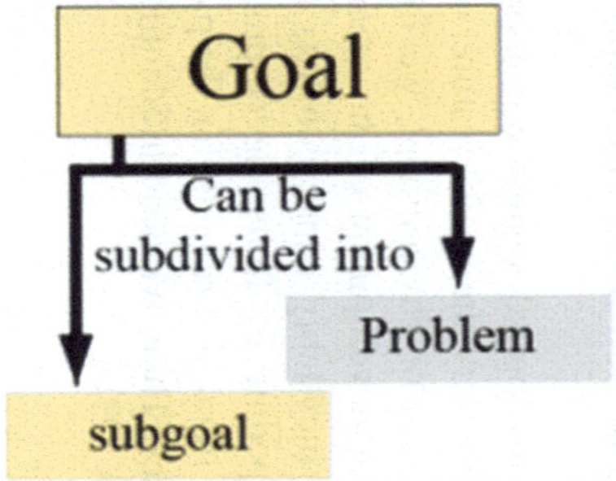
Problem → Partial solution	Can be solved by	A problem can be solved by a partial solution	

(continued)

Table 12.1 (continued)

Nodes	Relation	Description	Representation
			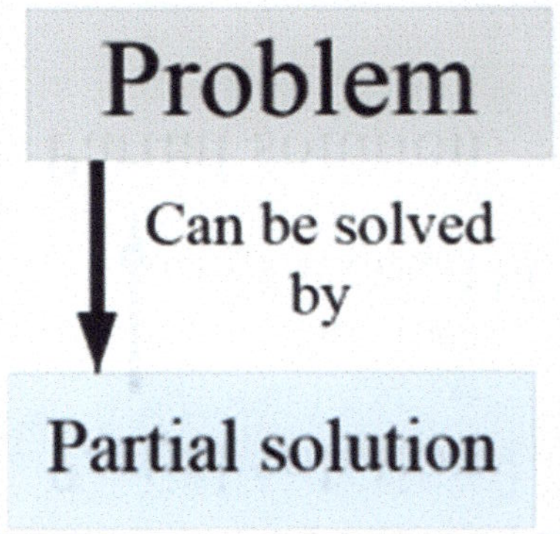
Problem → Problem	Provokes-causes or can be subdivided into	A Problem can cause another problem. A problem can be subdivided into sub-problems to facilitate its description	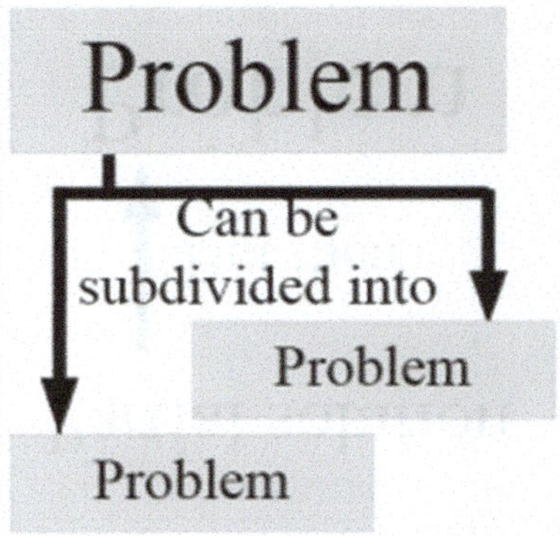

(continued)

Table 12.1 (continued)

Nodes	Relation	Description	Representation
Partial solution → Problem	But then	A partial solution can cause another problem	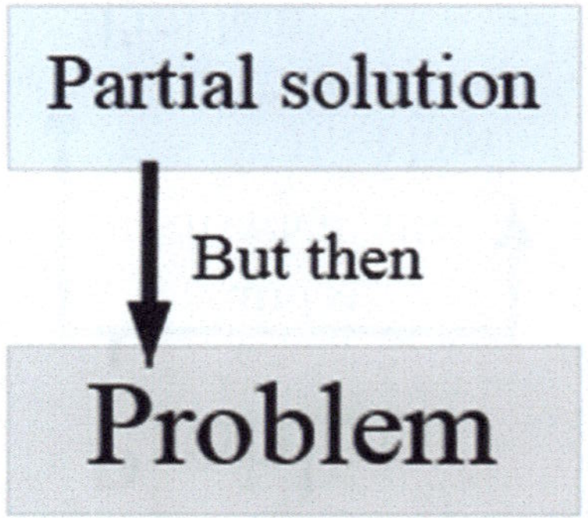
Partial solution → Partial solution	Can be implemented by	A general partial solution can be implemented by a specific solution. This specific solution gives more detailed information about how this solution will be applied	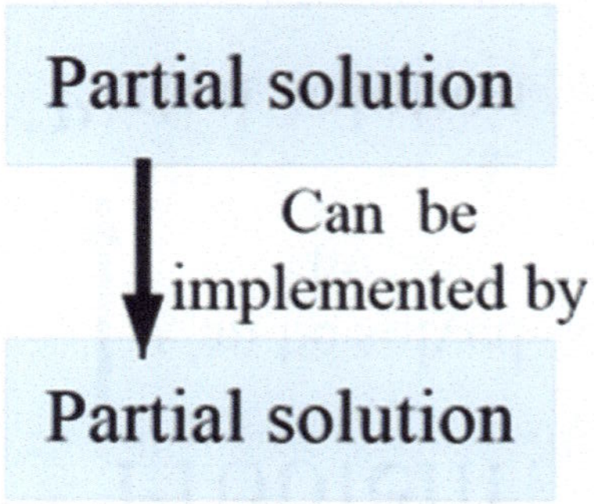

substance with completely free movement. This "division" can be applied to the system structure, materials, function, processes, and parameters (Zlotin and Zusman 2013). The dynamization of any technological system can be done by following these evolution lines (detailed in Annex 1 from Tables 12.2, 12.3, 12.4, 12.5, 12.6, 12.7, 12.8, 12.9, 12.10, 12.11 and 12.12): (1) Substance dynamization, (2) Field dynamization, (3) System dynamization.

3.2.3 Coordination Trend

The purpose of this trend is to ensure optimal performance of a technological system by coordinating or deliberately not coordinating the existing actions, interactions, processes, fields, and substances and their characteristics (speed, geometry, form, color, etc.). For instance, the coordination of frequency or resonance between product and instrument to reduce the use of energy. Also, the non-coordination of self-frequency to prevent harmful effects. This non-coordination is also used when two or more actions or processes are incompatible. In this case, one action is executed when the other is paused or an action is introduced during each paused interval of the other (Salamatov 1999). The line of evolution of this trend is provided in Annex 1.

3.2.4 Interactions and Controllability Trend

A technological system is composed of several parts that interact. This interaction is known in TRIZ as substance field interaction (su-field interaction) or substances interacting through a field. Examples of su-field interactions include the mechanical interaction, thermal interaction, acoustic interaction, electromagnetic interaction, chemical and biological reactions (Rantanen and Domb 2010). Thus, another option to enhance the functioning of the technological system is to improve the interactions, amplify deficient interactions, eliminate harmful interaction, or increase the controllability of the interactions. Finally, the evolution lines of this trend (in more detail in Annex 1) are: (1) Use of substance, (2) Use of field, and (3) Action introduction.

3.3 Evolution to Higher-Level System

Once all the possibilities of improvement at system levels have been exhausted (the limit of its own development has been reached), the evolution of technological system takes place at higher level or "macrosystem". That is, the system becomes a subsystem of a more general level or a higher-level system (Altshuller 1984). At this level, the system and its components, the neighboring systems, and the environment merge (into a whole), or bring together (temporarily or permanently) to

Table 12.2 Use of substance (resource) trends

Trend	Evolution line	Keywords
Material specialization[a] (Savransky 2000; Fey and Rivin 2005)	1. Homogeneous material 2. Separation of the homogeneous material into functional zones (parts, areas, layers, regions). Each part is modified (by changing the physical and/or chemical properties) to provide one function or specific property 3. Replacement each zone with new materials or composite material. If the existing resources cannot provide the desired output, then we can add other materials such as composites that exhibit advantageous properties 4. Transform the compound material into a specialized individual element 5. Convolution of the different individual elements into the ideal material with higher performance and adaptability	Allow variation in properties, optimum functionality, variation in characteristics
Evolution in space (Innovation Suite 2002)	Line → plane/surface → volume	Increase the magnitude and efficiency of useful effect
Linear evolution (Machine; Gadd 2011)	Single point → straight line → curved 2d line → complex curved 3d line	Improve structural strength, the surface area, aesthetics, and ergonomics. Identify/change component orientation and the ability to see objects
Surface evolution (Innovation Suite 2002)	Plane → cylindrical surface → spherical surface → complex surface	Increase surface area, improved use of space
Surface segmentation (Innovation Suite 2002)	Smooth surface → surface with protrusions in 2D → surface with protrusions in 3D → Rough surface → rough surface with active pores	Heat transfer, traction, drainage, aerodynamic, controllability, and insert a new useful function
Space segmentation (Innovation Suite 2002)	Solid → hollow → several hollow → pores → active pores	Reduce weight, multiple functions
Substance segmentation (Innovation Suite 2002)	Solid → segmented solids (with different degrees) → liquid → aerosol → gas → field	Surface area, arrange, separate different functions, isolate stresses/loads, transportability, reduced density, transmission of loads/energy, flexibility, controllability, complexity, maintainability

(continued)

Table 12.2 (continued)

Trend	Evolution line	Keywords
Segmentation of flow (segmenting the action) (Machine)	Continuous flow → flow separated into two parts → flow separated into several parts → flow separated into many parts	Flow distribution, flow control, increase the magnitude of the useful effect
Maximum use of plane/surface[b] (Petrov 2002)	Plane → backside of plane → Möbius strip	Increase surface area
Maximum use of a volume (Savransky 2000)	Volume → "nested doll"	Space utilization, save space, protect objects from damage, ease movement
Increase the degree of voidness (Innovation Suite 2002)	Solid substance → introduction of large void → solid with voids → solid with capillaries → porous substance → micro-porous substance → solid with dispersed micro-voids→	Weight reduction, reduced density, reduce waste
Increasing asymmetry (Savransky 2000)	Asymmetrical system → partial asymmetry → matched asymmetry	Ergonomics, component assembly (poka-yoke), ease of operation, compact installation, orientation visibility, ergonomics, aesthetic appearance, ease of operation
Increasing transparency (degree of transparency) (Innovation Suite 2002)	Opaque construction → partially transparent → transparent → active transparent elements	Improve the ability to see through the object, ability to make simple yes/no measurement, warning indicator, improved aesthetic appearance, increased natural illumination, save energy, increased visibility for safety, ease of inspection

[a]The study of the evolution of technological systems revealed that to improve the functioning of a system, sometimes, it is necessary to possess different properties, including contradictory ones, in various areas in the system (Fey and Rivin 2005). Therefore, it is required to divide the system into specialization areas or functional zones. After this segmentation, the substance's material is modified to be able to adapt to changing conditions

[b]This line of evolution aims to take full advantage of the existing plane using its backside

increase the number of functionalities, to share resources or to show new properties only observable in this configuration (Salamatov 1999; Fey and Rivin 2005). The development towards the macro system is divided into three transitional sub-trends: the first is the expansion sub-trend in which two independent systems (or more) are joined together. However, when this "new super-system" continues to evolve, the

Table 12.3 Evolution of field (resource) trend

Trend	Evolution line	Keywords
Evolution of field[a] (Zlotin and Zusman 2001)	Permanent field → asymmetrical field → alternating field → pulsed field → combination of fields with different parameters	Reduction of time, energy usage, energy loss

[a]The evolution line of substance dynamization follows the transition from stationary components to moving ones by changing its physical state or phase state

Table 12.4 Dynamization trend

Trend	Evolution line	Keywords
Substance dynamization (Salamatov 1999; Kwatra and Salamatov 2013)	Immobile solid substance → partial mobile part or decrease the degree of stability → two parts with one joint → multiple joints → flexible substance/add flexibility to a rigid part → liquid → gas → field (substitution of the substance). This also includes the use of optical copies (images)	Flexibility, self-organizing, adaptability, deformation capability, durability, variation in properties, fold to make more compact, deflection, improve response to external changes, increase user convenience, increased the ability to adapt to different use circumstances
Field dynamization (Kwatra and Salamatov 2013)	Constant action → pulse action → alternating action (change frequency, phase, wavelength)	Increase system efficiency convenience. Reduce the damage, and energy usage
Technological system dynamization (Fey and Rivin 2005)	Rigid system → modular system → programmable system → self-adapting system/ smart materials	Similar to substance dynamization

Table 12.5 Coordination and not coordination trend

Trend	Evolution line	Keywords
Coordination and not coordination (Gadd 2011)	Un-coordinated actions → partially coordinated actions → coordinated actions → realization of other actions by using the pause during intervals	Reduction of time, wear, and energy usage. Improve the response to external changes, user convenience

number of parts and operations increase, resulting in increased complexity to a level that the disadvantages produced by adding new elements are higher than the benefits obtained, thus reducing the ideality of the system. Then, the next transition is towards the recombination of elements (convolution) or the substitution (trimming). The lines of evolution of these three transitional sub-trends are described in detail in Annex 1.

Table 12.6 Su-field interaction and controllability (change substance) trend

Trend	Evolution line	Keywords
Introduction of modified substances (Salamatov 1999)	Change the aggregation state or the value of a parameter → use the mix of two or more existing substances including a void and derived substance → use a substance derived from the existing resources in a technological system	Improved user safety, ability to control function delivery to specified requirements, reduced user involvement, increase system efficiency, improve response to external changes, reduced energy loss
Introduction of a new substance[a] (Innovation Suite 2002)	Introduced internally → introduced externally → introduced in the environment	Similar to introduction of modified substances
Substance fragmentation (Innovation Suite 2002)	Solid → segmented solids (with different degrees) → liquid → aerosol → gas → field	Surface area, arrange, separate different functions, isolate stresses/loads, transportability, reduced density, flexibility, controllability, use of substance, complexity, manufacture cost-improved reliability, maintainability
Using voids (Rantanen and Domb 2010)	Use hollow structures → use of foam → use vacuum	Similar to Introduction of modified substances

[a]The substance field interaction can be improved with the mix, substitution or adding of substances that interact with the existing field. For instance, the introduction of ferromagnetic substances when an electromagnetic field is being used

3.4 Evolution to Micro-system

In the process of evolution of a technological system, the working unit evolves towards the use of micro-level structures. The transition to a micro level improves the efficiency of the system by the progressive division of the working unit into smaller and smaller parts. In the next evolution level, the working unit and its elements are replaced with a substance or field at a micro-level state (molecules, atoms, ions, electrons, etc.) that is easier to control and delivers the desired function in an efficient and effective manner (Salamatov 1999). Thus, this transition means that the functions carried out by macro-objects are gradually transferred to micro-objects, or that the micro-objects control the physical properties and behavior of the working unit (Fey and Rivin 2005). This trend consist in two sub-trends (described in Annex 1), which are: (1) Transition to micro-levels in materials, and (2) Transition to micro levels using physical states or combinations of states.

Table 12.7 Su-field interaction and controllability-change field

Trend	Evolution line	Keywords
Use of fields (Salamatov 1999)	Use of existing fields in the system and environment → use the interaction of two or more similar fields, but with different properties (permanent and variable field) → use the interaction of two or more fields of the same physical type → use the interaction of two or more fields of different type → use of field combined with a field-sensitive substance	Improved user safety, ability to control function delivery to specified requirements, reduced user involvement, increase system efficiency, improve response to external changes. Reduction in energy usage, complexity, energy loss
Transition to more controllable fields (Salamatov 1999)	Gravitational → mechanical → thermal → magnetic → electric → electromagnetic	Same to above
Introducing a new action using a field (Machine)	Mechanic, electric, thermal, magnetic, acoustic	Same to above

Table 12.8 Evolution to higher-level system trend

Trend	Evolution line	Keywords
Expansion (Kwatra and Salamatov 2013)	1. Mono system 2. Homogeneous bi-system: dual system from two similar elements 3. Homogeneous poly-system: several similar systems 4. Heterogeneous bi-system: a dual system from two different elements with shifting or unequal characteristics 5. Counteracting system: one system compensates the disadvantage of the other 6. Opposed bi-system: dual system from the union of two contradictory elements (opposite functions and/or properties) 7. Poly-system: combining systems with different characteristics into a common system (super-system) to perform multiple fu2nctions	Improve the amount of useful function deliverable, function distribution (e.g. load spreading), user convenience, synergy effects, system functionality, user convenience, adaptability, ability to achieve the opposite functions
Convolution (Kwatra and Salamatov 2013)	1. Expanded system 2. Overlapped systems: bi and poly-systems with characteristics and parts in common (various components are integrated into a single part) 3. Merged system: bi and poly-systems are combined into one unified system or the function of one system is transferred to another system 4. Alternative system: a new system replaces the existing system (dynamic or smart substance)	Reduced packaging, reduced number of systems, reduced net system size, reduced number of systems, reduced net system size
Trimming components (Machine; Kwatra and Salamatov 2013)	1. Delete one component and use another component in the system that provides the same desired function 2. Delete one component and use the resources in the environment to provide the same desired function 3. Delete entire system and add a new component/system to provide the main function and supplementary functions	Functionality, reliability, maintainability, user safety, complexity, manufacture cost, energy loss, waste/waste products, reduced number of systems, reduced net system size, system efficiency
Trimming energy conversion (Kwatra and Salamatov 2013)	Multiple energy conversions → reduce energy conversions → one energy conversion → no conversion of energy	Reliability, maintainability, user safety, complexity, manufacture cost, energy loss, system efficiency, waste/waste products, reduced number of systems

Table 12.9 Evolution to micro-system trend

Trend	Evolution line	Keywords
Line of transition to micro-levels in materials (Fey and Rivin 2005; Zlotin and Zusman 2013)	Solid substance → fragmented substance (powder, micro-spheres, etc.) → capillaries or porous substances → effects associated with substance structure (super-molecular or crystal level) → molecular phenomena → atomic phenomena → elementary particles → field actions instead of the part of the system	Improve the use of resources, overall size of system, performance, efficiency, response to external changes, user convenience. Reduction of the energy loss, system damage, wear, energy usage, and complexity
Transition to micro levels using various physical states or combinations of states (Zlotin and Zusman 2001)	Solid-state substances → elastic materials → gels → liquids → liquids in critical or supercritical phases gases → plasma → voids → mix of solid, liquid, gas and plasma states (foam, liquids saturated with gas, sprays, dust, etc.) → aggregate state transformations (melting, evaporation, condensation, solidification, etc.)	Same to line of transition to micro levels in materials

Table 12.10 Reduction of human involvement (additional trends)

Trend	Evolution line	Keywords
Decrease the human involvement (Zlotin and Zusman 2001)	1. Human 2. Mechanization: the function of the engine, working unit and transmission is performed by a technological system 3. Automation: the function in the control unit (guidance and control)is performed by the technological system such as amplifiers, reducers, gyroscopes 4. Computerization: decision making is performed by a technological system. For example, the use of sensors or system for processing information 5. Self-system: automated system performs actions without human intervention such as self-reproduction or self-driving 6. Fully-automated system: the system performs additional auxiliary functions for its own	Eliminate interference with human interfaces, self-learning systems, reduced human drudgery, reduced likelihood of 'human error' effects, increased accuracy, ability to deliver extremes of a function outside the human range—e.g., lifting, joining, throwing, etc.-reduction of fatigue effects, reduced cost

(continued)

Table 12.11 Trend to improve measurements

Trend	Evolution line	Keywords
Improve measurements (Machine; Altshuller 1984; Fey and Rivin 2005)	1. Direct measurement 2. Measuring using the existing resources 3. Measuring with the help of additives 4. Measuring with the help of field 5. Use of copies 6. Remove the need for measurements by changing the system in such a way that it is not necessary to carry out these measurements	Ability to make simple measurements, increased flexibility of measurement, eliminate interference with human interfaces, add new function through the employment of effects present, (e.g. use IR to achieve a heat-seeking capability), increased the range of measurement possibilities, ease of inspection, save energy/ energy management, self-repairing systems

Table 12.12 Product and service evolution

Trend	Evolution line	Keywords
Increasing the perception of the product[a] (Innovation Suite 2002)	Vision → touch → smell → taste → hear	Improved interaction control, increased human involvement, 'sensual immersion', improved realism of a simulation
Market evolution to more interactive experience for the customers (Innovation Suite 2002)	1. Product 2. Product-service: the product includes services such as assistance or maintenance 3. Product-service-experience: the product is replaced by a service that provides it such as the self-service bicycle sharing system	Customers desire, consumer perception, service development, competitive advantage, marketing, improve the capability to provide a service

[a]The evolution of a product evolves towards increasing use of senses to improve the consumer's level of perception (the order of sense may vary)

3.5 *Additional Trends*

3.5.1 Reduction of Human Involvement

Technological systems evolve in the direction towards the reduction of human intervention. Therefore, the aim of this trend is to replace the human action by technological systems to perform tasks that are boring, repetitive, dangerous, time-consuming, or even socially unacceptable. Thus, in this trend, the humans are excluded and transferred their responsibilities to machines through the mechanization, automation, and computerization, or the humans are replaced by machines as a source of energy (Savransky 2000). The evolution line of this trend is described in Annex 1.

3.5.2 Trend to Improve Measurements of the System

Another trend is the evolution in the process of measuring, detection and/or monitoring of information (property, characteristic, state, etc.) in a technological system and its components when cannot be performed directly. Annex 1 describes the evolution line of this trend.

3.5.3 Product and Service Evolution

The technological system can also be considered as a product or service and, consequently, the TRIZ theory has extended its analysis to study the trends of evolution of these two elements. As a result, evolution lines have been proposed that could be very useful for products and services in the food industry. These evolution lines (detailed in Annex 1) are: (1) Increasing the product perception, and (2) Market evolution.

4 TRIZ Evolution Trend Strategy: Approach Description

The TETs described before can provide a different vision to the traditional product development in the food industry. Thus, the aim of this section is to propose an approach for their implementation. Although there is no common agreement how to use the TETs, there are different ways to apply them:

1. Wang et al. (2016) and Park et al. (2013) propose to generate concepts by analyzing how similar technological systems (systems that perform the same function) evolve following a specific trend.
2. Other authors (Cavallucci and Weill 2001; Jianguang Sun et al. 2008; Berdonosov 2011; Yang and Chen 2012) propose to identify the trends in the actual state of the technological system and then generate concepts with those trends that are not applied at all.
3. The most common way is to brainstorm to generate ideas using the trends, as presented by Ishi and de Carvalho (2015).

In our approach, the TETs are combined with the technology push and/or market pulls strategies to analyze the actual state of the technological system. This association will enable us to anticipate future problem situations or generate ideas for future stages of the system. This approach involves five stages described below.

Stage 1: Collecting of information. In this stage, we can use the market pull strategy to identify the market needs and the main problems linked to the technological system, or use the technology push strategy to describe the new technology created. The information collected is described using goals, requirements, problems, ideas for improving the functioning of the existing system and its components. Then, a more extensive search about these goals, requirements, problems, ideas is carried out from different sources such as scientific research, patents, journal articles, internet website, textbooks, etc. Finally, this new information gathered together with the prior information are used in the next step.

Stage 2: Network of problems. The network of problems is a graphical formalism to link goals, problems, and solutions to create a semantic network to describe the actual state of a technological system. Also, this network is used for problem-solving methods in the TRIZ theory to improve any technological system (Khomenko et al. 2006; Khomenko and Guio 2007).Thus, the network of problems provides a well-structured set of sequential problem-solution situations that could simplify the identification of existing evolution trends as well as a means for identifying new opportunities for applying the trends.

To create the **network of problems** is necessary to translate the information collected in the previous stage into sentences describing the goals, problems, and partial solutions. Next, the sentences are transformed into nodes and linked between them by arrows to create the network structure. Table 12.1 describes the various types of possible links between the nodes. Finally, once the network of problems stops evolving, the network is analyzed to identify opportunities to use the TETs. This step is described in the next stage.

Stage 3: Assessment of the evolution of the system and opportunity identification. The network of problems describes the different ways to improve the technological system through problems and their partial solutions. Thus, in this stage, all these problems and partial solutions are analyzed to identify situations where the trends are only partially applied, situations in which the trends are ignored, or identify the trends that are not applied, but that can be potential opportunities for improvement. If this is the case, the relationship between the problem-solution situation and the trend observed is graphically represented (Fig. 12.1). This representation facilitates the visual analysis of the actual situation and identifies the possible evolutionary changes of the system. Also, when we can identify some opportunities where the TETs are not applied, they can be defined as new problematic situations. For example, if we found that the coordination trend was not applied, we can transform this situation into a problematic situation: how the system can be harmonized? With this new problem, a search of information is carried out to create a new part of the network.

Stage 4: Concept generation. The aim of this stage is to generate ideas using the trends to define the possible evolutionary changes of the system as potential future states of the system under study. Thus, once having identified the partial or full implementation of a trend in the actual state of the system or an opportunity where

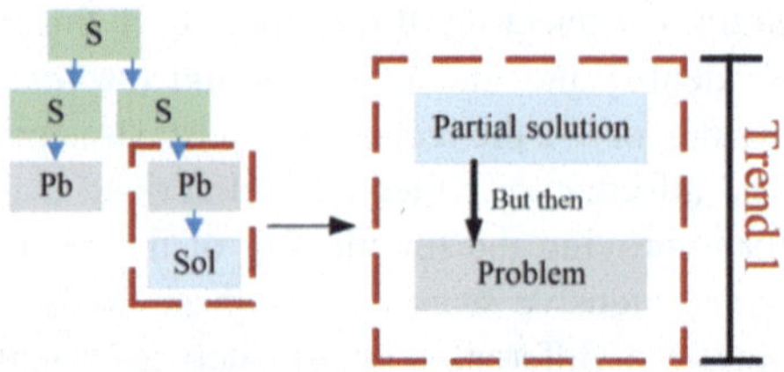

Fig. 12.1 The identification of an opportunity to apply the "trend 1"

the trends were not applied, ideas of how to apply these trends are generated which are later translated into workable concepts. Patent analysis and the TRIZ tools such as the Standards for the solutions of inventive problems and the Scientific-engineering effects can be helpful to facilitate the generation of ideas. The concepts created can describe how to apply one trend or trends in combination (Fig. 12.2).

Stage 5: Technological Evolution scenarios. The aim at this stage is to prioritize the different variants (concepts) of the evolution of the technological system. First, the concepts are evaluated considering the specific conditions and constraints (specifications, requirements and resource availability) and the possible obstacles facing their implementation. In order to consider a concept as a viable solution, it has to maintain the ideality ratio positive: the sum of useful outputs should always be larger than the sum of disadvantages. Next, the incompatible concepts are discarded, and the feasible concepts are classified as short-term, medium-term or long-term concepts. The concepts involving improvements with low complexity are defined as short-term concepts, the concepts that require more significant changes, and sometimes additional resources are defined as next-generation concepts or medium-term, and the concepts involving radical changes or the full replacement of the old system are defined as a long-term concept. However, when a concept with significant changes or radical changes may create an important strategic advantage for the organization, it can be considered as a short-term concept. After this classification, the concepts are organized in a list to create a technological system evolution scenario. This scenario is a kind of technology roadmap that organizes the concepts in a frame describing the possible technology pathways (scenarios) to follow. Each evolution path describes the probable states that the system could take. Finally, the most convenient pathway to explore is selected, and the concepts in this path are integrated into the new product development process as new solutions to explore, and integrated into the network of problems for further analysis.

5 Case Study

The membrane separation technology represents an environmental and economical alternative to conventional separation processes and as a means of obtaining high-value substances in industrial food production (Brazinha et al. 2015). Thus, we

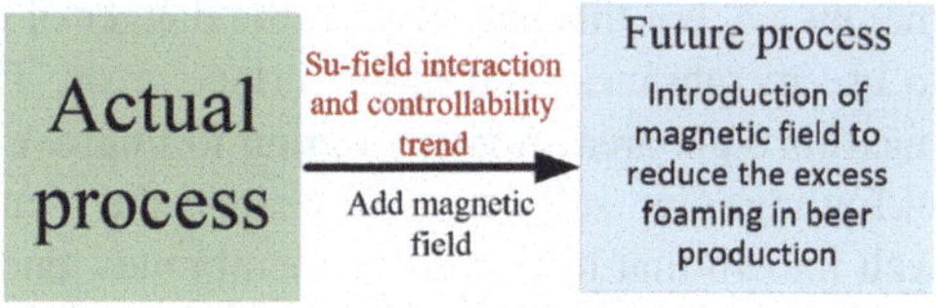

Fig. 12.2 Application of the su-field interaction and controllability trend to improve the beer production process (Shokribousjein et al. 2015)

aim to apply the TETs using our approach to identify opportunities for future improvement of this technology.

5.1 Collecting Information

The system to analyze is the cross-flow membrane separation for ultrafiltration with flat sheet modules. The main function is the separation of high-value substances or undesirable substances. Following the analysis of market needs (market pull), the main challenges of the membrane technology are: (1) easy maintenance, (2) improving the separation process, and (3) reducing fouling (an undesirable accumulation of material on the membrane that causes the blockage of the pores). Then, we searched and collected information about these main challenges from patents, scientific journals, and specialized books.

5.2 Network of Problems

The information collected was used to create the network of problems. This network shows the possible ways to improve the membrane technology. First, the membrane technology is divided into product and process. One way to improve the membrane (product) is to improve the maintenance. The possible alternatives to enhance this maintenance process are: (1) to implement a regular cleaning. Many procedures do exist for this purpose: back-pulsing, chemical cleaning, etc. (2) to eliminate the need for regular cleaning through preventing the accumulation of material on the surface. Regarding this latter alternative, we can use a disposable membrane and removable membrane, but this solution has a profound impact on the cost of the process. Another solution to prevent the accumulation of material is to create turbulence and shearing forces by using turbulence promoters that will require additional components (Popović and Tekić 2011). It is also possible to increase the cross-flow velocity of the flow to create turbulence, but this could result in a significant pressure drop from the inlet. Thus, rather than increasing the velocity, it could be possible use vibrating modules, but this system is complex. The new problem would be to find an alternative way to create turbulence.

Furthermore, the separation process can be improved by modifying the membrane. Increasing the packing density could improve the efficiency of the membrane, but this may lead to the danger of clogging. An additional alternative is to use membranes with large surface areas. Thus, it is necessary to find a way to maximize the area. Another option to improve the process is to reduce the problem of fouling. One solution is to change the membrane properties. For example, it is well known that hydrophobic membranes tend to increase fouling (Chang and Lee 1998), thus making the membrane hydrophilic could reduce it. However, this requires important changes in the membrane materials. The pretreatment of the feed could also reduce the fouling (Maartens et al. 1999) and depends on the specific characteristics of the feed. Finally, creating high turbulence and high cross-flow, and cleaning are two alternatives to eliminate fouling that were described above in the maintenance process. Figure 12.3 shows the network of problem.

5.3 Assessment of the Evolution of the System and Opportunity Identification

1. Su-field interaction and controllability trend. This trend can be applied to the problem of creating turbulence to generate shearing forces. Turbulence is difficult to control by increasing the cross-flow velocity of the flow. Also, a further problem caused by this high velocity is a significant pressure drop. Thus, this trend leads us towards the use of an easily controlled field combined with a field-sensitive substance.
2. Macro system evolution trend. The method used for regular cleaning show that this process has evolved through the use of more controllable fields. Thus, the next evolution step could be to the higher-level system by the combination (temporarily or permanently) of these methods or the use of one method to provide multiple functions or process.
3. Macro system evolution trend. This trend could be used to explore the possibility of merging the membrane and the turbulence promoters or the membrane takes the form of turbulence promoters.
4. Evolution of substance trend. This trend could be used to increase the surface area of the membrane.

5.4 Concept Generation

To facilitate the search for effects and phenomena and then generate conceptual solutions, the Goldfire Innovator database was used (Machine). The application of ultrasound was retrieved from the database using the query “create turbulence.” The ultrasound follows the trend su-field interaction and controllability because the process can be easily controllable, relatively cheap with low-energy consumption (Awad et al. 2012). This process is already used in food processing for elimination

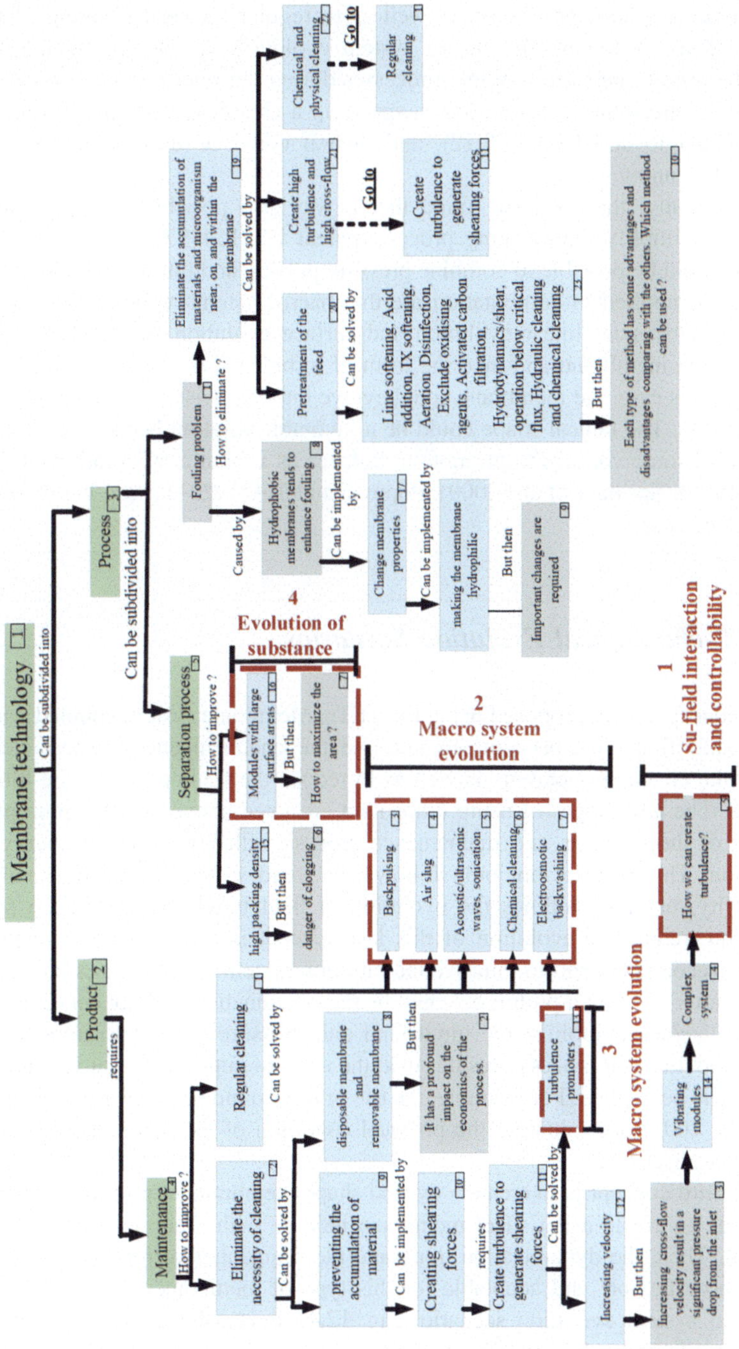

Fig. 12.3 Network of problems and identification of opportunities to apply the TETs

of bacteria and homogenization as well as to regular physical cleaning of membranes. Thus, following the macro system evolution, turbulence by ultrasound could be used to perform multiple processes during the filtering and also cleaning. Moreover, ultrasound can be used to speed up a chemical reaction (Cavani et al. 2009; Cravotto and Cintas 2012), and then it could be used to accelerate the chemical cleaning.

Additionally, the database found that it was also possible to apply pressure pulsing to simplify the cleaning process (patent US5154197) (Auld et al. 1991). Thus, it could be possible to combine pressure pulsing and ultrasound. Other trends that can be applied in combination are the macro system evolution and the evolution of substance (linear evolution and surface evolution) to make evolve the shape of the membrane and take the form of turbulence promoters with a twisted shape. In this case, the membrane could evolve into the twisted shape (Popović and Tekić 2011). The helical shape could be a favorable solution because it allows the creation of some vortices to promote turbulence as similar configurations in Gupta et al. (1995), Springer et al. (2009), Popović and Tekić (2011), and the increase in the surface area.

5.5 Technological Evolution Scenarios

Three concepts were proposed using the TETs. Membrane system coupled with the ultrasound (first concept) aims to improve the chemical and physical cleaning during the cleaning operation as well as to reduce the fouling during the filtration process. The advantages of using ultrasound, as noted above, are the more precise control of the filtration and maintenance process with low-cost and energy consumption. This concept not only follows the trend su-field interaction and controllability, but also follows the evolution of field trend and the macro system evolution trend. The evolution of this last concept is to add a pressure pulsing system (second concept) to improve the filtration system. In both cases, the trend to explore after creating a prototype could be the coordination and not coordination to synchronize/desynchronize the ultrasound and pressure pulsing to improve even further the filtration process. Besides that, the trend to improve measurements of the system can be applied. For example, the use of ultrasound for measuring the fouling (Li et al. 2003) or evaluating the physical condition of the membrane (Repacholi and Benwell 2012).

The third concept is to create a twisted shape membrane to promote turbulence. However, to create a flat sheet membrane module with a curved shape could be difficult. Then, the dynamization trend and the segmentation trend can be used to make more flexible and adaptable for this type of membranes. Finally, the technological system evolution scenario (Fig. 12.4) shows the membrane and ultrasound as a feasible and practicable alternative solution concept in a short-term. The solution system that integrates a membrane, a pressure pulsing and an ultrasound is categorized as a medium-term solution concept because it would only need to add

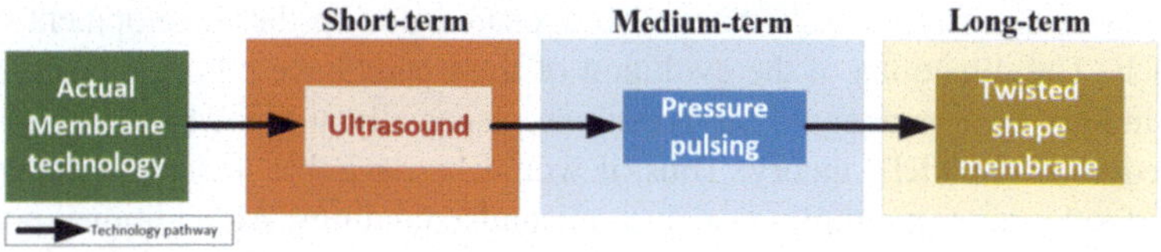

Fig. 12.4 Technological evolution scenario for the membrane technology

the pressure pulsing system to the short-term solution concept. Finally, the twisted shape membrane is considered as a long-term concept because it requires a more radical change. Furthermore, spiral surfaces usually involve sealing-related problems. Thus, a more in-depth investigation should be carried out about this issue.

6 Discussion

The TRIZ theory with their concepts, methods, techniques, and tools offer an alternative to traditional methods in the design stage of food product development. Following this premise, we proposed an approach that uses the network of problems, a tool of TRIZ, to analyze and organize a wide range of information about the system to improve into problems and partial solutions. This stage is followed by another stage that consists of identifying key opportunities to apply the TRIZ Evolution Trends (TETs).The approach ended in technological evolution scenario that organizes in a frame the feasible concepts that could improve the technological system under study.

One contribution of this research was the classification and adaptation of the existing TRIZ Evolution trends in food technology that were used in our approach. We structured the trends depending on the technological change of the system: general evolution trends, evolution trends at the system level, evolution trends at the higher-level system, evolution trends at micro-system, and other additional relevant trends. Another contribution was the analysis of the evolution of technological systems using the network of problems, which facilitates the identification of existing trends or opportunities to apply the TETs. This analysis cannot be only used in food technology, but it can be made available for all TRIZ practitioners.

The use of this approach in the improvement of membrane filtration technology showed its applicability by generating three conceptual solutions. Thus, this approach can be an alternative to traditional strategies (technology push and market pull strategies), or it can be combined with them to assist the development of new products and processes in the food industry. For instance, each evolution trend can help to improve the performance, quality, safety, and the cost reduction of food products bringing benefits to consumers and producers. Also, this approach can serve as an educational or training purpose to promote the creative thinking of future food engineers and scientists.

However, some drawbacks have also appeared during the development of our approach. The directions of the evolution of the trends have a broad meaning that could induce into erroneous interpretations, especially those who have no prior knowledge of the TRIZ theory. Thus, it would be desirable to have a knowledge base of food science examples to reduce misunderstandings. An additional problem is associated with the increase in complexity in the development of new food products that requires the access of knowledge from different fields of science. A knowledge management system to create a collaborative work environment as proposed in Barragan-Ferrer et al. (2017) could be very useful to share knowledge from different domains.

7 Conclusions

The inventive activity in the implementation of innovations in the food industry is typically governed by the technology push and market pull strategies, ignoring the evolution of the technological systems. From a different perspective, we observed that market (market pull) and technology (technology push) strategies could only influence the rate of evolution of the technological systems, but not modify its direction defined by the evolution trends. Thus, in this chapter, we presented an approach for promoting the development of the technological systems in the food industry from this evolutionary perspective using the TRIZ Evolution Trends as a strategy. These trends that have been used to anticipate the future development of the technological system were applied in the food industry to propose recommendation of the most favorable transformation to improve or develop technological (food) systems.

Finally, to extend our research, we would like to integrate the Web 2.0 collaborative technologies to create an open and collaborative innovation platform based on TRIZ theory for the food industry to facilitate the interdisciplinary collaboration of multiple domains of science such as biotechnology, biomedicine, chemical process, and mechanical engineering to develop innovative food technologies.

References

Altshuller, G. (1984). Creativity as an exact science: The theory of the solution of inventive problems. Gordon & Breach Science Publishing, New York.

Altshuller, G. S. (1999). The innovation algorithm: TRIZ, systematic innovation and technical creativity.

Arthur, W. B. (2007). The structure of invention. *Research Policy, 36*, 274–287. https://doi.org/10.1016/j.respol.2006.11.005.

Awad, T. S., Moharram, H. A., Shaltout, O. E., et al. (2012). Applications of ultrasound in analysis, processing and quality control of food: A review. *Food Research International, 48*, 410–427. https://doi.org/10.1016/j.foodres.2012.05.004.

Barragan-Ferrer, J.-M., Barragan-Ferrer, D., Lopez-Flores, R., et al. (2017). Knowledge management framework for integrating biomedicine into the new product development. *Sveik Moksl/Health Science, 27*, 78–79.

Auld G. D., Baum, A. J., Esposito, J. B., Stenger, J. (1991). Chemical cleaning method for steam generators utilizing pressure pulsing.

Becattini, N., Cascini, G., Rotini, F. (2015). An OTSM-TRIZ based framework towards the computer-aided identification of cognitive processes in design protocols. In J. Gero & S. Hanna (eds), *Design Computing and Cognition'14* (pp. 99–117). Cham: Springer.

Berdonosov, V. (2011). Application characteristics of the law of system completeness. *Procedia Engineering, 9*, 337–344. https://doi.org/10.1016/j.proeng.2011.03.123.

Bigliardi, B., Galati, F. (2013a). Innovation trends in the food industry: The case of functional foods. *Trends in Food Science and Technology, 31*, 118–129. https://doi.org/10.1016/j.tifs.2013.03.006.

Bigliardi, B., Galati, F. (2013b). Models of adoption of open innovation within the food industry. *Trends in Food Science and Technology, 30*, 16–26. https://doi.org/10.1016/j.tifs.2012.11.001.

Brazinha, C., Cadima, M., Crespo, J. G. (2015). Valorisation of spent coffee through membrane processing. *Journal of Food Engineering, 149*, 123–130. https://doi.org/10.1016/j.jfoodeng.2014.07.016.

Brem, A. (2008). Pull vs. Push—strategic technology and innovation management for a successful integration of market pull and technology push activities. In *The Boundaries of Innovation and Entrepreneurship* (pp. 39–71). Gabler, Wiesbaden.

Capitanio, F., Coppola, A., Pascucci, S. (2010). Product and process innovation in the Italian food industry. *Agribusiness, 26*, 503–518. https://doi.org/10.1002/agr.20239.

Cavallucci, D., Khomenko, N. (2007). From TRIZ to OTSM-TRIZ: Addressing complexity challenges in inventive design. *International Journal of Product Development, 4*, 4.

Cavallucci, D., Weill, R. D. (2001). Integrating Altshuller's development laws for technical systems into the design process. *CIRP Annals Manufacturing Technology, 50*, 115–120.

Cavallucci, D., Rousselot, F., Zanni, C. (2010). Initial situation analysis through problem graph. *CIRP Journal of Manufacturing Science and Technology, 2*, 310–317.

Cavani, F., Centi, G., Erathoner, S., Trifirò, F. (2009). Sustainable industrial chemistry: Principles, tools and industrial examples. Wiley.

Chang, I.-S., Lee, C.-H. (1998). Membrane filtration characteristics in membrane-coupled activated sludge system—the effect of physiological states of activated sludge on membrane fouling. *Desalination, 120*, 221–233. https://doi.org/10.1016/s0011-9164(98)00220-3.

Cravotto, G., Cintas, P. (2012). Harnessing mechanochemical effects with ultrasound-induced reactions. *Chemical Science, 3*, 295. https://doi.org/10.1039/c1sc00740h.

Del Giudice, T., Nebbia, S., Pascucci, S. (2012). The "Young" consumer perception of functional foods in Italy. *Journal of Food Products Marketing, 18*, 222–241. https://doi.org/10.1080/10454446.2012.666444.

Di Stefano, G., Gambardella, A., Verona, G. (2012). Technology push and demand pull perspectives in innovation studies: Current findings and future research directions. *Research Policy, 41*, 1283–1295. https://doi.org/10.1016/j.respol.2012.03.021.

Ferrer, J. B., Negny, S., Robles, G. C., Le Lann, J. M. (2012). Eco-innovative design method for process engineering. *Computer Chemical Engineering, 45*, 137–151.

Fey, V., Rivin, E. (2005). Innovation on demand: new product development using TRIZ. Cambridge University Press.

Gadd, K. (2011). TRIZ for engineers: Enabling inventive problem solving. Wiley.

Galanakis, K. (2006). Innovation process. Make sense using systems thinking. *Technovation*. https://doi.org/10.1016/j.technovation.2005.07.002.

Godin, B., Lane, J. P. (2013). Pushes and pulls: Hi(S)tory of the demand pull model of innovation. *Science, Technology, and Human Values, 38*, 621–654.

Gupta, B. B., Howell, J. A., Wu, D., Field, R. W. (1995). A helical baffle for cross-flow microfiltration. *Journal of Membrane Science, 102*, 31–42. https://doi.org/10.1016/0376-7388(94)00241-p.

Hirsch Kreinsen, H., Jacobson, D., Robertson, P. L. (2006). “Low-tech” industries: Innovativeness and development perspectives—a summary of a European research project. *Prometheus, 24*, 3–21. https://doi.org/10.1080/08109020600563762.

Innovation Suite (2002) CREAX.

Ishi, R. Y., de Carvalho, M. A. (2015). Using the trends of evolution (TEs) from the theory of inventive problem solving (TRIZ) to generate new product ideas for optical fiber distribution boxes. *Applied Mechanics and Materials, 741*, 814–822. https://doi.org/10.4028/www.scientific.net/AMM.741.814.

Khomenko, N., Guio, R. (2007). OTSM network of problems for representing and analysing problem situations with computer support. *IFIP International Federation for Information Processing, 250*, 77–88. https://doi.org/10.1007/978-0-387-75456-7_8.

Khomenko, N., Schenk, E., Kaikov, I. (2006). OTSM-TRIZ problem network technique: application to the history of German high-speed trains. In *TRIZ Future 2006*. Belgium, Kortijk,

Kleinknecht, A., Verspagen, B. (1990). Demand and innovation: Schmookler re-examined. *Research Policy, 19*, 387–394. https://doi.org/10.1016/0048-7333(90)90022-x.

Kwatra, S., Salamatov, Y. (2013). Trimming, miniaturization and ideality via convolution technique of TRIZ: A guide to lean and high-level inventive design. Springer.

Li, J., Hallbauer, D. K., Sanderson, R. D. (2003). Direct monitoring of membrane fouling and cleaning during ultrafiltration using a non-invasive ultrasonic technique. *Journal of Membrane Science, 215*, 33–52. https://doi.org/10.1016/s0376-7388(02)00600-2.

Maartens, A., Swart, P., Jacobs, E. P. (1999). Feed-water pretreatment: Methods to reduce membrane fouling by natural organic matter. *Journal of Membrane Science, 163*, 51–62. https://doi.org/10.1016/s0376-7388(99)00155-6.

Mäkimattila, M., Melkas, H., Uotila, T. (2013). Dynamics of openness in innovation processes—A case study in the finnish food industry. *Knowledge and Process Management, 20*, 243–255. https://doi.org/10.1002/kpm.1421.

Mann, D. L. (2002). Hands-on systematic innovation. CREAX Press.

Mowery, D. C., Rosenberg, N. (1991). Technology and the pursuit of economic growth. Cambridge University Press.

Negny, S., Belaud, J. P., Cortes Robles, G., et al. (2012). Toward an eco-innovative method based on a better use of resources: Application to chemical process preliminary design. *Journal of Cleaner Production, 32*, 101–113.

Nemet, G. F. (2009). Demand-pull, technology-push, and government-led incentives for non-incremental technical change. *Research Policy, 38*, 700–709. https://doi.org/10.1016/j.respol.2009.01.004.

Park, H., Ree, J. J., Kim, K. (2013). Identification of promising patents for technology transfers using TRIZ evolution trends. *Expert Systems With Applications, 40*, 736–743. https://doi.org/10.1016/j.eswa.2012.08.008.

Peters, M., Schneider, M., Griesshaber, T., Hoffmann, V. H. (2011). The quest for adequate technology-push and demand-pull policies: Country-level spillovers and incentives for non-incremental innovation. *SSRN Electron Journal*. https://doi.org/10.2139/ssrn.1752414.

Petrov, V. (2002). *The Laws of System Evolution* (pp. 1–22).

Popović, S., Tekić, M. N. (2011). Twisted tapes as turbulence promoters in the microfiltration of milk. *Journal of Membrane Science, 384*, 97–106. https://doi.org/10.1016/j.memsci.2011.09.016.

Rantanen, K., Domb, E. (2010). Simplified TRIZ: New problem solving applications for engineers and manufacturing professionals. CRC press.

Repacholi, M. H., Benwell, D. A. (2012). Essentials of medical ultrasound: A practical introduction to the principles, techniques, and biomedical applications. Springer Science & Business Media.

Salamatov, Y. (1999). TRIZ: the right solution at the right time: A guide to innovative problem solving. Hattem: Insytec.

Savransky, S. D. (2000). Engineering of creativity: Introduction to TRIZ methodology of inventive problem solving.

Shokribousjein, Z., Galan, D. R., Michiels, C., et al. (2015). Effect of a magnetic field on dispersion of a hop extract and the influence on gushing of beer. *Journal of Food Engineering, 145*, 10–18. https://doi.org/10.1016/j.jfoodeng.2014.08.008.

Springer, F., Carretier, E., Veyret, D., Moulin, P. (2009). Developing lengths in woven and helical tubes with dean vortices flows. *Engineering Applications of Computational Fluid Mechanics, 3*, 123–134.

Sun, J., Gao, J., Yang, B., Tan, R. (2008). Achieving disruptive innovation—forecasting potential technologies based upon technical system evolution by TRIZ. In *2008 4th IEEE International Conference on Management of Innovation and Technology* (pp. 18–22). IEEE.

Terninko, J., Zusman, A., Zlotin, B. (1998). Systematic innovation: an introduction to TRIZ (theory of inventive problem solving). CRC press.

Traitler, H., Coleman, B., Hofmann, K. (2014). Food industry design, technology, and innovation. Wiley.

Wang, C., Zhao, W., Wang, J., et al. (2016). An innovative approach to predict technology evolution for the desoldering of printed circuit boards: A perspective from China and America. *Waste Management Research, 34*, 491–501. https://doi.org/10.1177/0734242x16640330.

Machine, I. Goldfire Innovator.

Yang, C. J., Chen, J. L. (2012). Forecasting the design of eco-products by integrating TRIZ evolution patterns with CBR and simple LCA methods. *Expert Systems Applications, 39*, 2884–2892. https://doi.org/10.1016/j.eswa.2011.08.150.

Zlotin, B., Zusman, A. (2001). Directed evolution: Philosophy. *Theory and Practice*. Ideation International Inc.

Zlotin, B., Zusman, A. (2013). Patterns of technological evolution. In *Encyclopedia of Creativity, Invention, Innovation and Entrepreneurship* (pp. 1438–1446). New York, NY: Springer, New York.

A Service Design Process Based on the Business Model CANVAS and the C-K Theory

Sergio E. Castillo-López, Sheyla K. Flores Guirao, Eduardo Roldán-Reyes, Aidé A. Maldonado Macias and Diego Manotas Duque

Abstract Services are fundamental to any economy. They are responsible for most of the employment and collaborate with a substantial proportion of the Gross Domestic Product. Despite its importance, the tools and methodologies for service design in Latin America have a slow evolution. One of the reasons that explain this phenomenon is the lack of practical tools and methods to take into account the complexity of the service and the intrinsic dynamism of the design process. Another limitation of the adoption of new service design tools is a nontechnical view of the design process, situation that puts the common sense as the first resource for service design. Hence, the lack of a service design methodology avoids the learning process, the reproducibility of successful service models in a different context, and the assimilation of new design techniques. Consequently, this chapter proposes a framework to guide the conceptual design or redesign of services through the combination of the Business Model CANVAS with the C-K theory. The Business Model CANVAS is a graphical tool to represent, as simple as possible, the interaction of different elements to capture and deliver value to the market. On the other hand, the C-K theory proposes a graphical tool where Concepts (C) and Knowledge (K) interact dynamically to shape a process, without increasing complexity. The

S. E. Castillo-López (✉) · S. K. Flores Guirao · E. Roldán-Reyes
Instituto Tecnológico de Orizaba, Avenida Oriente 9 no. 852 Col. Emiliano Zapata, C.P. 94320 Orizaba, Veracruz, Mexico
e-mail: scastillo@ito-depi.edu.mx

A. A. Maldonado Macias
Department of Industrial Engineering, and Manufacturing—Institute of Engineering and Technology, Autonomous University of Ciudad Juarez, Avenida del Charro 450 Norte, Col. Partido Romero, Ciudad Juárez, Chihuahua, Mexico
e-mail: amaldona@uacj.mx

D. M. Duque
Universidad del Valle, Escuela de Ingeniería Industrial, Calle 13 no Edificio E56—Ciudadela Universitaria Meléndez, Universidad del Valle, 100-00 Cali, Colombia
e-mail: diego.manotas@correounivalle.edu.co

G. Cortes-Robles et al. (eds.), *Managing Innovation in Highly Restrictive Environments*, Management and Industrial Engineering,
https://doi.org/10.1007/978-3-319-93716-8_13

integration offers a design framework useful to deal with the challenging context of service development. A case study illustrates this practical approach.

Keywords CANVAS · C-K theory · Service design · Conceptual design

1 Introduction

Nowadays, the market moves beyond the conception of manufactured goods to propose experiences. Successful products are in fact one part of the network that involves services as a strategy to provide experiences. Mostly, the sell-point and the assistance for customers are typically the results of a user-centered design but when these services are not the result of a design effort, the service just "happens" and is not an integrated system. One of the service design objectives is to fix and repair the unbalanced composition between the human and technical side of service to propose new, attractive, congruent, and effective service systems. Another significant opportunity in the service design domain is the potential for fueling innovation at different levels (enterprises, government, and universities) condition that reveal new paths to solve complex problems such as the transportation, the sustainable development, the access to health services, among others (Reason et al. 2013).

Aboal et al. (2015) explain the relevance of the service activities in the economy of Latin America. Authors underline that the productivity of services is a key factor to promote growth in emerging economies. The service sector contributes approximately 60% of the total employment in Latin America and the Caribbean (LAC). Nevertheless, the productivity of services had not evolved at the same pace, particularly compared with manufacture activities and natural resources-based activities. The low productivity in the LAC economies is another reason to improve the performance of the service sector. A successful strategy to enhance the competitiveness of the LAC economies is to stimulate innovation and efficiency in services (Aboal et al. 2015). Services have a significant impact on any economy from different perspectives:

- Services have a direct effect on the manufacturing industry. Traditional services such as transportation, distribution, and communication are the fundamental blocks in the value chain. Hence, any improvement in the service productivity will produce a positive impact in the production of tangible products.
- The need to adapt services to fast-changing conditions generates new knowledge. Knowledge Intensive Based Services such as financial services, software design, and product design, among others, encourage the inventiveness, the new product development, and promote economic growth (Umaña-Aponte et al. 2013).
- The difficulty to patent service is another relevant condition to boost the technological development. This situation obligates enterprises to speed up the development pace and to propose new or adapted service faster, circumstances that move the frontiers of technology because of the apparent facility to copy

> and reproduce a service. Hence, new advances in services are crucial in the innovation process of other production systems. This statement contradicts traditional perspectives about service, which consider that services demand low assimilation of technology and produce almost no innovation. Consequently, in LAC, scientific literature does not focus their attention on services, which produces the lack of flexible and suitable methodologies to cover service design and the innovation management in this domain. Besides, at least in Mexico, the national initiatives to foster innovation underestimate the relevance of services to the economy. This situation is probably a consequence of an inadequate conception of the internal instruments to evaluate the impact of this industrial sector, and also, due to the misunderstanding of the technical requirements and needs in the service design and evolution (Álvarez et al. 2013).

Aboal et al. (2015) suggest that service companies aim to manage innovation under the same circumstance and with similar objectives that manufactured goods enterprises: To increase the market share, to improve the service quality, to expand the services portfolio, and to anticipate a shift in the market. Despite the recent interest in the service field in LAC, the research in this domain is almost inexistent. There are just a few isolated research projects about the most significant factors to encourage innovation and service productivity. Hence, it is desirable to develop new and more suitable methodologies to facilitate the service design with the goal to stimulate innovation. One of the most evident factors that promotes the change in the economy is the integration between products and services. This phenomenon pushes the economy to a different state, one where other economic activities envelop the growth of the service activities. Hence, it is now clear the presence of new services in global business. Services are, in fact, the strategy to compete and increase a market share, and simultaneously, a mechanism to differentiate a tangible product. Even in traditional domains like agriculture, where services are transforming the way a product is perceived in the market.

Consequently, it is necessary to develop new methodologies for service design but far from the influence of the product design. Hence, the differences between manufactured goods and services are another requirement to propose more suitable methodologies that also need to be functional in emergent markets.

Before underlying the relevance of services for any economy, this work proposes a framework to assist the new service design process. Witell et al. (2015) explain the urgency to introduce new methodologies to support the new service design. The authors underline that the service design process has several limits and restrictions that make the creation of new knowledge and its diffusion difficult. Another obstacle is that many successful methodologies are an adaptation of methodologies for tangible products. Hence, the context is more than adequate for the development of more appropriate methodologies for service design. This chapter sets as initial conditions that any methodology for service design should comply with the following criteria:

- **Simplicity**: The methodology must be simple and easy to apply. This requirement will facilitate the adoption of a potential user.
- **Intuitive**: The methodology needs to take into account that the people, which collaborate in the service design, come from different domains. The design process must be represented in such a way that people who did not participate in the conceptual design will understand all necessary steps and transformations to configure the service.
- **Flexible**: The methodology must apply to different contexts and responds to several restrictions.

An initial search for service design methodologies put in evidence that it is necessary to propose a methodology that will not demand a huge training effort, but that will be flexible enough to respond for a challenging contextual design. Two different approaches satisfy the three initial criteria: The Business Model CANVAS and the C-K theory.

The Business Model CANVAS or just a CANVAS model has a broad acceptation in the literature. The CANVAS offers a graphical structure that puts into relation nine blocks: The client, the relation, a distributions process, the value proposition for the market, some crucial activities and resources, the strategic partners, and finally, the blocks that are responsible from costs and revenues. The CANVAS model is simple and easy to use. It offers a graphic representation of a business model, which deals well with the differences between the service objective and the design process (Toro-Jarín et al. 2016). However, the CANVAS model has some opportunities for improvement, which according to Zolnowski et al. (2014), the next points are the most relevant.

- **It is a static modeling approach**: The CANVAS model does not have any mechanism to facilitate the continual evolution of each of their components. The user is responsible for updating any block according to the expert or the user's perspective. The service design is a dynamic process and demand frequent adjustments.
- **Lack of information**: The CANVAS model does not make evident the complex relationships that exist among their components. Hence, the user does not know how to deal with the complexity. A service is a complex system and any change in one of their components will have an impact on the system.
- **Not flexible enough**: Once the user has the graphic representation of the business model, it is not possible to observe the multiple alternatives that are available to produce the service.

Without damaging the CANVAS advantages but surmounting the above limitations, this chapter proposes to combine the CANVAS model with another approach capable of coping with these limitations: The C-K theory. This theory offers a versatile tool for modeling a product or a service (Hatchuel et al. 2011). Also, the C-K theory produces a framework where it is feasible to describe and represent the service design requirements and the more important relationships among the elements that interact in the design process. Hence, the hypothesis of this work aims to explain how to combine both approaches to assist the service design. Next section describes the complementarity between both methods.

2 Background

The complementarity between the Business Model CANVAS and the C-K theory is the most relevant reason to combine both approaches. The Business Model CANVAS provides a functional framework that synthesizes a huge volume of information to model a value proposition. On the other hand, the C-K theory creates a context where it is possible to explore and evaluate the relationships among concepts and the knowledge to implement them. Both approaches are relevant for the service design process, and next points briefly explain the reasons behind this combination.

- The Business Model CANVAS produces a design framework, which interacts with the elements of the value chain. The challenge is to transform these linear relations into proactive interactions (DaSilva and Trkman 2014).
- The C-K theory provides a graphical tool that CANVAS can use to model the most representative changes in any of the nine components. Hence, the C-K theory offers a dynamic point of view to the business model, a condition that improves the adaptability of the business model.
- The C-K theory can store any design intention under the form of a process. From the C-K theory perspective, any design intention is an iterative process. As a consequence, the C-K theory stores all design iterations to build a design memory, which facilitates the comprehension of the process to another person. It stores and preserves the logic behind the design intention with the goal to share and reuse the knowledge deployed during the design process (Hatchuel et al. 2011).
- The combined application of the Business Model CANVAS and the C-K theory make possible the visualization of the complex interaction that takes place within a service and the impact that the process has to fulfill the client or the market requirements. As a result, the graphical model of the service design enables the ability to evaluate the entire possible alternative to produce a service, which facilitates the cost analysis of the system.

Table 1 synthesizes the complementarity between the Business Model CANVAS, the C-K theory, and the combination of both approaches. The purpose is to depict the value that this synergy has.

Qualitative evaluation is a practical approach to compare performance among the Business Model CANVAS, the C-K theory, and their combination (Albarrak et al. 2010). In this evaluation strategy, a positive answer (Yes = 2) gives two points to one technique; a negative answer (No = 1) adds just one point. Table 2 classifies the alternatives.

Table 1 The advantages of the combined approach CANVAS-C-K

	Can model a business	Describe a process	Design oriented	Service oriented	Graphical tool	Simplicity	Service analysis	Store knowledge
CANVAS	Yes	No	No	No	Yes	Yes	No	No
C-K theory	No	Yes	Yes	Yes	Yes	No	No	Yes
Combined approach	Yes	Yes	Yes	Yes	Yes	Yes	Yes	Yes

Table 2 Comparison between the Business Model CANVAS and the C-K theory

Approach	Points
CANVAS	11
C-K theory	13
Combined approach	16

The Business Model CANVAS gets 11 points. The most significant features of this tool are that (1) CANVAS can synthesize the business logic and explain it easily through a graphical model. (2) It is an easy-to-use tool. Nevertheless, it demands a huge volume of information. (3) It is not an easy task to model a process, and it was not originally conceived as a tool for assisting or assessing design. On the other hand, the C-K theory has 13 points. Despite its usefulness, the C-K theory does not propose any specific tool or methodology to implement a business model and its use for assisting design demands some training, which in turn offers the capacity to represent processes and the design logic. Also, Table 2 underlines that a combined approach has several advantages:

- The combination of CANVAS and C-K theory can describe a business model and use the relations proposed by the CANVAS to get information to conceive the service production process.
- The C-K theory proposes a graphic tool that represents the service as a process. The resulting diagram offers a wide perspective about the different service configurations.
- The natural orientation of the Business Model CANVAS to services adds this competency to the combination of both approaches. Hence, it makes easier to model a service.
- Given the fact that both approaches have a graphical tool to represent a process or a business model, it is possible to manage the information of the service design.
- The simplicity of the Business Model CANVAS is another inherited characteristic of the combined approach.
- The combination of both approaches enables a new resource: The capacity to model and analyze a service. The C-K theory models a service through a sequence of activities, but the business model is not explicitly considered in this sequence. Thus, the designer uses their past experiences to finish the modeling stage and guarantee the coherence between the design process and the business model.
- Finally, the C-K theory offers the capability to store the logic behind a model. Thus, the reutilization of some processes or the information to make decisions become accessible to other users, conditions that facilitate the knowledge transfer. Hence, if several models are available, and there is a common structure, then, in theory, it will be possible to reuse partially or entirely the service design.

Once the advantages of combining the Business Model CANVAS and the C-K theory were briefly explored, next section describes the state of the art that shapes this chapter.

3 State of the Art

This section offers a succinct description of some related works about the use of the Business Model CANVAS and the C-K theory.

3.1 The Business Model CANVAS

A business model is a set of principles that envelop the global strategy of an organization to create, deliver, sustain, and improve value to the market or the society. The term value refers to economic, social, cultural, technological, or environmental benefits and even a combination of these different perspectives. Hence, a business model is a term used to represent the unique characteristics of an organization or enterprise. It includes—but is not limited to—the purpose, products, services, strategies, infrastructure, practices, processes, and politics (Muhtaroglu et al. 2013). The literature about business models explains and classifies how an enterprise deploys their business mission. About this subject, the work of DaSilva and Trkman (2014) offers a clear and objective perspective. A business model includes a conceptual model, depicts some fundamental components, and sometimes, it proposes some crucial indicators.

Zott and Amit (2010) define a business model as the content, the structure, and the operation management organized to create value through the exploitation of some business opportunities. Chesbrough (2010) underlines the functions of a business model as to identify the market share, specify the mechanisms that create an income, determine the role of the enterprise inside the value chain, and finally, to formulate the competitive strategy that will preserve a competitive advantage. Research about the role of the business model is not something new (Drucker 1995), but it is recently that it attracts the attention of the academic world, which does not accept the current definition of a model business (Zott et al. 2011). It was Osterwalder and Pigneur (2010) who offer a clear and simple definition of a business model: The representation of how an enterprise creates, captures, and delivers value from a product or service. According to Osterwalder (2004), a business model must involve three aspects: (1) It needs to explain how the integration of components and crucial functions delivers value to customers; (2) It depicts how all components and functions are interconnected within the enterprise, how they interact in the supply chain or in the networks of other stakeholders, and finally, (3) To describe how these interconnections produce value or profit for the enterprise. Osterwalder and Pigneur (2005) argue that the research in the field of business models is an emergent but relevant topic. The objective of this research is to propose or conceive the concepts and tools that help to capture, understand, communicate, analyze, design, and change the business logic of an organization. Once the business model is understood, it provides the vision, and the coherence to direct the strategies and actions to impel the competitiveness (Masanell and

Ricart 2010). It is important to underline that the interconnections of the business model are frequently understood at the tacit level, it is essential to move this tacit perception to the explicit level. The transition from tacit to the explicit level encourages the innovation process by discovering non-seen opportunities (Johnson et al. 2008). The work of Osterwalder and Pigneur (2010) proposes a basic format to describe a business model. This format, known as CANVAS, has the purpose to help users to declare the logic relations among each component of the business model. The CANVAS takes into account the four essential management perspectives: The product, the client, the infrastructure, and some financial indicators. The relations among these components become evident through nine building blocks that transform an idea into some interconnections (see Fig. 1). Table 3 briefly describes the role of each element in the CANVAS.

The practical point of view of the Business Model CANVAS was rapidly accepted for many companies, for instance, Proctor & Gamble, Nestle, Airbnb, to mention but a few. The Business Model CANVAS produces a change in the business to move from a product-centered perspective to a business-centered thinking. In Mexico, an association between Mexico and USA proposes the TechBA program. In Mexico is the Secretary of Economy and the US–Mexico Foundation for Science (FUMEC) that provides access to a dynamic business ecosystem for small- and medium-sized technology-based Mexican businesses. The purpose is to expand the market share and activities to global markets. This program uses the Business Model CANVAS to represent the value proposition of Mexican enterprises (TechBA 2017). Other successful programs in Mexico that use the Business Model CANVAS is the National Institute for Entrepreneurs (INADEM), the National Council of Science and Technology in the Program for

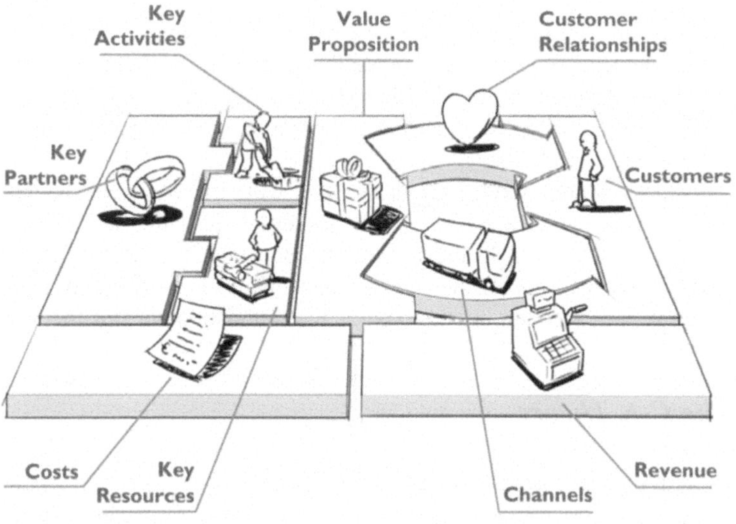

Fig. 1 The Business Model CANVAS (Osterwalder and Pigneur 2010)

Table 3 Description of the CANVAS building blocks

CANVAS component	Description
Customers	The enterprise focuses on at least one market segment. The model can take into account multiple segments; each one will have their strategy
Value proposition	This block describes how the enterprise solves problems and produces something that satisfies a need or activates a different need. It is the set of features, products, services, or a combination that produces value for the market segment
Channels	It is the interface between the enterprise and the customer. This element contains the strategies, infrastructure, distribution process, and communication to deliver the value proposition
Customer relationship	The politics, protocols, mechanism, among other elements that enable and sustain a relationship with a particular customer. It also searches to maintaining this relationship in time
Revenue stream	This block describes the mechanisms to produce the profit. It is the answer to the question of what are customers willing to pay for
Key resources	This block contains all necessary resources to produce and configure the value proposition. It includes physical resources, financial, intellectual, and human resources. Also, this block contributes to maintain the customer relation and the channels efficiency
Key activities	This element contains those activities that are crucial for the value proposition and also that make operational the business model
Key partners	This section depicts the network behind the value proposition. This network mobilizes knowledge, competencies, and resources, among other relevant elements to support the model
Cost structure	This block accounts all the costs that imply to deliver value to customers. It also includes the acquisition of key resources, the execution of key activities, and the relationship with crucial partners

Technological Innovation (FIT), just to mention a few examples. Hence, it is relevant in the Mexican context to assimilate this approach for designing a business model (Molina-Morejón et al. 2015). The use of the CANVAS model is undoubtedly well accepted in Latin America (Pereira et al. 2016); (Montenegro-Marín et al. 2017).

The Business Model CANVAS deals well with the logic of the business, however, it cannot model a process. Thus, to complement this lack, this chapter proposes the use of the C-K theory, which can assist the design process. As mentioned before, both approaches are complementary. Due to the importance of the C-K theory, next section briefly exposes the foundations of this theory.

3.2 *The C-K Theory*

Hatchuel and Weil (1999, 2002) propose the C-K theory as part of a movement oriented to a unified design theory. In the C-K theory, the letter C refers to Concepts, while K stands for Knowledge. Both terms are necessary to identify design singularities, situations that are complex to identify, particularly compared with problem-solving approaches or other reasoning processes. The C-K takes their foundation on other design approaches like the Axiomatic Design, General Design Theory, Systematic Design, diverse creativity theories to mention some relevant sources (LeMasson et al. 2017). From the C-K point of view, the design process is a continual interaction between Concepts and Knowledge. In this interaction, the concept space modifies and recreates the available knowledge. In turn, the knowledge space transforms and gives shape to concepts. This complex relationship is thus useful to model how a concept and the knowledge evolve in the design process, revealing the recursive nature of design. The theory has the next initial conditions:

- K represents the Knowledge space. It is in this space where a designer finds some propositions that have a logical condition. This space is underestimated or assumed available in most of the cases.
- A proposition with a logical status or condition represents the degree of confidence that a designer determines for a certain proposition. The logic could be as simple as a true or false evaluation or a more complex structure like the fuzzy logic. Whatever the designer logic, all proposition have a logical status.
- A concept is at least one proposition or a set of propositions that do not have a logical status in K. Hence, when a concept matches with this condition, it is extremely complex to validate that it is a proposition of K. The absence of a concept design makes reference to past experiences, and in consequence, the challenge is to conceive the best way to mobilize the available knowledge. Finally, a concept contains a set of properties that apply to one or several entities.
- Under the logic of the interaction of concepts and knowledge, the authors define the design process as the continual transformation process where a concept generates other concepts or is transformed into knowledge. This definition of design is particularly useful in the new service design process, due to the continual and even simultaneous interaction of concepts and knowledge.

The C-K theory has a tool called a C-K diagram that helps to visualize the design process, which puts into evidence the opportunities to propose new solutions, and hence, to lead toward a potential innovation. Figure 2 depicts the interaction between the concept and knowledge spaces.

The C $\leftarrow$ K (K-C) relation means that the knowledge space modifies the concept space. In this process, some properties available in K are added, transformed, or eliminated in C. The effect is that K creates disjunctions in C, which transform one proposition into a concept.

In the C $\rightarrow$ K (C-K) relation, C searches to add, transform, or eliminate some properties of K to reach a logical status. This relation verifies the design process. Also, this operation expands the knowledge space through the interaction with the concept space. The interaction also takes into account the natural relation that exists between C to C (C-C) and K to K (K-K).

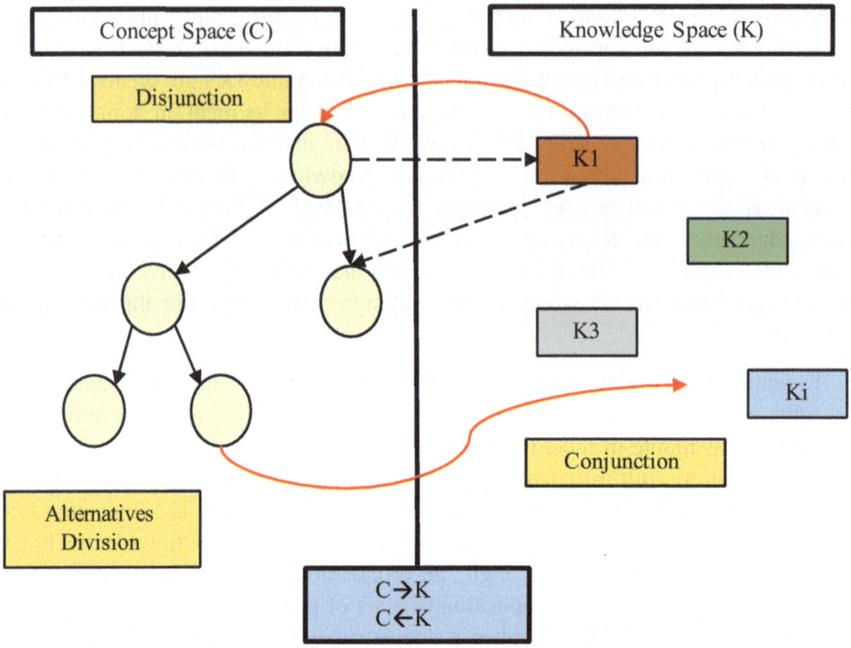

Fig. 2 The C-K diagram Hatchuel and Weil (1999)

According to Hatchuel and Weil (2012), the C-K diagram can organize and guide the innovation process, due to a global perspective about crucial processes and the new knowledge resulting from the interaction between C and K. According to (Ullah et al. 2012), a significant characteristic of the C-K theory is that it recognizes creativity as a crucial component in the design process. The authors explain that the creative dimension of the design process promotes the generation or recreation of the knowledge space.

Benguigui (2012) analyze the influence of the C-K theory on a period from 2002 to 2012. The author underlines that the C-K theory elucidates the initial stages of the innovation process and helps to propose new management tools to assist innovation. The work of Reich et al. (2012) has a similar orientation. In this work, the authors propose to combine the C-K theory with the Advanced Systematic Inventive Thinking (ASIT) method. This combination has the goal to improve the modeling process through the incorporation of some tool for inventive problem-solving. Another successful case of the implementation of the C-K theory is the Thales Alenia Space Corporation, an enterprise that uses the C-K theory as a design guideline. Besides, according to Defour et al. (2010), the C-K theory enables a team to organize the inventive design process. The work of Agogue and Kazakçi (2014) evaluate the impact of the C-K theory and propose some guidelines to manage a portfolio of innovation projects. This work also describes the possibility to develop new tools based on the C-K theory.

Ondrus and Pigneur (2009) evaluate the use of the C-K theory outside the design context and demonstrate the usefulness of this approach for solving a multiplicity of different conflicts, for instance, the application of the C-K theory to design information systems, data mining structures, knowledge management processes, in the evaluation of projects (Lenfle 2012), psychology, and cognition are other examples (Hatchuel et al. 2011).

Next points describe some of the C-K theory advantages:

- A clear definition of design, independently of the domains even in nontechnical fields.
- A design theory that has the same rigor and modeling capacities than other approaches related to decision-making.
- Concepts such as inventive thinking and innovation are at the core of this approach.
- It proposes a reusable framework to capture any design effort, thus, it stores knowledge about the design process.

4 Methodological Approach

The conceptual design of new services asks for an approach capable of combining in the same structure the capability to implement the business model with the right service production process. The integration between the Business Model CANVAS and the C-K theory uses the simplest possible approach: To transform the nine building blocks of the Business Model CANVAS in a set of classes in the knowledge space of the C-K theory. The Business Model CANVAS contains the service dynamics while the C-K theory provides the ability to model the service production process. The complex interaction between the CANVAS building blocks and the knowledge space is captured as a simultaneous interrelationship where any decision in one building block will affect the rest, which depicts the systemic point of view if the proposal. Figure 3 shows the logic of the integration.

Next points describe each phase in the integration:

- Phase one: Service design CANVAS. This stage uses only the CANVAS methodology to configure the business model. According to Osterwalder and Pigneur (2010), this process has five activities: (1) It is necessary to mobilize knowledge, competencies, and resources to obtain valuable information about the market and the client. (2) In this activity, the team analyzes all valuable information to create an initial concept of the business model. (3) The conceptualization of the business model involves creativity, a time horizon, and the anticipation of some changes in the market. (4) The implementation stage, in this case, does not mean to put into operation the service, but validate the basic assumptions. Finally, the activity (5) concentrates its attention to get feedback that will produce valuable information to update the business model.

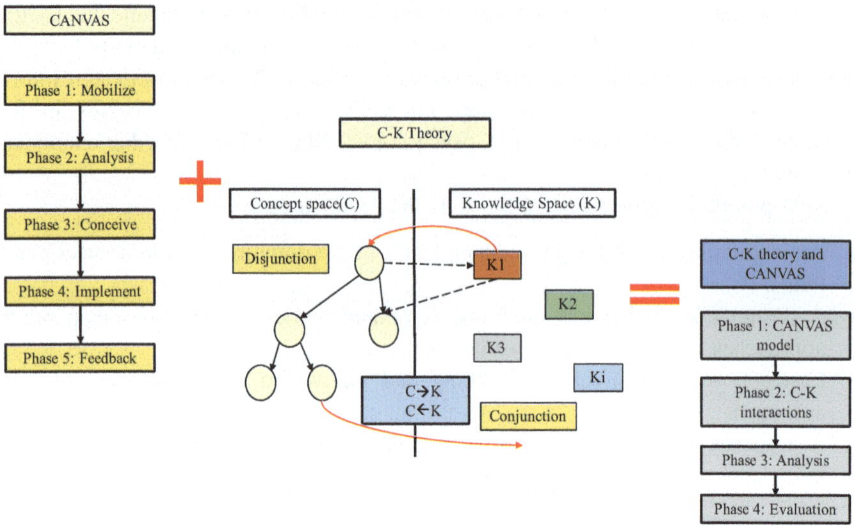

Fig. 3 The integration between the C-K theory and the Business Model CANVAS (Osterwalder and Pigneur 2010), (Hatchuel et al. 2017)

- Phase two: Combining the CANVAS building blocks in the Knowledge space (K). In this activity, the nine building blocks of CANVAS divide the knowledge space. A designer can start at any point. However, this chapter suggests a bottom-up approach, which commences in the customer block to move then to the value proposition and then explore the rest of the building blocks. The cost structure is the last block that takes shape. Figure 4 describes the integration as a continual process where the knowledge space and the concept space interact dynamically.
- Phase three: Once the designer or team has an initial configuration of the service production process, the next step is to deploy an analysis of the arborescence that takes shape in the concept space. This activity aims to verify that the process captures the service logic and that the concept and knowledge space are coherent. A designer can go back to any previous step to add, eliminate, or modify anything and to observe the impact of any given decision in the service design process. Another benefit in phase three arrives when the designer uncovers new opportunities for developing other aspects of the service, and hence, to propose a more flexible service production system or simply to conceive an entirely new service. Figure 5 describes an example of the arborescence.
- Phase four: In this stage, the designer proposes some indicators to evaluate the most relevant parameters in the concept space with the purpose to validate the conceptual design. If this activity offers a satisfactory result, then the team tests the concept to obtain feedback. In the other case, the team or designer needs to make the necessary changes in the process. This is an iterative process that ends when the designer or the team has a valid design.

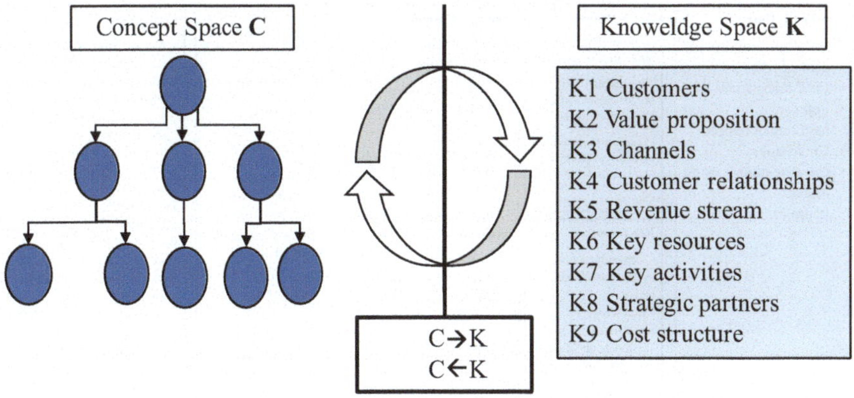

Fig. 4 The interaction between the CANVAS knowledge space and the concept space

Fig. 5 Division of the concept space that forms the arborescence

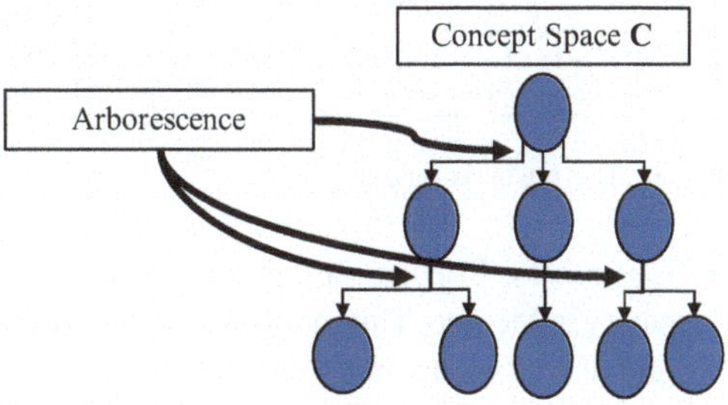

5 Case Study

The case study describes the conceptual design of a new service. The implementation of this service will be coded in a piece of software that is consumed through a website and also by smartphones. The service offers a system to supervise the water level of a river with the goal to alert the population about increased risk. The software also recommends what the nearest refuge facility is. Next points describe how the methodology is applied in the case study.

Phase one: In this first activity, the team produces the first conceptualization of the Business Model CANVAS. Next points describe each building block:

- Value proposition: A system that offers a continual supervision of a flow to prevent population about an increasing risk. The software notifies to users what are the nearest facilitates to take shelter. The system also preserves the knowledge deployed by the local authorities to manage risks. It is possible to reuse the successful strategies applied in the past and solves future incidences. In the future, when enough cases will be stored in the memory, the team aims to propose a simulation of a river with two goals: To increase the interest of younger people, and for learning about the behavior of a river or flow.

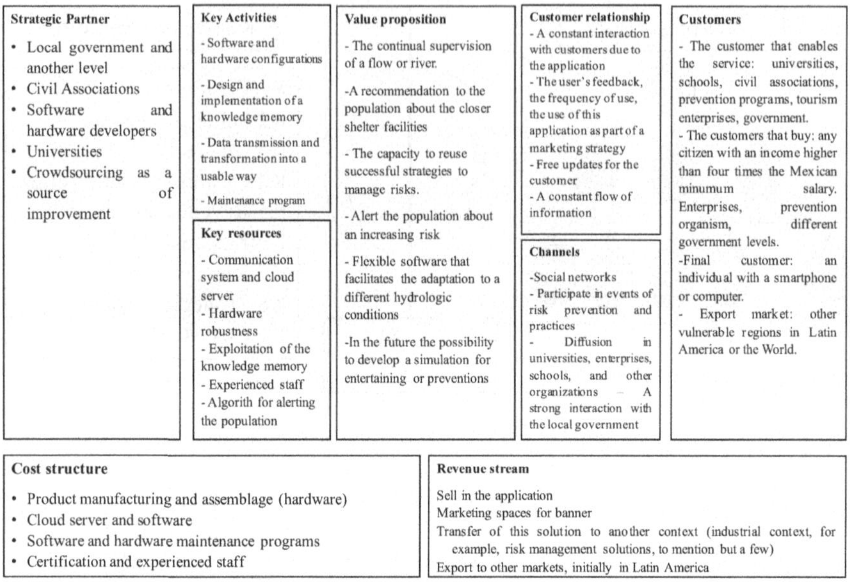

Fig. 6 The business model

- Customers: Population that has an income higher than four times the minimum salary in Mexico. This class includes prevention organizations, touristic enterprises, and of course, any individual having a smartphone. The piece of software offers the possibility to hire a space or temporal banners as part of a marketing strategy.
- Channels: Social networks are a critical resource to put the service in the customer's preferences. A flexible infrastructure to adapt the service to the particularities of any given region. The enterprise will participate in some events of prevention and also in risk training events in collaboration with local authorities. The use of different software application to distribute the piece of software is part of the distribution strategy.
- Customers relationship: The piece of software will produce a constant interaction with the customer. This CANVAS block uses a piece of software that offers a constant interaction with the customer. The software stores and index information about the user's feedback, the frequency of use, the customer announces, and the evaluation of the most relevant parameters of the application.
- Revenue stream: Direct sells of the product is consumed through a monthly payment or an annual subscription. The piece of software also offers to other companies the possibility to advertise their trademarks. Finally, in the midle term, it is expected to transfer the business model to other domains and export the application to Latin American countries.
- Key resources: The capacity of the cloud server, the continued operation of the network of sensor installed in a river or flow. The system that transmits information to the server, experienced staff for problem-solving, and the right criteria for launching an alert to the population.

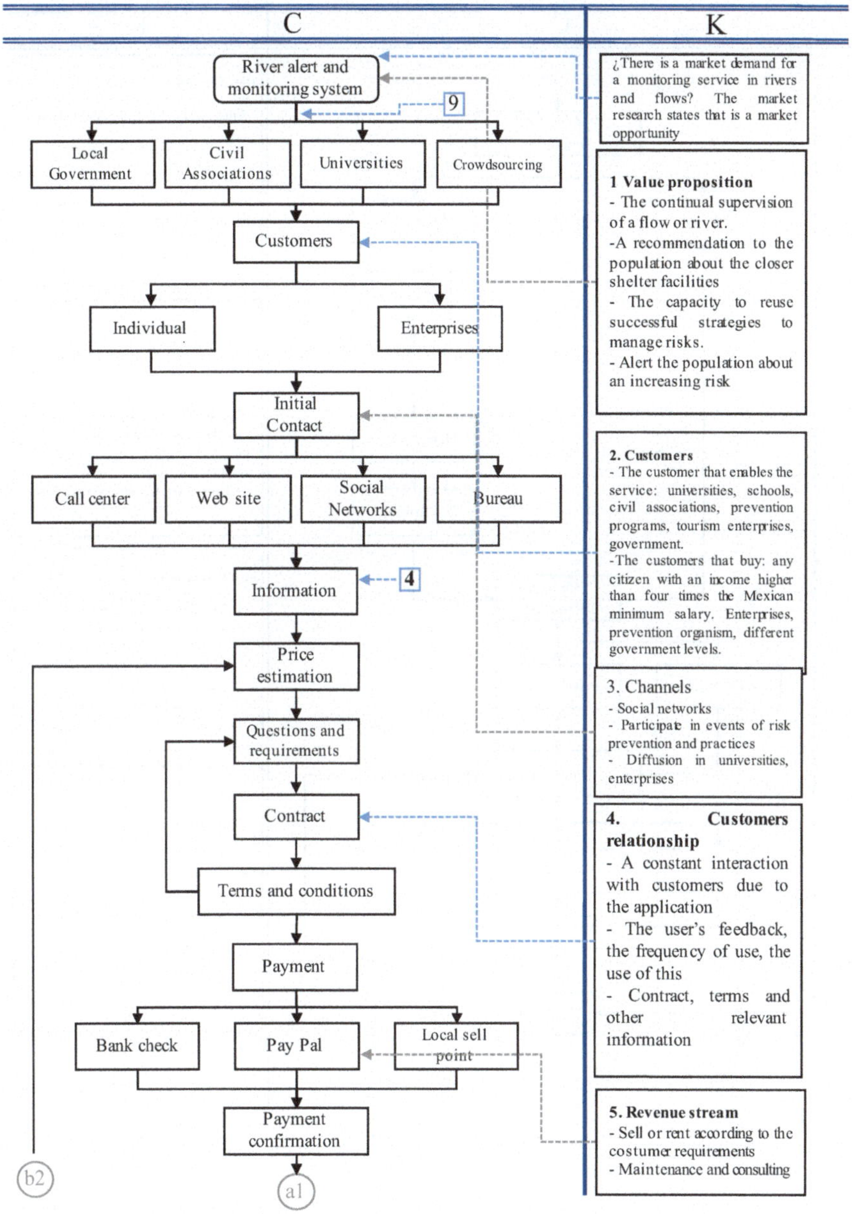

Fig. 7 The initial configuration of the service

- Key activities: The frequency for updating the application and an effective transmission of data from the source to the algorithm that transforms this information in a usable form. The maintenance of the network of sensors and the

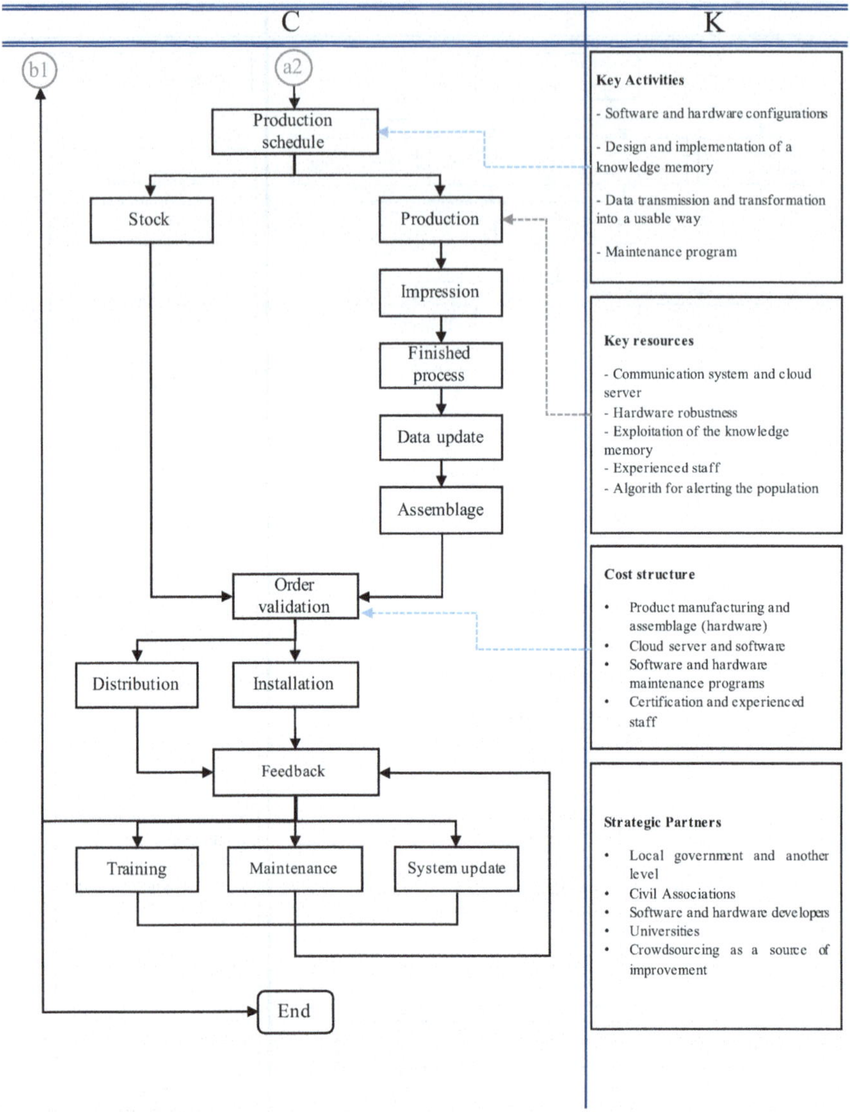

Fig. 7 (continued)

coherence between the primary measuring elements and the mechanism to transform data into information. The collaboration with the local authorities is crucial in this case.

- Strategic partners: Different government levels, civil associations, software and hardware developers, universities as a source of knowledge, and of course, the collective intelligence of the population.

- Cost structure: The manufacturing, installation, and maintenance of the sensors network are a relevant part of the cost structure. The construction, indexation, and maintenance of the knowledge memory. The conception and implementation of a simulation process that the user will exploit. Figure 6 schematizes the business model.

Phase two: Figure 7 describes the design process as a dynamic interaction between the concept and knowledge spaces. The first step in the process is to express questions or disjunctive, which cannot be answered with the knowledge in the current state. Hence, this disjunctive obligates the concept space to react. Then, new knowledge is necessary to satisfy the lack of knowledge. What comes next is a creation and modification of both spaces, which produce new concepts and recreate, transform, or generate knowledge. In the next step, the designer chooses the CANVAS building block that fits the best design process. Figure 4 proposes a basic structure flexible enough to fit several requirements. When other CANVAS building blocks get involved in the process, then different alternatives to produce a service emerge. It is important to underline that behind a C-K diagram, there is a design memory where are exposed and documented significant decisions. Each alternative in the arborescence illustrated in Fig. 7 represents a different configuration to produce and deliver the service. Consequently, the relation C- > K searches for some properties in K that can be added or eliminated to reach a logical status. In this process, the designer validates the process. The inverse relation, C < -K modifies the concept space, and in this process, it adds or eliminates some properties of K toward C (Hatchuel et al. 2004).

It is important to underline that the C-K diagram is an iterative process. Figure 7 shows only one of the first models. The enterprise is launching the service at present; hence, the final model is out of the scope of this chapter.

Phase three: This stage explores what is the impact of each alternative in the production process. The analysis has the purpose of estimating the cost of each solution, the time in the system, and even reveals the possibility to model the process to answer questions like what happens if…? This phase produces information that makes possible to identify a potential failure, and then, to fix it directly from the source. The analysis in this stage also determines if the infrastructure, resources, and capital are enough to assure the production of the service. This information enables a cost–benefit relation that makes possible to estimate the cost and facilitate the decision-making process.

Phase four: The designer determines what are the more practical parameter to evaluate the pertinence and correctness of the service. A qualitative, quantitative, or hybrid approach is possible. In the case study, the production time was a crucial parameter. This information gives the enterprise the opportunity to explore what will happen if the market demand is as expected, slightly higher or under the forecast, and thus, modify the marketing strategy in the channels or other building blocks of the business model.

6 Conclusions and Future Work

Despite the importance of the service domain for any economy, in Latin America, the methodologies and tools for service design are underestimated. There are just a few universities that cover this topic, and even less postgraduate programs oriented to the service design. The subjacent statement of this initial paragraph explains that common sense is the dominant approach for service design. However, it is necessary to explore more in depth the context to uncover the roots of this design lack in Latin America.

The state of the art of both approaches, the Business Model CANVAS and the C-K theory underlines their complementarity, and creates the right context to propose a simple but useful integration. The goal of this combination is to provide a mechanism to transform a service business model into a knowledge space to assist the service design process. The chapter describes four stages, and the methodology to combine the C-K theory and the Business Model CANVAS. The first phase or stage focuses on the business model. Then, in the next stage, this information is useful to divide the knowledge space, and the interaction between the concept and knowledge space take place. At the end of this stage, the team or designer obtain a C-K diagram, which contains the most significant alternative to produce the service. In stage three, the team or designer elaborates a detailed analysis about each alternative to extract the necessary information to estimate the cost, the production time, and the probable conflicts among different paths, to mention some of the information that this stage produces. The last stage is the validation of the process by the team or designer. In this step, the expert can use qualitative, quantitative, or a combination of indicators to validate that the most relevant parameters for the customers are satisfactorily implemented in the process. It is important to underline that each phase is connected with the next one; so, a problem or the lack of knowledge in one phase will affect the next one and the precedent. For this reason, in this chapter, it is accepted that the service design is a simultaneous process, and not a sequential one.

As a future work, the authors consider that it is necessary to develop a knowledge memory to facilitate the service design. From our point of view, it is possible to propose a flexible and practical approach if the knowledge memory uses the Business Model CANVAS as a fundamental structure. Hence, the research team is developing an ontology to guide the service design process.

References

Aboal, D., Crespi, G., & Rubalcaba, L. (2015). La innovación y la nueva economía de servicios en América Latina y el Caribe: Retos e implicaciones de política. *Centro de Investigaciones Económicas*, 9–205.

Agogué, M., & Kazakçı, A. (2014). 10 years of C-K theory: A survey on the academic and industrial impacts of a design theory. In *An anthology of theories and models of design:*

Philosophy, approaches and empirical explorations (pp. 219–236). Springer. Obtenido de https://hal-mines-paristech.archives-ouvertes.fr/hal-00983009.

Albarrak, A., Aboalsamh, H., & Abouzahra, M. (2010). Evaluating learning management systems for University medical education. *International Conference on Education and Management Technology*, 672–677. https://doi.org/10.1109/ICEMT.2010.5657569.

Álvarez, A., Bravo, C., & Zahler, A. (2013). *Innovation in services: Case study analysis for Chile*. Chile: CINVE Working Paper.

Benguigui, J. (2012). Les 10 ans de la théorie C-K : Revue de littérature. *XXIth International French Conference on Strategic Management (AIMS)*.

Boland, R., & Fred, C. (2004). *Managing as designing*. Stanford: Stanford University Press. Obtenido de http://www.sup.org/books/title/?id=1448.

Burgi, P., & Victor, B. (2004). Case study: Modeling how their business really works prepares managers for sudden change. *Strategy & Leadership*, *32*(2), 28–35. https://doi.org/10.1108/10878570410525106.

Casadesus-Masanell, R., & Ricart, J. (2010). From strategy to business models and onto tactics. *Long Range Planning*, *43*(2–3), 195–215. https://doi.org/10.1016/j.lrp.2010.01.004.

Chesbrough, H. (2010). Business model innovation: Opportunities and barriers. *Long Range Plan*, 43, 354–363. https://doi.org/10.1016/j.lrp.2009.07.010.

Cortés Robles, G. (2006). *La innovación en los servicios. El conocimiento como elemento central en la creación de nuevos servicios*. Orizaba, México: Instituto Tecnológico de Orizaba. División de Estudios de Postgrado e Investigación.

DaSilva, C. M., & Trkman, P. (2014). Business model: What it is and what it is not. *Long Range Planning*, *47*, 379–389. http://dx.doi.org/10.1016/j.lrp.2013.08.004.

Defour, Delaveau, & Dupas. (2010). *Avionique. Des technologies innovantes au services des plus belles réussites aéronautiques*. Paris.

Drucker, P. (1955). *The practice of management*. Allied Publishers.

Fritscher, B., & Yves, P. (2014a). Computer aided business model design: Analysis of key features adopted by users. *Proceedings of the 47th Annual Hawaii International Conference on System Sciences*. Computer Society Press. https://doi.org/10.1109/HICSS.2014.487.

Fritscher, B., & Yves, P. (2014b). Visualizing business model evolution with the business model canvas: Concept and tool. *IEEE 16th Conference on Business Informatics*, 151–158. https://doi.org/10.1109/CBI.2014.9.

Hatchuel, A., & Weil, B. (1999). *Pour une théorie unifiée de la conception*. Lyon: Conception et Cognition, GIS Cognition CNRS.

Hatchuel, A., & Weil, B. (2002). *C-K theory: Notions and applications of a united design theory*. Lyon: Ecole des Mines de Paris.

Hatchuel, A., & Weil, B. (2012). La théorie CK, Fondements et usages d'une théorie unifiée de la conception. *Proceedings of the Sciences of Design: The Scientific Challenge for the 21st Century In Honour of Herbert Simon*. Obtenido de https://www.techniques-ingenieur.fr/base-documentaire/procedes-chimie-bio-agro-th2/innovations-en-genie-des-procedes-42487210/theorie-c-k-j8115/.

Hatchuel, A., Le Masson, P., & Weil, B. (2004). C-K theory in practice: Lessons from industrial applications. *International Design Conference*, 1–13.

Hatchuel, A., Le Masson, P., & Weil, B. (2011). Teaching innovative design reasoning: How C-K theory can help to overcome fixation Effects. *Artificial Intelligence for Engineering Design, Analysis and Manufacturing*, *25*, 77–92. https://doi.org/10.1017/S089006041000048X.

Hatchuel, A., Weil, B., & Le Masson, P. (2017). *Design theory: Methods and organization for innovation*. París: Springer. https://doi.org/10.1007/978-3-319-50277-9_4.

Johnson, M., Christensen, C., & Kagermann, H. (2008). Reinventing your business model. *Harvard Business Review*, *86*(12), 50–59.

Le Masson, P., Hatchuel, A., & Weil, B. (2017). Designing in an innovative design regime—introduction to C-K design theory. *Design Theory*, 125–185. https://doi.org/10.1007/978-3-319-50277-9_4.

Lenfle. (2012). Exploration, project evaluation and design theory: A rereading of the Manhattan case. *International Journal of Managing Projects in Business*, *5*(3), 486–507. https://doi.org/10.1108/17538371211235335.

Martin, R. (2009). *The design of business: Why design thinking is the next competitive advantage*. Harvard Business Press.

Molina-Morejón, V. M., López, S., & Casio Agüero, V. (2015). Flexibilización del modelo de negocio de las PYME en el marco de sus componentes financieros. *XX Congreso Internacional de Contaduría, Administración e Informática*.

Montenegro Marín, C., Acosta Contreras, L., Barreto López, A., & Gaona-García, P. (2017). Fuzzy logic system based on canvas model to evaluate the initial viability of a business proposal. *WEA 2017: Applied Computer Sciences in Engineering*, 365–377. https://doi.org/10.1007/978-3-319-66963-2_33.

Muhtaroglu, F., Demir, S., & Obali, M. (2013). Business model canvas perspective on big data applications. *IEEE International Conference on Big Data*, 32–37. https://doi.org/10.1109/BigData.2013.6691684.

OCDE (Organización para la Cooperación y el Desarrollo Económicos). (2010). *Sciencie, Technology and Industry Outlook 2010*. París: OCDE.

Ondrus, J., & Pigneur, Y. (2009). C-K design theory for information systems research. *4th International Conference on Design Science Research in Information Systems and Technology*, 7–8. https://doi.org/10.1145/1555619.1555656.

Osterwalder, A. (2004). *The business model ontology-a proposition in a design science approach*. University of Lausanne.

Osterwalder, A., Pigneur, Y. (2010). *Business model generation: A handbook for visionaries, game changers, and challengers*. Hoboken: John Wiley & Sons.

Osterwalder, A., Pigneur, Y., & Tucci, C. (2005). Clarifying business models: Origins, present, and future of the concept. *Communications of the Association for Information Systems*, *16*(1), 1–25.

Pereira, S., Medina, F., Gonçalves, R., & da Silva, M. (2016). System thinking and business model canvas for collaborative business models design, 461–468. Sao Paulo, Brazil: Springer International Publishing. https://doi.org/10.1007/978-3-319-51133-755.

Reason, B., Polaine, A., & Lovlie, L. (2013). *Service design: From insight to implementation*. Rosenfeld Media.

Reich, Hatchuel, Shai, & Subrahamanian. (2012). A theoretical analysis of creativity methods inengineering design: Casting and improving ASIT within C–K theory. *Research in Engineering Design*, *23*(2), 137–158. https://doi.org/10.1080/09544828.2010.493505.

Teece, D. (2010). Business models, business strategy and innovation. *Long Range Planning*, *43*(2–3), 172–194. https://doi.org/10.1016/j.lrp.2009.07.003.

Toro-Jarín, M., Ponce Jaramillo, I., & Güemes Castorena, D. (2016). Methodology for the of building process integration of Business Model Canvas and Technological Roadmap. Elsevier, 1–13. https://doi.org/10.1016/j.techfore.2016.01.009.

Ullah, S., Rashid, M., & Tamaki, J. (2012). On some unique features of C–K theory of design. Elsevier, *5*(1), 55–66. https://doi.org/10.1016/j.cirpj.2011.09.001.

Umaña-Aponte, M., Estupiñan, F., & Duque, C. (2013). *Innovation and productivity in services: An impact evaluation of Colciencias funding programs in Colombia*. Colombia: CINVE Working Paper.

Witell, L., Snyder, H., Gustafsson, A., Fombelle, P., & Kristensson, P. (2015). Defining service innovation: A review and syntehsis. *Journal of Business Research*, 1–10. https://linkinghub.elsevier.com/retrieve/pii/S0148296315006967.

Zolnowski, A., Weiß, C., & Böhmann, T. (2014). Representing service business models with the service business model canvas - The case of a mobile payment service in the retail industry. *IEEE Computer Society*, 718–727. https://doi.org/10.1109/HICSS.2014.96.

Zott, C., & Amit, R. (2010). Business model design: An activity systems perspective. *Long Range Planning*, *43*, 216–226. https://doi.org/10.1016/j.lrp.2009.07.004.

Zott, C., Amit, R., & Massa, L. (2011). The business model: Recent developments and future research. *Journal of Management*, *37*(4), 1019–1042. https://doi.org/10.1177/0149206311406265.

Integration of Design Thinking and TRIZ Theory to Assist a User in the Formulation of an Innovation Project

Hugo Domingo García-Manilla, Jesús Delgado-Maciel, Diego Tlapa-Mendoza, Yolanda Angélica Báez-López and Leonardo Riverda-Cadavid

Abstract The principal limitations of the Design Thinking (DT) model reside in its high subjectivity, which generates some rigidity called psychological inertia and it is observed in the search for a solution within a very well-defined space. Also, DT does not offer strategies or techniques for the detailed solution of the intrinsic problems of the design process. This means that the procedure of the problem-solving that arises during the design process depends on the experience of a team or an individual. The Theory of Inventive Problem Solving (TRIZ) may assist the requirements of the Design Thinking model. TRIZ contains among its tools, a set of techniques that allow modeling and solving inventive problems. At the same time, TRIZ does not propose any tool or technique to identify the user's requirements, a fundamental aspect of the Design Thinking model. As a consequence, this chapter describes a strategy to combine both approaches and presents a basic structure to balance the best characteristics of both the approaches. The purpose is to guide the thinking and the creative efforts during the development of

H. D. García-Manilla (✉) · J. Delgado-Maciel
Instituto Tecnológico de Orizaba, Avenida Oriente 9 no. 852 Col.
Emiliano Zapata, C.P. 94320 Orizaba, Veracruz, Mexico
e-mail: hgarciam@ito-depi.edu.mx

J. Delgado-Maciel
e-mail: jdelgadom@ito-depi.edu.mx

D. Tlapa-Mendoza · Y. A. Báez-López
Universidad Autónoma de Baja California, Carretera Transpeninsular Ensenada - Tijuana 3917, Zona Playitas, C.P 22860 Ensenada, Mexico
e-mail: diegotlapa@uabc.edu.mx

Y. A. Báez-López
e-mail: yolanda@uabc.edu.mx

L. Riverda-Cadavid
Universidad Del Valle, Escuela de Ingeniería Industrial, Edificio E56—Ciudadela Universitaria Meléndez, Universidad Del Valle, Calle 13 no, 100-00 Cali, Colombia
e-mail: leonardo.rivera.c@correounivalle.edu.co

G. Cortes-Robles et al. (eds.), *Managing Innovation in Highly Restrictive Environments*, Management and Industrial Engineering,
https://doi.org/10.1007/978-3-319-93716-8_14

an innovation project and offer to a user a set of tools to solve the problems that invariably the design process contains. Finally, this chapter illustrates the application of both techniques in a case study.

Keywords Design thinking · TRIZ · Product design · Inventive problems

1 Introduction

Nowadays, the basis for the development of products and services is the satisfaction of the society needs and the resolution of its problems. There is also an industrial effort to introduce new products to the market that are consistent with the client's requirements, but above all, at a pace that allows a company to maintain its competitive position. The product (a manufactured good or a service) has success in the market, depending on the correct transformation and appropriation of the user's needs and the functionalities that it incorporates. Also, the precise and effective identification of the market requirements may lead to the development of successful new products.

As a result, one of the starting points to create a new product or modify an existing product is in the identification of the present and future market demands. The process of the Design Thinking (DT) model makes use of the empathy as a central process. Thus, the DT is a useful tool for developing products and services that focus on the balance between the human needs, the business needs, and technical needs, linked to the corresponding liability issues (Pavie and Carthy 2015).

Despite the usefulness of the DT, it does not propose some specific tools to deal with problems that are always associated with the design process. A method based on the evolution of technology offers a series of concepts and tools to solve problems: the Theory of Inventive Problem Solving (TRIZ). In its structure, TRIZ does not have a mechanism to identify the market needs. This task is based on an individual's experience and also on the use of other tools to extract this information. Therefore, both methods are complementary: DT produces valuable information to guide the design effort, while TRIZ can provide a repeatable and efficient toolbox for problem-solving. Thus, this chapter proposes a framework for combining the advantages of both methods and demonstrates the utility of this combination to provide new resources with the potential to employ them in the design of new products.

This chapter has five sections; the first one describes the role of the Design Thinking model in the innovation process, mentioning their basic concepts and the process to deploy this methodology, followed by an analysis of the state of the art to observe how the DT has been used in both industry and academy.

The next section illustrates the function of the Theory of Inventive Problem Solving (TRIZ), referring to their basic concepts, the benefits that provide to the innovation process and the basis that substantiate it. In the same way, this section offers a literature review to analyze how it has been used during product design and problem-solving.

Subsequently, the following section explains the reason for the integration, analyzing, and evaluating both the methodologies separately, thus discovering their strengths and opportunities. In this part is also described the potential of using both jointly.

Immediately, the integration of design thinking with the TRIZ theory is described by explaining that elements of TRIZ are incorporated in specific phases of the DT.

Then, this chapter presents a case study on the innovation of a laptop cover by employing the proposed framework (DT + TRIZ). Finally, the conclusion section refers to the feasibility of integrating DT and TRIZ, and its potential to assists the conceptual design of new products and later in the innovation process.

2 The Design Thinking (DT) Role in the Innovation

Innovation is the introduction of a new or significantly improved product, a process, a new method of marketing, or a new method of organization for the market and society (OCDE 2005). Therefore, innovation covers both the production and implementation of new ideas (Schmidt and Rosenberg 2015). The Oslo Manual (OCDE 2005) identifies four types of innovation: Product Innovation, Process Innovation, Marketing Innovation, and Organizational Innovations.

This chapter focuses on product innovation, which corresponds to the introduction of a new or significantly improved product or a service about its characteristics or the use for which it is intended. It includes a significant improvement in technical characteristics, components, and materials, integrated computing, ease of use and other functional aspects (OCDE 2005).

Product innovations can use new knowledge, technologies or rely on new uses and combinations of existing knowledge or technologies. Also, it is important to note that the Oslo Manual explains that the development of a new method of a product and the slight modification of their technical specifications is an innovation of the product. Making some changes in materials, components or other features that improve the performance of these products makes significant improvements in existing products. Consequently, to develop a new product or significantly improve one already existing in the market, it is essential to implement a design methodology. Several authors quote the use of DT to assist the innovation process. The following section describes briefly the most relevant works in the context of this chapter.

2.1 Design Thinking (DT): Basic Concepts and Process

A systematic innovation process that prioritizes deep empathy for desires, needs, and challenges to fully understand a problem in the hope of developing more comprehensive, and practical solutions (Roberts et al. 2016). DT transforms the way organizations develop products, services, processes, and strategies. The reason behind this argument is that DT brings together what is desirable from the human point of view, with characteristics that are technologically and economically viable. DT is also a flexible approach to address a wide range of challenges. It is a method for designing products, using the way of thinking of a designer, who uses a type of unconventional reasoning in the business world, that is, the deductive thinking. Therefore, the designer seeks the formulation of questions through the understanding of the phenomena. In other words, express questions that must be answered based on the information gathered during the observation of the universe surrounding the problem. Therefore, when thinking deductively, the solution does not derive from the obstacle: more precisely, it fits into it (Vianna 2016). The Design Thinking process covers five stages:

1. Empathize: Empathy is the basis of a people-centered design process. To empathize, it is necessary to display three functions: (1) observe the behavior of users in the context of their lives or the use of the product. (2) Interact with users and interview them through scheduled and "intercepted" meetings. (3) Submerge or experience what the user experiences or lives. Designers must understand people to transform this observation into knowledge and later into a product. Designers interpret and infer information about the meaning of the data obtained to discover ideas or perceptions. These perceptions channel the creative effort to generate solutions that meet their needs, requirements, and desires.
2. Define: This stage summarizes the findings collected through empathy and defines a specific and relevant problem. The objectives are to identify the problem or problems that the team must address and establish the evaluation criteria. It is also at this stage where a team formulates some relevant design restrictions and forecast the evaluation metrics to validate a concept.
3. Ideate: This stage calls for a creative effort. The team needs to focus their creativity to propose concepts that will solve the design problem. Traditionally, this stage utilizes psychological tools like brainstorming, six thinking hats, SCAMPER, among other available techniques. The ideation stage is the transition from problem-identification to the exploration of potential solutions. The team takes advantage of the collective perspectives, strengths, and creative effort.
4. Prototype: The prototype stage consists of transforming ideas in some way to the physical world. A prototype can be anything that has a physical shape, including a wall full of Post-it to represent a process or a service, a role-play activity, space, an object, an interface, or even a storyboard. The quality of the prototype increases with the project maturity. A prototype then shows to users what will be the product. Finally, a prototype is a way to test functionality.

Frequently, this stage unveils new problems not detected before, increasing the pertinence of the product.

5. Evaluate: This stage is the opportunity to refine the solutions and improve the concept and the prototype. The user interacts with the concept to learn more about their needs and requirements and provide useful feedback for the product and the previous stages in the methodology.

2.2 Design Thinking: State-of-the-Art Review

Table 1 presents a review of the literature about the benefits that Design Thinking give on the innovation process, its ability to combine with other techniques and its importance in the development of products and understanding of the client.

By analyzing the state of the art of the Design Thinking, the following conclusions rise about this design approach:

According to Volkova and Jakobsone (2016), one of the most effective strategies to improve competitiveness is achieved by creating new products, with meaning, with added value and through the personalization of products taking into account the following aspects:

(a) The understanding of the identity and culture of the user.
(b) The requirements currently requested.
(c) Future trends.

One of the main strengths of design thinking is the ability to communicate ideas and create value proposals with multidisciplinary teams (Geissdoerfer et al. 2016; Brown 2009a, b). Design Thinking has been increasingly recognized as a promising asset for fields other than design. It gained attention in the sectors of business, leadership, and management, as an option to deal with the growing complexity of the market and to be used as a source of innovation and commercial success (Davis 2010; Dorst 2011; Fraser 2007).

Design Thinking allows anyone without design experience to create solutions to everyday challenges. These solutions may be products, services, environments, organizations, and modes of interaction. The design team converges to transform the collected information into meaningful ideas (Shapira 2015). Principles that characterize DT according to Glen (2015):

- It is centered in the being. It tries to develop new knowledge about the problems faced by users in everyday life. It is necessary that the designer puts aside preconceived ideas and be willing to observe to define which problems deserve further investigation. It develops empathy.
- Observation. To understand human needs, the DT emphasizes observation that at the same time leads to empathy, understanding, and analysis. The purpose is to obtain an understanding of the user's requirements.

Table 1 Some works related to DT

#	Author (s)	Problematic	Technique	Contributions
1	Volkova and Jakobsone (2016)	Turbulent business environments, scarce resources, hyper-competition, and globalization exert great pressure on companies. Providing value to the customer is vital for the organization to remain in the market	Design thinking	Design thinking (DT) becomes the main tool to create value for customers. It highlights the potential of DT to develop new organizational abilities, improve the welfare of society and maintain competitiveness in challenging business conditions
2	Geissdoerfer et al. (2016)	Innovation in sustainable business models is an emerging issue. There are few tools available to model sustainable businesses	• Design thinking • Value mapping	To employ Design Thinking to improve the creative process during the development of value proposals and to improve the modeling of sustainable businesses
3	Geissdoerfe et al. (2017)	It is necessary a guide and tools to facilitate the design of sustainable business models	• Business model • Design thinking	A process to guide innovation efforts when developing business models, which integrates Design Thinking elements within its phases
4	Davis (2010)	There is a great opportunity to improve the value offer of a company. Traditional products must incorporate value. It is very important to teach Design Thinking in the academy	• Design thinking.	Tips to develop the ability to think like a designer. Design Thinking provides creativity to the decision-making process. It gives value to traditional products. The importance of teaching DT to business students

(continued)

Table 1 (continued)

#	Author (s)	Problematic	Technique	Contributions
5	Shapira (2015)	Design Thinking as a tool to respond to the pressing challenges of global sustainability	• Design thinking • Sustainable development	Enriched process of Design Thinking with sustainability strategies
6	Glen (2015)	To provide students with a framework to deal with unstructured problems and to manage the innovation process.	• Design thinking	Orientation to manage activities and challenges using Design Thinking. Principles that characterize DT
7	Mosely et al. (2018)	To explore how the public that does not belong to the design area uses Design Thinking	• Design thinking	Strategies to teach Design Thinking to business students
8	Pavie and Carthy (2015)	To understand and explore the problems in responsible innovation	• Design thinking.	A process to develop responsible products and services. To stimulate the creative capacity of innovation teams.
9	Chou (2017)	To generate optimal results during the realization of a social enterprise	• Design thinking	Detailed review of social entrepreneurship theories and Design Thinking methodology
10	Camacho (2016)	To know the way to perform design thinking in two institutions icon in this technique.	• Design thinking	Description of how Design Thinking came about in Stanford and IDEO
11	Frisental (2016)	Importance of improving the way organizations manage their businesses, and their relationships	• Design thinking	Redefine and expand business analysis through DT
12	Abrell (2016)	Studies show that corporations that use DT take better advantage of their opportunities	• Design thinking	Link between DT and corporate entrepreneurship

- Display. The approach used to make information meaningful. It takes advantage of empathy and intuition.
- Prototyping. The idea is represented with the purpose of obtaining feedback and learning. To generate many disposable prototypes in the fastest and cheapest way possible.

DT involves solving complex problems that require curiosity, imagination, and creativity to generate, explore, and develop possible solutions, with value for the final user (Dorst and Cross 2001). It offers to the companies the ability to transform products, services, processes, and strategies, where the last one determines new forms of value (Brown 2008).

DT defends the importance of empathy with the consumer. To develop good solutions, teams need to understand their users; how they think and what they feel about the problem that the group seeks to solve. DT offers a set of techniques about how to develop the ability to become empathetic with the users. It suggests to the innovation teams to immerse themselves in the lives of their users and observe how they interact with the products they want to improve. By watching, listening, and collecting stories, they can capture unexpected ideas or longings. In the line of DT's approach, innovation teams need to know their users and care about them and their lives to create meaningful products (Brown 2009a, b).

DT has become an integral part of the innovation process. In fact, it plays a strategic role in the generation of value through the creation of ideas that respond better to the expectations and needs of consumers.

Design thinking with its broad and generic applicability provides an efficient method for the creation of value-added products effectively (Brown 2008); justifying in this way their ability to be integrated to other techniques, such as TRIZ.

The flexible application of design thinking provides an effective method for creating value-added products effectively (Brown 2008). However, its flexibility of application implies that the expert who applies the DT is in need of solving problems outside his area of expertise. Consequently, it is necessary to aggravate the DT deployment process with a method capable of formulating and solving problems in multiple domains. The Theory of Inventive Problem Solving (TRIZ) has this competence. The next section explores the use of TRIZ in the innovation process.

3 The Role of the Theory of Inventive Problem Solving (TRIZ) to Help the Innovation Process

TRIZ is the Russian acronym for the Theory of Inventive Problem Solving, developed by Genrich Altshuller. Its objective is to solve inventive problems and the intensification of the creative and technical thinking. An inventive problem is the one that contains at least one contradiction (technical or physical), in which the conditions of the problem do not allow negotiation or compromise between the different parties to the conflict (Cortés 2003). TRIZ is a systematic, knowledge-based and human-oriented methodology of inventive problem solving (Savransky 2000).

3.1 Basic Concepts of TRIZ

Some basic concepts of TRIZ are described below:

Contradictions: They arise when in a system, the attempt to improve a specific parameter causes the unacceptable degradation of another parameter also useful or when a system requires two mutually exclusive states to reach an objective (Cortés 2003). There are two types of contradictions: (1) Physical contradiction: It requires the simultaneous existence of two mutually exclusive states or conditions in a single component or parameter in a system. (2) Technical contradiction: The improvement, modification, or transformation of a useful characteristic of a system causes the degradation of another useful characteristic or a blockage in the system.

Ideal system: Systems evolve toward an ideal behavior. The evolution goes in the direction of increasing the benefits and reducing costs (Domb 2017). The extreme result of this evolution is the ideal result, where the benefits are present, and the costs are not. It is the ideal solution to a technical problem (Fiorineschi et al. 2018).

Resources: According to TRIZ, the evolution of any technical system depends on available resources. A resource is, for example, an element available in a system or environment that remains inactive or underestimated, but with the potential to generate a useful effect or a valuable action at no additional cost or minimal cost. Some examples of resources are substances, energy fields, space, time, information, and practical resources.

Trends of Evolution: Evolutionary tendencies, or patterns of evolution, are a direct consequence of the laws of the evolution of the system. There are eight trends. The studies of Altshuller showed that if a system begins to follow one of these tendencies, during its evolutions, it will arrive at the predicted end. Therefore, analyzing the current state of a system and its history, it is possible to identify which trends have been undertaken and how the system can evolve shortly (Gadd 2011).

3.2 TRIZ Benefits

According to the Mexican TRIZ Association, the following points describe some of the advantages of this approach (AMETRIZ 2016).

- It creates inventive solutions and practices faster.
- It simplifies the products and processes technically, gaining in cost, reliability, and average life.
- It resolves conflicts and technical contradictions without the need of trade-off solutions.
- It quickly conceives the next generations of products and processes.
- It reduces the development cycle starting from a correct concept.

3.3 TRIZ Basic Assumptions

The research directed by Genrich Altshuller is based on the hypothesis that there are universal principles with the potential to guide the creative process. These principles are based on technological evolution and not on psychology. After, his investigation, these principles were identified and codified to give a more predictable creative process. There are three fundamental premises for solving a problem using TRIZ:

- The problems and the solutions are repeated in different technological domains. This implies that knowledge of one domain is transferable to others. The modeling of a problem using physical or technical contradictions is an adequate strategy to facilitate the transfer of knowledge among technical domains.
- The evolution of a technical system follows some trends that are repeated in different technological domains.
- The resolution of an inventive problem requires scientific knowledge from different domains.

3.4 TRIZ: State-of-the-Art Review

Table 2 presents a review of the literature about the benefits that the Theory of Inventive Problem Solving confers to the innovation process, its ability to be combined with other techniques, and its role in the New Product Development process.

Creativity is the ability to generate new and useful ideas. To overcome limitations in the generation phase of solution proposals, (Eppinger and Ullman 2007) as well as (Ilevbare et al. 2013) suggests the application of specific methods and/or tools to support the idea generation stage. One of these "assistants" to creativity is TRIZ, which is considered as support for designers in the generation of creative solutions.

TRIZ is a different method to conceive new products and processes that use various tools that propose several ways to propose inventive solutions (Vaneker and

Table 2 TRIZ: Review of your state of the art

#	Author (s)	Problematic	Technique	Contributions
1	Fiorineschi et al. (2018)	To complement FDM with some creative method to improve its functionality	• Functional and morphological analysis. • TRIZ	Integration of functional and morphological analysis with TRIZ for problem-solving
2	Vaneker and Van Diepen (2016)	To optimize maintenance activities during the design process	• Maintenance design • TRIZ	TRIZ is a useful addition to the set of guidelines for maintenance. Roadmap to solve maintenance problems with the support of TRIZ tools
3	Bertoncelli et al. (2016)	How to learn to be creative?	• TRIZ	Arguments that guarantee that creativity can be learned and trained
4	Chechurin and Borgianni (2016)	The scope of TRIZ is interpreted differently in industry and academia	• TRIZ	Study of the best 100 indexed publications related to TRIZ
5	Chang (2016)	To promote the skills of engineering students to solve problems	• TRIZ	Employment of TRIZ for product design, technology forecasting, and business management
6	Gronauer and Naehler (2016)	To guide the company's personnel in the development of products		TRIZ method and tools that show the advantages when developing innovative and creative products
7	Wang et al. (2016)	It strengthens highly important skills during creative development	• TRIZ	TRIZ as a means to be creative. It begins to be more used by university students
8	Ekmekci and Koksal (2015)	Companies need to create new products or improve existing ones	•TRIZ	Product design and improvement of existing ones through TRIZ
9	Albers et al. (2011)	Search for the best method in the development of innovative products	• TRIZ	TRIZ Toolbox to support the inventor in the development of a product

(continued)

Table 2 (continued)

#	Author (s)	Problematic	Technique	Contributions
10	Spreafico and Russo (2016)	To simplify the TRIZ method to make innovation	•TRIZ	Collection, analysis and processing of case studies of ETRIA TRIZ Future Conference, and TRIZ Journal
11	Chechurin (2016)	The development of new products and their commercialization require efficient and quality processes	• TRIZ	ImModern TRIZ. Berlinprovement of lean products through TRIZ
12	Becattini et al. (2015)	The time used in the analysis of design protocols is delayed	• OTSM	OTSM-TRIZ coding scheme Set of rules for the identification of cognitive processes. To reduce the time spent analyzing design protocols
13	Delgado et al. (2017)	Limitation of TRIZ to observe the progress of an inventive problem over time	• Dynamic of systems • TRIZ	System dynamics integrated with TRIZ for the resolution of inventive problems

Van Diepen 2016). It is a problem-solving approach strongly based on the concept of "Contradiction" using specific techniques to solve them (Fiorineschi et al. 2018).

TRIZ is compatible with the solution of problems, prevention of failures, incidents, management, the creation of new products/services, the definition of commercial concepts, and the resolution of administrative conflicts. Therefore, it contains a set of tools to be conveniently selected, according to specific needs (Vaneker and Van Diepen 2016).

4 Reason for DT + TRIZ Integration

The literature review showed that there is an area of opportunity to combine DT with TRIZ. The purpose of the integration is to assist users in the formulation of innovation projects to combine the best features of both methods.

Consequently, it is possible to propose a framework which makes possible to develop products such as designers do, through its 5 stages (empathy, definition, ideation, prototype, and evaluation) and the fundamental concepts of TRIZ, which might guide a team or an individual in the process of solving the intrinsic problems generated by designing, in that way, channel the creativity, imagination, and thought processes of the users. After reviewing the main contributions of each approach, Table 3 summarizes the advantages of each approach and its complementarity.

As the table shows, both approaches complement each other to cover certain deficiencies, TRIZ benefits from DT to identify the user's requirements and wishes for whom it is being designed, and guides the design process so that people without experience in design can carry out this process and equip themselves with products with added value and meaning. While DT benefits from TRIZ to identify, define, and model a problem, provides an efficient toolkit for the accurate resolution of

Table 3 Complementarity between TRIZ and DT

Advantages	DT	TRIZ	DT + TRIZ
It identifies a user's requirements	Yes	No	Yes
It models an innovation problem	No	Yes	Yes
It proposes tools to solve problems	No	Yes	Yes
It offers a framework to guide the design	Yes	No	Yes
It subjects the proposed solution to evaluation	Yes	Yes	Yes
It contemplates the possible evolution of a product in the future	No	Yes	Yes
It can be used by users without design experience	Yes	No	Yes
It gives meaning to the products that are being designed	Yes	No	Yes
It prototypes to evaluate the proposed solutions	Yes	No	Yes

these problems, and in this way, it channels the thinking and creative effort towards the development of solutions.

It may be determined that the complementarity of both approaches allows the design of a framework with highly technical and humanistic characteristics, suitable for designing new products that meet the requirements of users following a defined process, where the thinking and creative efforts are efficiently directed.

5 Integration of Design Thinking and TRIZ Theory (DT + TRIZ)

Figure 1 presents which elements of the TRIZ theory will be integrated into a certain phase of the Design Thinking model:

Description of the framework:

Stage 1: Empathy
Empathy is the basis of the Design Thinking process, to empathize it is necessary to: (a) Observe the behavior of users in the context of their lives. (b) Interact with the users and interview them. (c) Experience what the user experiences.

It is suggested the incorporation of the concept of an ideal system into the basis of the Design Thinking model: the empathy phase. The objective is to try to understand the users and the way they describe how the product or service should work to satisfy their needs, requirements, or desires.

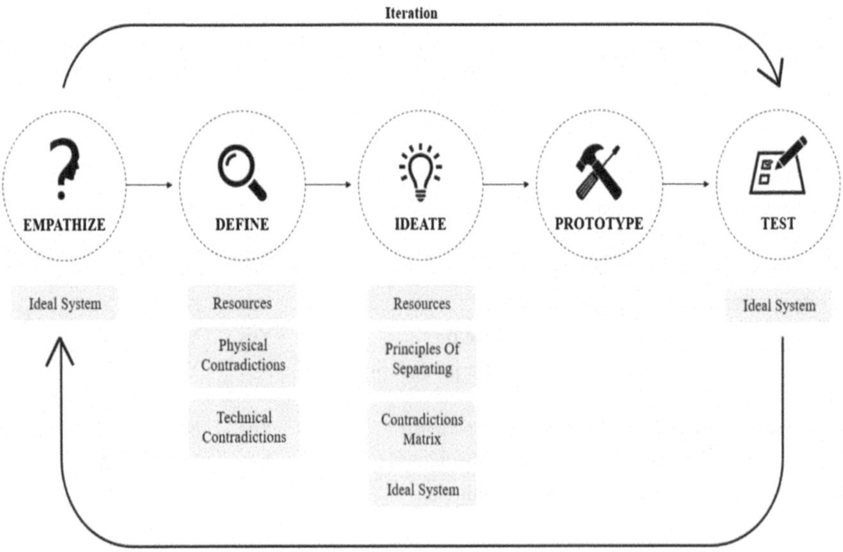

Fig. 1 DT + TRIZ

The type of information that they allow to collect may classify the methods used in the empathy stage: qualitative methods (observation, interviews, and group dynamics) and quantitative methods (surveys and questionnaires) (Garreta and Mor, 2010).

Results and Discussion of Stage 1

The objective of this stage is to understand the users for whom it is being designed. Understand them as a person and identify what is important to them. Observe what they do, how they interact with their environment, to infer what they think and feel as well as what they need. Interacting directly with people reveals the way of thinking and the values they have.

In this first stage of the DT, the ideal system concept of TRIZ will be used, which consists of the user describing the ideal behavior of his product and its benefits. In this way, the wishes and requirements that the user looks for in a product are possible to obtain which will become design parameters for the following stages.

Empathy is the fundamental stage of the whole design process; TRIZ does not consider this aspect when designing a product. As a result, it is the primary contribution that DT makes to this theory. Knowledge about users, their contexts of use, their needs, objectives, and attitudes are essential for user-centered design (Garreta and Mor 2010).

Stage 2: Define

The second phase of Design Thinking focuses on the definition of the problem. The analysis of the product aims to identify at least one physical or technical contradiction (s) in the product, process, or service.

To support this DT phase, some elements of TRIZ will be identified, such as (a) resources, which are elements available in a system or environment that remains inactive or underestimated, but with the potential to generate a useful effect or a valuable action at no additional cost or minimal cost. Some examples of resources are substances, energy fields, space, time, information, and practical resources (Orloff 2012). (b) Physical contradictions, which require the simultaneous existence of two states or conditions mutually exclusive in a single component or parameter in a system, and (c) Technical contradictions, which arise when the improvement, modification, or transformation of a useful characteristic of a system causes the degradation of another useful characteristic or a blockage in the system (Savransky 2000).

Results and Discussion of Stage 2

In the second stage of the DT, the findings obtained through empathy are synthesized. A specific problem has to be defined, to address it, therefore make solution proposals in the next stage. The contribution of TRIZ to the second stage of the DT consists in identifying and defining the problem in the form of physical or technical contradictions, in this way, the separation principles and the contradiction matrix of the TRIZ toolbox can be used in stage number 3 as support to solve the problem. The analytical tool of TRIZ to be used in stage number 2 is that of resources, where the substances, energy fields, space, time, information, and other resources present in the system will be identified (Orloff 2012). Thus, it is determined how the body of the system can be used to make a suggestion of improvement (wooden board or some other material).

Stage 3: Ideate

The resources, the separation principles, and the contradiction matrix (which contains the 40 inventive principles and the 39 technical parameters) and the ideal system are the elements of TRIZ that will be integrated into phase number three of Design Thinking: Ideate. The purpose is to channel thought and creative efforts toward the search for an inventive solution.

This stage of the DT model consists in generating ideas to solve the problem identified in the previous phase. The objective of this stage is to propose solutions by taking advantage of the different perspectives and strengths of the design team.

The role of TRIZ in the stage of ideation is to channel thought and creative efforts during the development of solutions through the use of concepts such as resources (described in stage 2), separation principles used to solve physical contradictions, there are four types:

I. The separation in space. If something is physically contradictory regarding space, assigning the contradictory characteristics to various components of the system will separate the requirement.
II. The separation in time. Condition the system so that contradictory requirements are executed in different moments of time
III. The separation between the whole and its parts. A characteristic can have a value that is only observed at the system level and an opposite value and even nonexistent at the component or subsystem level.
IV. The separation according to conditions. It consists in separating the contradictory requirements so that it is the medium that surrounds the system that executes the necessary changes that allow reaching an objective.

The next tool of TRIZ is the Contradiction Matrix; it is one of the first TRIZ instruments proposed by Altshuller to solve technical contradictions. The vertical elements of the Matrix are the engineering parameters that must be improved, and the horizontal columns contain the engineering parameters that may be affected and/or negatively degraded as a result of the improvement of the parameters, which are used to solve the technical contradiction (Savransky 2000). Then, it is considered to make use again of the ideal system (explained in the empathy stage) to analyze if the solution proposals generated in this stage are suitable results to satisfy the user's requirements.

Results and Discussion of Stage 3

The DT uses techniques such as brainstorming, the trial and error method and the six hats to think of Edward de Bono, which have a high psychological bias, generating the design team to ramble during the generation of ideas and consuming time. TRIZ provides in this stage the Contradictions Matrix (which contains the 40 principles of the inventive solution and the 39 technical parameters) and the principles of separation which will provide a guideline to the design team to start developing new ideas; these strategies channel the creative effort of the designer.

Stage 4: Prototype

In this stage of the DT, the proposals generated in the ideation stage will materialize, that is to say, the product with the relevant improvements will be taken to a physical plane, to experiment and analyze its functionality. The product is replicated in the material available (cardboard, plastic, wood, etc.), represent it by means of the storyboard, through an interface, a role-play, etc.

The importance of this stage is that through a prototype it is possible to learn in a better way, design disagreements are stipulated and resolved, a conversation with the user is initiated, if the proposal is wrong the design team will realize before launching it to the market, saving resources. This interaction with the prototype helps to drive empathy and in this way develop successful solutions.

Results and Discussion Stage 4

TRIZ does not explicitly consider the need to prototype in its resolution process, although it is assumed that it does it. Therefore, DT covers this limitation with its fourth stage. By making tangible the proposed solution improves the understanding of a product and its respective manipulation, they are made to find areas of opportunity to improve it. The design team and the user interact with the prototype to empathize with it in its everyday context.

Stage 5: Test

Finally, in the last phase of design thinking, the prototype will be tested with the user to receive feedback. In this final stage, the ideal system concept of TRIZ is used, so that the user can determine if the prototype meets the requested requirements, that is to say, the ideal behavior of the product. If necessary, relevant changes will be made. Thus, it is possible to return to previous phases of the model to rethink or improve the proposal.

The process described above is iterative, that is, it allows returning to the previous stages of Design Thinking if necessary, to rethink the proposed solutions and even analyze if the problem was established correctly or if there is some characteristic that was omitted in the empathy phase.

Results and Discussion Stage 5

In the last stage of the DT, there is an interaction with the prototype to find areas for improvement and propose solutions. If necessary, the design team is referred to previous stages of DT to solve the problem. TRIZ brings to this stage the concept of an ideal system with the purpose of comparing the developed proposal with the requirements and wishes stipulated by the user in the first stage of DT.

6 Results and Discussion of DT + TRIZ Integration

The integration of DT and TRIZ is feasible. After analyzing the advantages and limitations of each tool, its complementarity was determined. On the one hand, DT provides TRIZ with accurate information about the requirements, wishes, and desires of the users for whom it is being designed. It offers a framework to guide the design process, including to be carried out by users without design experience. Also, it gives to the product a meaning to make it more attractive to the user. It considers prototyping to materialize the ideas proposed and in this way facilitate interaction with the user. It submits the proposals generated for evaluation, to obtain feedback and if it is necessary to return to previous stages of DT to improve the prototype.

On the other hand, TRIZ gives DT the problem in the form of contradiction, by performing this strategy there is a possibility of having access to the TRIZ toolbox to solve this problem effectively based on scientific principles and proven over time.

Therefore, DT marks a process to design products and establishes the basis, which is the description of the user for whom the work is developed, while TRIZ provides the necessary tools to channel the creative effort of the design team during the generation of solution proposals.

7 Case Study

The following case study shows how the Design Thinking model was applied in combination with the Theory of solving inventive problems (TRIZ) to propose a conceptual design of a system that supports a laptop:

Stage 1: Empathy

In this first stage, it began with the technique of observation, which consisted of analyzing laptop users, to know what they do with their equipment and under what conditions they do it. The information obtained by performing this technique supports us to determine their behavior and to know how the user uses his laptop.

It was observed that users employ their laptop in different situations: in closed and external environments, static or moving. It is common for the user to place the computer equipment on their legs. When the user becomes distracted or exalted by interference from outside, drop the equipment to the ground, which receives the damage in the corners most of the time.

Another technique used in this phase is the interview, which aims to obtain qualitative information, a deep understanding of the needs, preferences, and past experiences of users regarding a product is pursued (Garreta and Mor 2010), in this case, the system that supports the laptop.

The laptop users were interviewed, obtaining more important data that as the time passes, the equipment begins to heat up more and more and to expel hot air, causing in the user a sensation of heat.

During the interview, the ideal system is used (See Sect. 3.1 Basic concepts of TRIZ), the user is asked what are the features and benefits that you are looking for in a laptop. It obtained security and portability as the most important requirements. Regarding safety, it is mentioned that the laptop's support system does not slip off the legs, and that the laptop does not slip from the support object and that in case of falling, the system receives the damage so that the laptop suffers the least possible damage. Regarding portability, it is desired that the support system of the laptop (average size 13 and 15 inches) can be stored in an average backpack (45–55 cm high, 30–35 cm long, 15–25 cm wide).

Observing and interviewing laptop users the aforementioned information were obtained, the difficulties encountered were experienced by the designers of this laptop to have empathy with the user. Another aspect found through this exercise was that when transporting the laptop, the user can be pushed and release the laptop, including doing that by its distraction.

After interacting and empathizing with the user, it is determined that the ideal system for the laptop is the one where the object can be assembled by itself, that prevents the laptop from sliding, in case of falling, that protects equipment from impact, comfortable to keep it in the legs for a long time and that becomes portable to carry it in a backpack.

Results Stage 1:

In summary, the most important aspects collected during this stage is that system: (I) Do not allow the equipment to slide, (II) In case of fall, you must protect from equipment damage, (III) It must be able to be used on the legs, (IV) It must not slide off the legs and (V) It must be portable (fit in an average backpack).

Stage 2: Definition of the problem

The objective of this stage is to identify which are the main problems when using a laptop on the legs, both outdoors and in closed or external environments, in static or moving way and the inconvenience to transport it.

It is determined that the main problem is that the system requires new functions to satisfy the user. Using functional analysis, it was defined that the system cannot be portable and to provide security to the laptop.

The main resource (See Sect. 3.1) with which the system has space, since it is that through the modification of its original form it becomes multipurpose, that is, with the minimum number of movements, the system offers different functions.

Once the functions that must be incorporated into the object are known (that the system changes form to provide different functions), we analyze the TRIZ tools based on knowledge, specifically, of the Technical Contradictions Resolution Matrix (MRCT) which is a relation between the 40 principles of inventive solution and the 39 generic parameters, arranged in a 39×39 matrix (Cortés 2003).

Then the parameter to be improved is determined, in this case, the stability, as well as the characteristic that degrades or worsens, which is the form. At the intersection of these two axes are contained principles of inventive solution that will serve as support in the search for solutions.

The inventive principles are strategies and heuristics, able to guide the reasoning of an individual toward the solution of a problem, without causing minimal harmful effects (Cortés 2003). The principles are classified according to their relevance, which means that the principle that occupies the first position is the most appropriate to resolve the contradiction. The technical contradiction asks for two parameters in conflict: stability versus shape. The solution principles found in the intersection of these parameters are 1, 4, 35, 17, 7, and 3, which correspond to the principles of segmentation, asymmetry, the transformation of the physical and chemical states of an object, movement to a new dimension, nesting, and local quality, respectively. Therefore, these are the guides to make improvements to the system that is being designed.

Results Stage 2:

The main problem to solve is that the system must be secure and portable. Space is the resource that should be used when designing the new system. The team chooses to model the conflict as a technical contradiction, due to its facility to understand this tool. The aspect of improving is the stability of the system for its use in exteriors and interiors, for this it is necessary to modify the form, being the aspect to get worse. The first inventive principle to modify the system is the principle of segmentation.

Stage 3: Ideation

The design team develops a series of proposals for solving the problem and the contradictions identified in the previous phase. It makes use of techniques such as brainstorming, six hats to think, and brain writing to interpret the principle of inventive solution (segmentation) and propose how it will be applied to the new system.

To carry out the improvement of the system, it is desired to improve the stability; on the other hand, the characteristic that can be affected is the form. Therefore, the solution principle that MRCT recommends to use is the segmentation principle, which consists of creating a sectioned object or increasing the degree of segmentation that it already has.

The proposed solution consists of segmenting the body of the system, which is rectangular so that when making a move the system goes from being a support structure to a transport structure for the laptop. During the experimentation, it was discovered that by severing the middle part of the body of the system it was possible for the walls of the system to move, in this way, it could be manipulated if the walls surround the computer equipment (to protect and transport it) or if they were joined to form a base that supports the computer equipment.

Regarding security, it is proposed to add a handle in the system that will act as insurance in case the user releases the system since by doing so, the equipment will be hanging from the wrist, giving the user enough time to reincorporate the system. To the system does not slip from the legs, it was proposed to incorporate some small legs, which will prevent the equipment from sliding when moving the legs, providing greater stability and security to the system. Garters will be placed in the corners of the system so that the laptop does not come off when transported. The proposal generated contains some elements of the principle of universality since the system can be used as a cover, support, and means to transport the equipment.

Results Stage 3

The principle of inventive solution was used to propose the new system is that of segmentation, which consists of creating a sectioned object or increasing the degree of segmentation that it already has. A handle will be added to the structure of the system to act as insurance in case the user releases this object. Legs will be added to improve the stability and safety of the system when it is placed on the legs.

Stage 4: Prototype

The objective of the fourth phase is the materialization of the proposed solution, in such a way that the new proposed functions, movements, portability, design, etc., can be analyzed. A prototype was made based on cardboard, with the purpose of replicating the "ideal" laptop support system requested by the user (see Fig. 2), complying with the segmentation principle, adding the handle and legs to the structure.

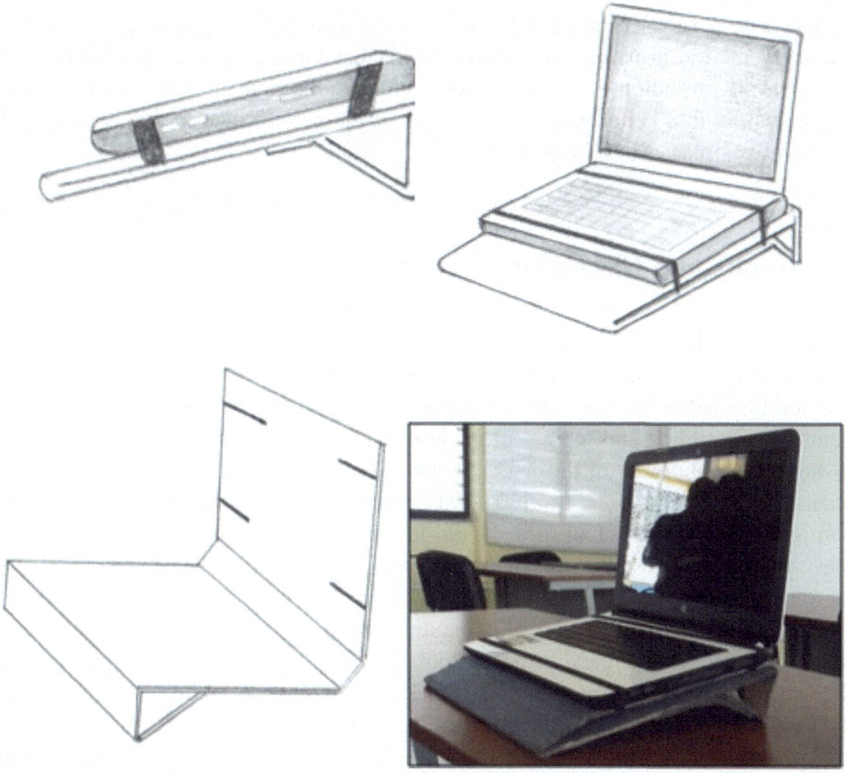

Fig. 2 Prototype

Results Stage 4:

A functional prototype of the proposed laptop system made of cardboard, wood, and elastic strips (see Fig. 2).

Stage 5: Test

Submit to evaluation the prototype developed by the design team and the user to obtain feedback. Take into account at all times the ideal system stipulated by the user in the first stage (See Stage 1 of the case study).

When interacting with the prototype, it was verified that the laptop is subject to the system, that is, it does not slip. It was found that the legs provide stability and safety, ensuring that even if the user moves the legs, the system does not fall off, but stays on the same legs. The handle was moved to a corner of the system, due to when the design team was manipulating the accessory determined that the equipment is held at an angle and not the middle part where the handle had previously been placed and that it should not be exposed when it is not used, so it is proceeded to nest it to the structure of the system. The system can be used as a support and as a means of transporting the laptop, according to the user's need.

Results Stage 5:

The goal is to verify the functionality of the designed system, determining that the segmentation principle was appropriated, that the garters and legs fulfill their objective of providing security, avoiding that the equipment slides and that the handle works in a better way in the corner of the system than in the average part where it was placed in a beginning.

8 Results and Discussion: Case Study

When DT + TRIZ was applied during the innovation to a laptop support system, a final product was obtained, which is an object that provides stability, security, and portability. Garters and legs give security to the computer equipment since it prevents it from sliding and falling off the legs; the structure of the system is designed to hold the equipment in the legs. At the same time, the user is prevented from receiving heat in his legs, by making a simple movement, the system covers the computer equipment, acting as a sheath, in this way, if the equipment goes down, the system receives the impact, avoiding or reducing the damage to the laptop. Another feature in the object that the user does not expect, but that increases the usefulness, is the addition of a handle. The user inserts the hand. If by distraction or accident he releases the equipment, the system will remain hanging from the wrist avoiding the computer to fall. The requested portability requirement is met, designing the system in such a way that it can be transported in an average backpack (45–55 cm high, 30–35 cm long, 15–25 cm wide) and for laptops from 13 to 15 inches. Technical parameters such as mechanical resistance, flexion points, and the number of kilograms that the laptop holder supports are evaluated through laboratory tests.

9 Conclusion

Potential/Feasibility of integration

It can be concluded that the characteristics of DT centered on the human being and the highly technical and logical attributes of the theory of solving inventive problems benefit each other. Consequently, they can be combined to obtain a model that uses the best features of both. By combining both approaches it is possible to understand the problems and needs of the user using the approach of Design Thinking, and will have a greater chance of success in formulating a solution to these requirements through the use of TRIZ theory, making use of its tools, to channel the creativity and innovation efforts of the design team.

Integration needs

- Integrating design thinking with the theory TRIZ seeks to increase the efficiency of the ideation process. Both in TRIZ and in DT, thought is channeled through considering specific elements, being a support to be in continuous generation of ideas, and that these, are argued, that is, that each idea has a rationale and is based on a specific principle or parameter, demonstrating in a better way why it would be the best option to solve the problem of the client/user.
- Decrease the psychological bias during the ideation phase, wander to generate ideas and even creative blocks, that is, when there are no more ideas to contribute, or it is not known how to start the approach, formulation or description of an idea.
- It is a proposal for those entrepreneurs who wish to innovate an existing product or want to formulate an innovation project starting from scratch, providing a framework to follow to have enough empathy to identify needs in the market and have sufficient tools to develop their respective proposals for solutions properly argued.

10 Future Work

- It is contemplated to incorporate the trends of evolution to try to explain what will be the stages of development of a prototype. In this way, a family of products could be planned, and the transitions of the product managed more appropriately.
- It is necessary to propose a strategy to systematize the use of DT + TRIZ through a software platform or an application of easy access and consumption.
- It is necessary to make many more case studies to improve the methodology and test it in different contexts and with different restrictions.

References

Abrell, T. (2016). *Design thinking and corporate entrepreneurship: An integration and avenues for future research*. Switzerland.

Albers, A., Deigendesch, T., Schmalenbach, V. (2011). TRIZ-box–improving creativity by connecting TRIZ and artifacts. *Procedia Engineering 9*:214–221. https://doi.org/10.1016/j.proeng.2011.03.113.

AMETRIZ (2016). Entrenamiento en la metodología TRIZ. Asociación Mexicana de TRIZ.

Becattini, N., Cascini, G., & Rotini, F. (2015). *Design computing and cognition*. Switzerland: Springer International Publishing.

Bertoncelli, T., Mayer, O., & Lynass, M. (2016). Creativity, learning techniques and TRIZ. *Procedia CIRP, 39,* 191–196. https://doi.org/10.1016/j.procir.2016.01.187.

Brown (2008). *Design thinking*. Harvard Business Review. New York

Brown, T. (2009a). *Change by design. How design thinking can transform organizations and inspire innovation*. New York: Harper Collins.

Brown, T. (2009b). *Change by design: How design thinking transforms organizations and inspires innovation*. New York: Harper Business.

Camacho, M. (2016). David Kelley: From design to design thinking at Stanford and IDEO. *She Ji: The Journal of Design, Economics, and Innovation, 2*(1), 88–101.

Chang, Y. S. (2016). Effect of TRIZ on the creativity of engineering students. *Thinking Skills and Creativity, 19,* 112–122. https://doi.org/10.1016/j.tsc.2015.10.003.

Chechurin, L. (2016). *TRIZ events increase innovative strength of lean product development processes*. Switzerland.

Chechurin, L., & Borgianni, Y. (2016). Understanding TRIZ through the review of top cited publications. *Computers in Industry, 82,* 119–134. https://doi.org/10.1016/j.compind.2016.06.002.

Chou, D. (2017). Applying design thinking method to social entrepreneurship project. *Computer Standards & Interfaces, 55,* 73–79. https://doi.org/10.1016/j.csi.2017.05.001.

Cortés, G. (2003). Gestion de l'innovation: application de la théorie TRIZ (in French). Touluse.

Davis, B. (2010). Creativity & innovation in business 2010 teaching the application of design thinking to business. *Elsevier, 2*(4), 6532–6538. https://doi.org/10.1016/j.sbspro.2010.04.062.

Delgado-Maciel, J., Cortes-Robles, G., Sánchez-Ramírez, C., Alor-Hernández, G., García-Alcaraz, J., & Negny, S. (2017). *System dynamics modeling and TRIZ: A practical approach for inventive problem solving*. In: D. Cavallucci (Ed.), TRIZ – The Theory of Inventive Problem Solving. Springer: Cham.

Domb, E. (2017). TRIZ Journal. https://triz-journal.com/ideal-final-result-tutorial/. Accessed 12 Dec 2017.

Dorst, K. (2011). The core of design thinking and its application. *Design Studies, 32*(6), 521–532. https://doi.org/10.1016/j.destud.2011.07.006.

Dorst, K., & Cross, N. (2001). Creativity in the design process: Co-evolution of problem–solution. *Design Studies, 22*(5), 425–437. https://doi.org/10.1016/S0142-694X(01)00009-6.

Ekmekci, I., & Koksal, M. (2015). Triz methodology and an application example for product. *Procedia Social and Behavioral Sciences, 195,* 2689–2698. https://doi.org/10.1016/j.sbspro.2015.06.481.

Eppinger, S., Ullman, D. (2007). Product design and development. Massachusetts.

Fiorineschi, L., Saverio, F., & Rotini, F. (2018). Enhancing functional decomposition and morphology with TRIZ: Literature review. *Computers in Industry, 94,* 1–15. https://doi.org/10.1016/j.compind.2017.09.004.

Fraser, H. (2007). The practice of breakthrough strategies by design. *Journal of Business Strategy, 4,* 66–74. https://doi.org/10.1108/02756660710760962.

Frisental, T. (2016). Design thinking business analysis: Business concept mapping applied. Switzerland.

Gadd, K. (2011). *TRIZ for engineers: Enabling inventive problem solving*. New York: Wiley.

Garreta, D., & Mor, E. (2010). *Diseño centrado en usuario*. Catalunya: Universitat Oberta de Catalunya.

Geissdoerfer, M., Bocken, N., & Hultink, E. (2016). Design thinking to enhance the sustainable business modelling. *Journal of Cleaner Production, 135,* 1218–1232. https://doi.org/10.1016/j.jclepro.2016.07.020.

Geissdoerfer, M., Savaget, P., & Evans, S. (2017). The Cambridge business model innovation process. *Procedia Manufacturing, 8,* 262–269. https://doi.org/10.1016/j.promfg.2017.02.033.

Glen, R. (2015). Teaching design thinking in business schools. *The International Journal of Management Education, 13*(2), 182–192. https://doi.org/10.1016/j.ijme.2015.05.001.

Gronauer, B., & Naehler, H. (2016). TRIZ as an amplifier for corporate creativity and corporate innovation ability. *Procedia CIRP, 39,* 185–190. https://doi.org/10.1016/j.procir.2016.01.186.

Ilevbare, I., Probert, D., & Phaal, R. (2013). A review of TRIZ, and its benefits and challenges in practice. *Technovation, 33*(2–3), 30–37. https://doi.org/10.1016/j.technovation.2012.11.003.

Mosely, G., Wright, N., & Wrigley, C. (2018). Facilitating design thinking: A comparison of design expertise. *Thinking Skills and Creativity, 27,* 177–189. https://doi.org/10.1016/j.tsc.2018.02.004.

OCDE. (2005). *Oslo manual: Guidelines for collecting and interpreting innovation data*. Grupo Tragsa: Third Edition.

Orloff, M. (2012). Modern TRIZ. Berlin.

Pavie, X., & Carthy, D. (2015). Leveraging uncertainty: A practical approach to the integration of responsible innovation through design thinking. *Procedia—Social and Behavioral Sciences, 213,* 1040–1049.

Roberts, J., Fisher, T., Trowbridge, M., & Bent, C. (2016). A design thinking framework for healthcare management and innovation. *Healthcare, 4*(1), 11–14. https://doi.org/10.1016/j.hjdsi.2015.12.002.

Savransky, S. (2000). *Engineering of creativity: Introduction to TRIZ methodology of inventive problem solving*. Florida: CRC Press.

Schmidt, E., & Rosenberg, J. (2015). *Cómo trabaja Google*. D.F.: Aguilar.

Shapira, H. (2015). The integration of design thinking and strategic sustainable development. *Journal Cleaner Production, 140*(1), 277–287. https://doi.org/10.1016/j.jclepro.2015.10.092.

Spreafico, C., & Russo, D. (2016). TRIZ industrial case studies: A critical survey. *Procedia CIRP, 39,* 51–56. https://doi.org/10.1016/j.procir.2016.01.165.

Vaneker, T., & Van Diepen, T. (2016). Design support for maintenance tasks using TRIZ. *Procedia CIRP, 39,* 67–72. https://doi.org/10.1016/j.procir.2016.01.167.

Vianna, M. (2016). *Design thinking: Innovación en los negocios*. Rió de Janeiro: MVJ Press.

Volkova, T., & Jakobsone, I. (2016). Design thinking as a business tool to ensure continuous value generation. *Intellectual Economics, 10*(1), 63–69. https://doi.org/10.1016/j.intele.2016.06.003.

Wang, F., Yeh, C., & Chu, T. (2016). Using the design for Six Sigma approach with TRIZ for new product development. *Computers & Industrial Engineering, 98,* 522–530. https://doi.org/10.1016/j.cie.2016.06.014.

The manufacturer's authorised representative in the EU is Springer Nature Customer Service Centre GmbH, Europaplatz 3, 69115 Heidelberg, Germany. If you have any concerns regarding our products, please contact ProductSafety@springernature.com

Printed and bound by CPI Group (UK) Ltd, Croydon, CR0 4YY
05/10/2026
02228141-0010